Phytochemicals: Dietary Sources, Innovative Extraction and Health Benefits

Phytochemicals: Dietary Sources, Innovative Extraction and Health Benefits

Editors

Yolanda Aguilera Gutiérrez
Vanesa Benítez García

MDPI • Basel • Beijing • Wuhan • Barcelona • Belgrade • Manchester • Tokyo • Cluj • Tianjin

Editors
Yolanda Aguilera Gutiérrez
Autonomous University of
Madrid
Spain

Vanesa Benítez García
Autonomous University of
Madrid
Spain

Editorial Office
MDPI
St. Alban-Anlage 66
4052 Basel, Switzerland

This is a reprint of articles from the Special Issue published online in the open access journal *Foods* (ISSN 2304-8158) (available at: https://www.mdpi.com/journal/foods/special_issues/ Phytochemicals_Dietary_Sources_Innovative_Extraction_Health_Benifits).

For citation purposes, cite each article independently as indicated on the article page online and as indicated below:

LastName, A.A.; LastName, B.B.; LastName, C.C. Article Title. *Journal Name* **Year**, *Volume Number*, Page Range.

ISBN 978-3-0365-3000-0 (Hbk)
ISBN 978-3-0365-3001-7 (PDF)

Contents

Yolanda Aguilera and Vanesa Benítez
Phytochemicals: Dietary Sources, Innovative Extraction, and Health Benefits
Reprinted from: *Foods* **2021**, *11*, 72, doi:10.3390/foods11010072 . . . . **1**

Manoj Kumar, Maharishi Tomar, Ryszard Amarowicz, Vivek Saurabh, M. Sneha Nair, Chirag Maheshwari, Minnu Sasi, Uma Prajapati, Muzaffar Hasan, Surinder Singh, Sushil Changan, Rakesh Kumar Prajapat, Mukesh K. Berwal and Varsha Satankar
Guava (*Psidium guajava* L.) Leaves: Nutritional Composition, Phytochemical Profile, and Health-Promoting Bioactivities
Reprinted from: *Foods* **2021**, *10*, 752, doi:10.3390/foods10040752 . . . . **5**

Mercedes M. Pedrosa, Eva Guillamón and Claudia Arribas
Autoclaved and Extruded Legumes as a Source of Bioactive Phytochemicals: A Review
Reprinted from: *Foods* **2021**, *10*, 379, doi:10.3390/foods10020379 . . . . **25**

Cecilia Moreno, Luis Mojica, Elvira González de Mejía, Rosa María Camacho Ruiz and Diego A. Luna-Vital
Combinations of Legume Protein Hydrolysates Synergistically Inhibit Biological Markers Associated with Adipogenesis
Reprinted from: *Foods* **2020**, *9*, 1678, doi:10.3390/foods9111678 . . . . **59**

Chaowanee Chupeerach, Amornrat Aursalung, Thareerat Watcharachaisoponsiri, Kanyawee Whanmek, Parunya Thiyajai, Kachakot Yosphan, Varittha Sritalahareuthai, Yuraporn Sahasakul, Chalat Santivarangkna and Uthaiwan Suttisansanee
The Effect of Steaming and Fermentation on Nutritive Values, Antioxidant Activities, and Inhibitory Properties of Tea Leaves
Reprinted from: *Foods* **2021**, *10*, 117, doi:10.3390/foods10010117 . . . . **81**

Mioara Slavu (Ursu), Iuliana Aprodu, Ștefania Adelina Milea, Elena Enachi, Gabriela Râpeanu, Gabriela Elena Bahrim and Nicoleta Stănciuc
Thermal Degradation Kinetics of Anthocyanins Extracted from Purple Maize Flour Extract and the Effect of Heating on Selected Biological Functionality
Reprinted from: *Foods* **2020**, *9*, 1593, doi:10.3390/foods9111593 . . . . **97**

Miguel Rebollo-Hernanz, Silvia Cañas, Diego Taladrid, Vanesa Benítez, Begoña Bartolomé, Yolanda Aguilera and María A. Martín-Cabrejas
Revalorization of Coffee Husk: Modeling and Optimizing the Green Sustainable Extraction of Phenolic Compounds
Reprinted from: *Foods* **2021**, *10*, 653, doi:10.3390/foods10030653 . . . . **111**

Piotr Kuropka, Anna Zwyrzykowska-Wodzińska, Robert Kupczyński, Maciej Włodarczyk, Antoni Szumny and Renata M. Nowaczyk
The Effect of *Ilex* × *meserveae* S. Y. Hu Extract and Its Fractions on Renal Morphology in Rats Fed with Normal and High-Cholesterol Diet
Reprinted from: *Foods* **2021**, *10*, 818, doi:10.3390/foods10040818 . . . . **135**

Editorial

Phytochemicals: Dietary Sources, Innovative Extraction, and Health Benefits

Yolanda Aguilera [1,2,*] **and Vanesa Benítez** [1,2,*]

1 Institute of Food Science Research, CIAL (UAM-CSIC), 28049 Madrid, Spain
2 Department of Agricultural Chemistry and Food Science, Universidad Autónoma de Madrid, 28049 Madrid, Spain
* Correspondence: yolanda.aguilera@uam.es (Y.A.); vanesa.benitez@uam.es (V.B.)

Citation: Aguilera, Y.; Benítez, V. Phytochemicals: Dietary Sources, Innovative Extraction, and Health Benefits. *Foods* **2022**, *11*, 72. https://doi.org/10.3390/foods11010072

Received: 28 October 2021
Accepted: 22 November 2021
Published: 29 December 2021

Publisher's Note: MDPI stays neutral with regard to jurisdictional claims in published maps and institutional affiliations.

Plants are the main natural source of numerous phytochemicals, although only a certain amount have been isolated and identified. Nutritional epidemiology has investigated the relation between diet and human health, reporting positive evidence on the role of phytochemicals. The studies carried out to date affirm that these compounds can reduce the incidence of several chronic diseases, including cardiovascular, obesity, diabetes, and cancer diseases, as well as high blood pressure and inflammation. Vegetables, fruits, pulses, chocolate, and teas are rich sources of phytochemicals; however, the wide diversity of these compounds requires optimized extraction methodologies to further characterization.

This Special Issue addresses interdisciplinary research about phytochemicals, highlighting their dietary sources, innovative extraction methodologies and their effect on human health. Seven papers have been selected to contribute further to phytochemical studies. Guava (*Psidium guajava* L.) leaves are studied by Kumar et al. [1] due to their health benefits. In this review, the authors critically discussed the presence of several bioactive compounds and the biological activities of guava leaf extracts, emphasizing the potential use of this plant's leaves as an ingredient in the development of functional foods and pharmaceuticals. The authors reported the existence of a high level of antioxidants and phenolic compounds responsible of biological activities, including antioxidant, hypoglycemic, anticancer, antimicrobial, and anti-obesity activities. The addition of guava leaves did not cause any change in the texture properties, did not alter the rheological and sensory properties, and showed no interaction with medicines or drugs.

Two papers about legumes have been included in this Special Issue due to the importance of this food product as source of bioactive compounds. On the one hand, the study performed by Pedrosa et al. [2] is focused on the changes produced in the bioactive phytochemical content of pulses processed by domestically or industrially processing techniques to develop new pulse flour ingredients. The review describes the processing effects on bioactive phytochemicals, concluding reductions in galactosides, phenolic compounds, protein inhibitors, and myo-inositol content. Contrarily, protein digestibility and mineral absorption was improved in processed samples. This studied food-processing technology focused on the use of autoclaving, extrusion/cooking, and cold extrusion that would allow the inclusion of pulse flours in new food products. On the other hand, Moreno et al. [3] investigated the anti-adipogenesis potential of selected legume protein hydrolysates and their combinations by biochemical assays and in silico predictions. The protein hydrolysates showed high antioxidant activity and capacity to inhibit pancreatic lipase and HMG-CoA reductase, this capacity being higher when the hydrolysates of different legumes were combined. Therefore, the authors concluded that the combination of legume protein hydrolysates from different protein sources could synergistically enhance their anti-adipogenic potential.

Chupeerach et al. [4] carried out studies on the effect of steaming and fermentation on tea leaves, analyzing nutritive values, phenolics, antioxidant activities, and in vitro health properties. The results suggest that tea preparations play a significant role in the amount of nutrients and bioactive compounds. After the steaming process, most nutrients decreased and, after fermentation, energy, fat, sodium, potassium, and iron contents increased, while calcium and vitamins decreased. However, the total phenolic content and antioxidant activities increased significantly after steaming and fermentation. With respect to health properties, steamed tea exhibited a high inhibition against lipase, α-amylase, and α-glucosidase, while fermented tea possessed high anti-cholinesterases and anti-angiotensin-converting enzyme activities. Therefore, these results indicate that tea preparations could be used as a source of nutrients and bioactive compounds to develop other functional foods and drinks.

Anthocyanins, apart from their health promoting effects, are being intensely studied as natural colors. These compounds can change their structure due to chemical and enzymatic reactions that can lead to the loss of color and its beneficial properties. In this regard, Slavu et al. [5] have studied the thermal degradation kinetics of anthocyanins extracted from purple maize flour and the impact of heating on antioxidant activity and anthocyanin release during in vitro digestion. The results showed the thermostability of anthocyanins between 80 and 120 °C, following a first-order kinetic model at higher temperatures. The thermal degradation of anthocyanins reduced their antioxidant activity, and heating decreased their stability and release after in vitro digestion.

The Special Issue also incorporates a study about the revalorization of industry by-products, in particular coffee husks, by optimizing the extraction of phenolic compounds using green sustainable techniques. Rebollo-Hernanz et al. [6] used artificial neural networks and response surface methodology to model the effect of time, temperature, acidity, and the solid-to-liquid ratio on phenolic compounds. The results validated the model and the phenolic aqueous extract obtained (100 °C, 90 min, no acid, 0.02 g of coffee husk per mL) could be utilized as nutraceutical or sustainable food ingredient.

Finally, a preliminary verification of the safety of a European hybrid of *Ilex* for use as a substitute for yerba mate is included in this Special Issue. Kuropka et al. [7] studied the effect of *Ilex x meserveae* S. Y. Hu extract and its fractions on the renal morphology of rats fed with a normal and a high cholesterol diet compared to *Ilex paraguariensis* (yerba mate). The results indicated that, at the administered concentrations, a saponin fraction from *Ilex x meserveae* seems not to influence kidney status. However, polyphenols and terpenoids existing in dry extracts and the fresh infusions from both *Ilex x meserveae* and *Ilex paraguariensis*, together with co-extracted compounds in a normal diet, cause a nephrotoxic effect that is less pronounced when a high-cholesterol diet is administered.

We are pleased to present this Special Issue that contributes to the growth of this research area. We would like to thank all the authors who have shared their scientific knowledge through their contributions and the reviewers of the papers published for their great contributions. We are also grateful to the editorial board members for their support during the preparation of this Special Issue. We sincerely hope that the readers will find this Special Issue motivating and informative.

Funding: This research received no external funding.

Conflicts of Interest: The authors declare no conflict of interest.

References

1. Kumar, M.; Tomar, M.; Amarowicz, R.; Saurabh, V.; Nair, M.S.; Maheshwari, C.; Sasi, M.; Prajapati, U.; Hasan, M.; Singh, S.; et al. Guava (*Psidium guajava* L.) Leaves: Nutritional Composition, Phytochemical Profile, and Health-Promoting Bioactivities. *Foods* **2021**, *10*, 752. [CrossRef] [PubMed]
2. Pedrosa, M.; Guillamón, E.; Arribas, C. Autoclaved and Extruded Legumes as a Source of Bioactive Phytochemicals: A Review. *Foods* **2021**, *10*, 379. [CrossRef] [PubMed]
3. Moreno, C.; Mojica, L.; González de Mejía, E.; Camacho Ruiz, R.M.; Luna-Vital, D.A. Combinations of Legume Protein Hydrolysates Synergistically Inhibit Biological Markers Associated with Adipogenesis. *Foods* **2020**, *9*, 1678. [CrossRef] [PubMed]
4. Chupeerach, C.; Aursalung, A.; Watcharachaisoponsiri, T.; Whanmek, K.; Thiyajai, P.; Yosphan, K.; Sritalahareuthai, V.; Sahasakul, Y.; Santivarangkna, C.; Suttisansanee, U. The Effect of Steaming and Fermentation on Nutritive Values, Antioxidant Activities, and Inhibitory Properties of Tea Leaves. *Foods* **2021**, *10*, 117. [CrossRef]

5. Slavu, M.; Aprodu, I.; Milea, S.A.; Enachi, E.; Gabriela Râpeanu, G.; Bahrim, G.E.; Stanciuc, N. Thermal Degradation Kinetics of Anthocyanins Extracted from Purple Maize Flour Extract and the Effect of Heating on Selected Biological Functionality. *Foods* **2020**, *9*, 1593. [CrossRef] [PubMed]
6. Rebollo-Hernanz, M.; Cañas, S.; Taladrid, D.; Benítez, V.; Bartolomé, B.; Aguilera, Y.; Martín-Cabrejas, M.A. Revalorization of Coffee Husk: Modeling and Optimizing the Green Sustainable Extraction of Phenolic Compounds. *Foods* **2021**, *10*, 653. [CrossRef]
7. Kuropka, P.; Zwyrzykowska-Wodzińska, A.; Kupczyński, R.; Włodarczyk, M.; Szumny, A.; Nowaczyk, R.M. The Effect of Ilex meserveae S. Y. Hu Extract and Its Fractions on Renal Morphology in Rats Fed with Normal and High-Cholesterol Diet. *Foods* **2021**, *10*, 818. [CrossRef] [PubMed]

Review

Guava (*Psidium guajava* L.) Leaves: Nutritional Composition, Phytochemical Profile, and Health-Promoting Bioactivities

Manoj Kumar [1], Maharishi Tomar [2], Ryszard Amarowicz [3,*], Vivek Saurabh [4], M. Sneha Nair [5], Chirag Maheshwari [6], Minnu Sasi [7], Uma Prajapati [4], Muzaffar Hasan [8], Surinder Singh [9], Sushil Changan [10], Rakesh Kumar Prajapat [11], Mukesh K. Berwal [12] and Varsha Satankar [13]

[1] Chemical and Biochemical Processing Division, ICAR—Central Institute for Research on Cotton Technology, Mumbai 400019, India; manoj.kumar13@icar.gov.in

[2] ICAR—Indian Grassland and Fodder Research Institute, Jhansi 284003, India; Maharishi.tomar@icar.gov.in

[3] Institute of Animal Reproduction and Food Research, Polish Academy of Sciences, Tuwima 10 Str., 10-748 Olsztyn, Poland

[4] Division of Food Science and Postharvest Technology, ICAR—Indian Agricultural Research Institute, New Delhi 110012, India; vivek_11593@iari.res.in (V.S.); uma_11103@iari.res.in (U.P.)

[5] Department of Nutrition and Dietetics, Faculty of Allied Health Sciences, Manav Rachna International Institute of Research and Studies, Faridabad 121004, Haryana, India; sneha.fas@mriu.edu.in

[6] Department of Agriculture Energy and Power, ICAR—Central Institute of Agricultural Engineering, Bhopal 462038, India; Chirag.Maheshwari@icar.gov.in

[7] Division of Biochemistry, ICAR—Indian Agricultural Research Institute, New Delhi 110012, India; minnusasi1991@gmail.com

[8] Agro Produce Processing Division, ICAR—Central Institute of Agricultural Engineering, Bhopal 462038, India; Muzaffar.Hasan@icar.gov.in

[9] Dr. S.S. Bhatnagar University Institute of Chemical Engineering and Technology, Panjab University, Chandigarh 160014, India; ssbhinder@pu.ac.in

[10] Division of Crop Physiology, Biochemistry and Post-Harvest Technology, ICAR—Central Potato Research Institute, Shimla 171001, India; Sushil.Changan@icar.gov.in

[11] School of Agriculture, Suresh Gyan Vihar University, Jaipur 302017, Rajasthan, India; rakesh.prajapat@mygyanvihar.com

[12] Division of Crop improvement, ICAR—Central Institute for Arid Horticulture, Bikaner 334006, India; Mukesh.kumar4@icar.gov.in

[13] Ginning Training Centre, ICAR—Central Institute for Research on Cotton Technology, Nagpur 440023, India; satankarvarsha@gmail.com

* Correspondence: r.amarowicz@pan.olsztyn.pl

Citation: Kumar, M.; Tomar, M.; Amarowicz, R.; Saurabh, V.; Nair, M.S.; Maheshwari, C.; Sasi, M.; Prajapati, U.; Hasan, M.; Singh, S.; et al. Guava (*Psidium guajava* L.) Leaves: Nutritional Composition, Phytochemical Profile, and Health-Promoting Bioactivities. *Foods* **2021**, *10*, 752. https://doi.org/10.3390/foods10040752

Academic Editors: Yolanda Aguilera and Vanesa Benítez García

Received: 6 March 2021
Accepted: 30 March 2021
Published: 1 April 2021

Publisher's Note: MDPI stays neutral with regard to jurisdictional claims in published maps and institutional affiliations.

Abstract: *Psidium guajava* (L.) belongs to the Myrtaceae family and it is an important fruit in tropical areas like India, Indonesia, Pakistan, Bangladesh, and South America. The leaves of the guava plant have been studied for their health benefits which are attributed to their plethora of phytochemicals, such as quercetin, avicularin, apigenin, guaijaverin, kaempferol, hyperin, myricetin, gallic acid, catechin, epicatechin, chlorogenic acid, epigallocatechin gallate, and caffeic acid. Extracts from guava leaves (GLs) have been studied for their biological activities, including anticancer, antidiabetic, antioxidant, antidiarrheal, antimicrobial, lipid-lowering, and hepatoprotection activities. In the present review, we comprehensively present the nutritional profile and phytochemical profile of GLs. Further, various bioactivities of the GL extracts are also discussed critically. Considering the phytochemical profile and beneficial effects of GLs, they can potentially be used as an ingredient in the development of functional foods and pharmaceuticals. More detailed clinical trials need to be conducted to establish the efficacy of the GL extracts.

Keywords: phenolic extracts; essential oils; polysaccharides; valorization; crop residue

1. Introduction

Plants are a predominant natural source of numerous bioactive compounds [1,2]. Several diseases have been cured using a variety of plant preparations in folk medicine since

5

ancient times [3] and, presently, cosmetic, pharmaceutical, and nutraceutical industries are paying more attention to plant preparations and pure phytochemicals. The projected growth of the plant preparation market is around USD 86.74 billion by 2022, with the largest market share belonging to the pharmaceutical sector, followed by the nutraceutical industry. Interestingly, the utilization of plant preparations for cosmetics, beverages, food, and medicine is mainly dependent on plant leaves. Among all plant organs, leaves are the largest accumulators of bioactive compounds, such as secondary metabolites. Several recent studies reported phytochemical profiles and biological activities of leaf extracts of various cultivated plants [2,4–6]. Hence, although plant leaves are considered as agricultural waste, they are a rich source of high-value nutra-pharmaceutical compounds.

The guava (*Psidium guajava* L.) tree (Figure 1), belonging to the Myrtaceae family, is a very unique and traditional plant which is grown due to its diverse medicinal and nutritive properties. Guava has been grown and utilized as an important fruit in tropical areas like India, Indonesia, Pakistan, Bangladesh, and South America. Different parts of the guava tree, i.e., roots, leaves, bark, stem, and fruits, have been employed for treating stomachache, diabetes, diarrhea, and other health ailments in many countries. Guava leaves (*Psidii guajavae folium*; GL) are dark green, elliptical, oval, and characterized by their obtuse-type apex. Guava leaves, along with the pulp and seeds, are used to treat certain respiratory and gastrointestinal disorders, and to increase platelets in patients suffering from dengue fever [7]. GLs are also widely used for their antispasmodic, cough sedative, anti-inflammatory, antidiarrheic, antihypertension, antiobesity, and antidiabetic properties [8]. Studies on animal models have also established the role of GL isolates as potent antitumor, anticancer, and cytotoxic agents [9,10].

Figure 1. (**A**) Guava fruit and leaves, (**B**) bunch of guava leaves with dorsal view on the left and ventral view on the right, (**C**) guava leaf with dorsal view on the left and ventral view on the right.

GLs are widely employed for treating diarrhea and digestive ailments, while the fruit pulp is utilized to enhance the platelet count for treating dengue fever. The potential of guava leaf extracts for diarrhea treatment was also studied [11,12]. The flavonoids present in guava leaf extract chiefly determine their antibacterial activity, while quercetin, which is the most predominant flavonoid of guava leaves, exhibits strong antidiarrheal activities. The antidiarrheal activity of quercetin is ascribed to its relaxing effect on the intestinal muscle lining which prevents bowel contractions. Guava leaf polysaccharides (GLPs) can be utilized as an antioxidant additive in food and for diabetes treatment.

The presence of a unique variety of bioactive polyphenolic compounds, like quercetin and other flavonoids, and ferulic, caffeic, and gallic acids, present in guava leaves primarily determine their bioactive and therapeutic properties [8,13]. These phenolic compounds are known as secondary metabolites which exhibit strong antioxidant and immunostimulant activities. This review aims to discuss the various nutritional and bioactive compounds present in guava leaves and decipher the molecular basis of their pharmacological and medicinal properties concerning human health, nutrition, and as complementary medicine.

2. Chemical Composition

2.1. Proximate Composition

Guava leaves (GLs) are a rich source of various health-promoting micro- and macronutrients as well as bioactive compounds. They contain 82.47% moisture, 3.64% ash, 0.62% fat, 18.53% protein, 12.74% carbohydrates, 103 mg ascorbic acid, and 1717 mg gallic acid equivalents (GAE)/g total phenolic compounds [14]. The overall proximate profile of GLs is presented in Table 1.

2.1.1. Polysaccharides

Polysaccharides are macromolecules that are ubiquitously present in nature. They are made of long polymeric chains, which are composed of monosaccharide units. These polysaccharides demonstrate various physicochemical, biological, and pharmacological properties, such as antioxidant, anti-inflammatory, antidiabetic, immunomodulatory, and antitumor activities [15]. Guava leaf polysaccharides (GLPs) can be isolated using ultrasound-assisted extraction (UAE) (time: 20 min, power: 404 W, temperature: 62 °C). These GLPs contain about 9.13% uronic acid and 64.42% total sugars, out of which 2.24% are reducing sugars. GLPs are soluble in water, while insoluble in organic solvents like ethanol, diethyl ether, ethyl acetate, acetone, and chloroform. Extracted GLP with a concentration of 100 μg/mL exhibits good antioxidant capacity with 56.38% and 51.73% 2,2-diphenyl-1-picrylhydrazyl (DPPH) radical- and 2,2′-azino-bis(3-ethylbenzothiazoline-6-sulfonic acid (ABTS) radical cation-scavenging capacity, respectively [16]. Similar results were also reported by Kong et al. [17]. They obtained up to 0.51% GLP using UAE that exhibited good DPPH$^\bullet$- and $^\bullet$OH-scavenging activity (72–86% and 42.94–58.33%). GLPs can be categorized into two groups: unsulfated and sulfated GLPs. Sulfated GLP contains about 18.58% sulfate content. Sulfated GLP exhibited good antioxidant activity in terms of DPPH, hydroxyl, and alkyl radical-scavenging activity (0.10, 0.02, and 0.17 IC_{50}, mg/mL, respectively). Studies showed that guava leaves extracts (GLE) effectively reduced the oxidative stress and toxicity caused by hydrogen peroxide in mammalian cell lines (Vero cells) [18]. GLPs are also found to be beneficial in treating diabetes mellitus symptoms. Acarbose (an antidiabetic drug) is commonly used for the treatment of type 2 diabetes [16]. It acts as an inhibitor of glycoside hydrolases like α-glucosidase and α-amylase and thus prevents rapid glucose release from complex carbohydrates [19]. This activity causes some of the incompletely digested complex carbohydrates to remain in the intestine and be transported to the colon. The intestinal microflora digests these complex carbohydrate fractions, causing gastrointestinal problems like diarrhea and flatulence. A study reported that GLP inhibited α-glucosidase more efficiently than acarbose without significantly blocking the α-amylase activity [19]. Moreover, it also caused a substantial drop in fasting blood sugar, total cholesterol, total triglycerides, glycated serum protein, creatinine, and malonaldehyde in diabetic mice without causing any major side effect [15]. Therefore, GLP can be used as a replacement of acarbose for managing diabetes mellitus and also as an antioxidant additive in foods.

Table 1. Nutritional profile of guava leaves.

Compounds	Content/Composition	References
Elements and ascorbic acid		
Potassium	1.11%	
Phosphorus	0.23%	[20]
Nitrogen	1.02%	
Ascorbic acid	142.55 mg/100 g	
Carbohydrates/phenols/sulfates		
Fucose	1.44%	
Rhamnose	3.88%	
Arabinose	22.6%	
Galactose	29.41%	
Glucose	33.79%	
Mannose	0.59%	[18]
Xylose	7.71%	
Phenol	15.28%	
Sulfate	18.58%	
Carbohydrate	48.13%	
Sulfate polysaccharide	66.71%	
Protein		
Association of Official Analytical Chemists (AOAC) method	22.98 ± 0.036% [dry weight (DW) basis]	[21]
AOAC method	9.73%	[22]
Lowry's method	16.8 mg/100 g	[23]
Ninhydrin method	8.0 mg/100 g	

2.1.2. Proteins

Guava leaves contains 9.73% protein on a dry weight basis [22]. Proteins are large biomolecules composed of amino acids and act as building blocks of cells. Proteins play a major role in growth and maintenance, enzyme regulation, and cell signaling, and also as biocatalysts [24]. Recently, plant-based nutrients have gained potential because of the high demand for nutritionally rich food, particularly protein. A great effort is now being made to find highly sustainable nutritionally rich food sources [25]. Thomas et al. [23] reported 16.8 mg protein/100g and 8 mg amino acids/100g in guava leaves as estimated according to Lowry's and ninhydrin methods, respectively. Jassal et al. [21] reported that guava leaves can be utilized as a novel and sustainable dietary source as they are a rich source of proteins, carbohydrates, and dietary fibers.

2.1.3. Minerals and Vitamins

Guava leaves are the rich source of minerals, such as calcium, potassium, sulfur, sodium, iron, boron, magnesium, manganese, and vitamins C and B. The higher concentrations of Mg, Na, S, Mn, and B in GLs makes them a highly suitable choice for human nutrition and also as an animal feed to prevent micronutrient deficiency [26]. Thomas et al. [23] reported the concentration of minerals such as Ca, P, K, Fe, and Mg as 1660, 360, 1602, 13.50, and 440 mg per 100g of guava leaf dry weight (DW), respectively. The concentration of vitamins C and B was 103.0 and 14.80 mg per 100g DW, respectively. Consumption of Ca- and P-rich GLs reduces the risk of deficiency-related diseases like hypocalcemia, hypophosphatemia, and osteoporosis. The study also reported that the concentration of Ca, P, Mg, Fe, and vitamin B in GLs was higher than that in guava fruit. The higher vitamin C content in GLs may help in improving the immune system and maintain the health of blood vessels, whereas vitamin B plays an important role in improving blood circulation, nerve relaxation, and cognitive function stimulation.

2.2. Phytochemical Profile

2.2.1. Essential Oil Profile

GLs are a rich source of essential oils (Table 2). The major constituent of GL essential oil includes 1,8-cineole and *trans*-caryophyllene [27]. Chen et al. [8] identified 50 compounds in GL essential oil using gas chromatography (GC) and gas chromatography/mass spectrometry (GC–MS), where they found β-caryophyllene, α-pinene, and 1,8-cineole to be the major ones. GL essential oil from the Philippines was found to contain a different profile, with limonene, α-pinene, β-caryophyllene, and longicyclene as major compounds [28]. Ecuadorian GL essential oil contained a higher content of monoterpenes (limonene and α-pinene) whereas Tunisian guava leaf oil displayed a higher content of veridiflorol and *trans*-caryophyllene [29,30]. Soliman et al. [31] reported a larger amount of monoterpenes, contrary to the other studies, where sesquiterpenes constituted the major compound in GL essential oil. El-Ahmady et al. [32] reported 4 α-selin-7(11)-enol, α-selinene, β-caryophyllene, and β-caryophyllene oxide as the major constituents of GL essential oil. In another study, sixty-four different compounds were determined in essential oil extracted from GLs by gas chromatography–mass spectrometry (GC–MS). Among them, caryophyllene (24.97%) was found to be predominantly present, which acts as an antioxidant, anticancer, anti-inflammatory, and antimicrobial agent [21]. This study reported the concentration of non-oxygenated sesquiterpenes, oxygenated sesquiterpenes, and monoterpenes as 73.67, 12.94, and 8.55%, respectively.

Table 2. Essential oil components of guava leaves.

Compounds	Content/Composition	References
Essential oil components		
α-Pinene	1.53%	
Benzaldehyde	0.83%	
p-cymene	0.52%	
Limonene	54.7%	
1,8-Cineole	32.14%	
β-*cis*-Ocimene	0.28%	[31]
γ-Terpinene	0.38%	
α-Terpineol	1.79%	
β-Caryophyllene	2.91%	
α-Humulene	0.77%	
Total identified constituents	95.85%	
Caryophyllene, copaene, nerolidol, caryophyllene oxide, humulene, limonene, eucalyptol, beta-bisabolene, cadin-4-en-10-ol, *trans*-cadina-1,4-diene, sesquiterpenes, eugenol, isoeugenol, cevadine, emetine (extracted from guava leaves, Ludhiana, India using hydro-distillation by Clevenger-type apparatus)	-	[21]

2.2.2. Phenolic Compounds

GLs are widely popular as a traditional source of medicine in Asian countries due to their antihyperglycemic effect. As mentioned in the previous sections, they contain superior quality bioactive polysaccharides, proteins, lipids, essential oils, vitamins, and minerals. The various secondary metabolites present in GLs include phenolic acids, flavonoids, triterpenoids, sesquiterpenes, glycosides, alkaloids, and saponins. Phenolic compounds (PCs) serve as key bioactive compounds which provide antioxidant and hypoglycemic properties to GLs. Generally, these PCs play a major role in managing various metabolic and physiological activities in the human body. About seventy-two different phenolic compounds have been determined in GLs using high-performance liquid chromatography–diode array detector–quadrupole time-of-flight tandem mass spectrometry [33]. Generally, five quercetin glycosides are present in GLs. The presence of two new benzophenone galloyl glycosides (guavinosides A and B) and one quercetin galloyl glycoside (guavinoside C) was also reported [34]. Seventeen types of triterpenoids, thirty types of flavonoids, and nineteen types of sesquiterpenoids in GLs have also been reported [10]. Moreover, diphenylmethane [35] sesquiterpenoid-diphenylmethane meroterpenoids (psiguadials A and B) [36] and psiguanins A–D (1–4) [37] were also found in GLs. Epidemiological studies have established the roles of polyphenolic compounds against chronic diseases, such as diabetes, cancer, and neurodegenerative and cardiovascular diseases [38]. Phenolic compounds modulate numerous physiological processes like cell proliferation, enzymatic activity, cellular redox potential, and signal transduction pathways to fight against chronic pathologies [39]. Various phenolics reported in GLs are summarized in Table 3 and the structures can be seen in Figure 2.

Table 3. Phenolic compounds of guava leaves.

Origin of Guava Leaves	Extract/Fraction	Bioactive Compounds	References
Leaves from Guangzhou (China)	Ethyl acetate-soluble fraction, n-butanol-soluble fraction, 75% ethanol extract, residual fraction, dichloromethane-soluble fraction	Quercetin, avicularin, apigenin, guaijaverin, kaempferol, hyperin, myricetin	[40]
Leaves from Jing-cin Farm (Tianzhong Township, Changhua County, Taiwan)	Aqueous extract	Gallic acid, catechin, epicatechin, quercetin, chlorogenic acid, epigallocatechin gallate, caffeic acid	[41]
Leaves from Motril (Spain)	Acetone, water, and acetic acid extract	Proanthocyanidins (PAs)	[33]
Leaves from Jiangmen (China)	Methanol extract	Gallic acid, chlorogenic acid, epicatechin, mono-3-hydroxyethyl-quercetin-glucuronide, rutin, isoquercitrin, quercetin-3-O-α-L-arabinofuranoside, quercetin-3-O-β-D-xylopyranoside, avicularin, quercitrin, kaempferol-3-arabofuranoside, quercetin, kaempferol	[42]

Among phenolic compounds, quercetin is a major bioactive phenolic compound in GLs. Diets enriched with bioactive compounds have been gaining much attention in recent years due to their potential to lower the risk of the development of numerous chronic diseases. Seven pure compounds, quercetin, avicularin, apigenin, guaijaverin, kaempferol, hyperin, and myricetin, were separated from the ethyl acetate (EtOAc)-soluble GL fraction using Sephadex LH-20 column chromatography with reversed-phase thin layer chromatography (RP-TLC) to monitor separation. Mass spectrometry and nuclear magnetic resonance spectroscopy were used to elucidate the compound structures [40]. Wang et al. [43] extracted and analyzed phenolic compounds from non-fermented guava leaves (NFGLs) and fermented guava leaves (FGLs) using high-performance liquid chromatography coupled to electrospray ionization quadropole–time-of-flight mass spectrometry (HPLC–TOF–ESI/MS). The authors reported the presence of gallic acid, rutin, chlorogenic acid, avicularin, isoquercitrin, quercitrin, and kaempferol in NFGL and FGL samples. Among them, quercetin, rutin, gallic acid, avicularin, and isoquercitrin occupied about 65% of the total peak area on the chromatogram. Another study reported higher concentrations of catechin (2.25%) and epicatechin (1.45%), whereas gallic acid, chlorogenic acid, quercetin, caffeic acid, and epigallocatechin gallate were present in lower concentrations in GL extract [41]. Additionally, phenolic compounds (eugenol and isoeugenol) and alkaloids (cevadine and emetine) were detected. Díaz-de-Cerio et al. [44] optimized the extraction of proanthocyanidins, as antidiabetic and antiobesity agents [45], from GLs by

HPLC–fluorimetric detector (FLD)–ESI–MS and studied their degree of polymerization in different oxidation states. Thus, the phytochemical profile of GL extract depicts the presence of numerous phytochemicals with distinct medicinal properties, suggesting its application to cure human diseases.

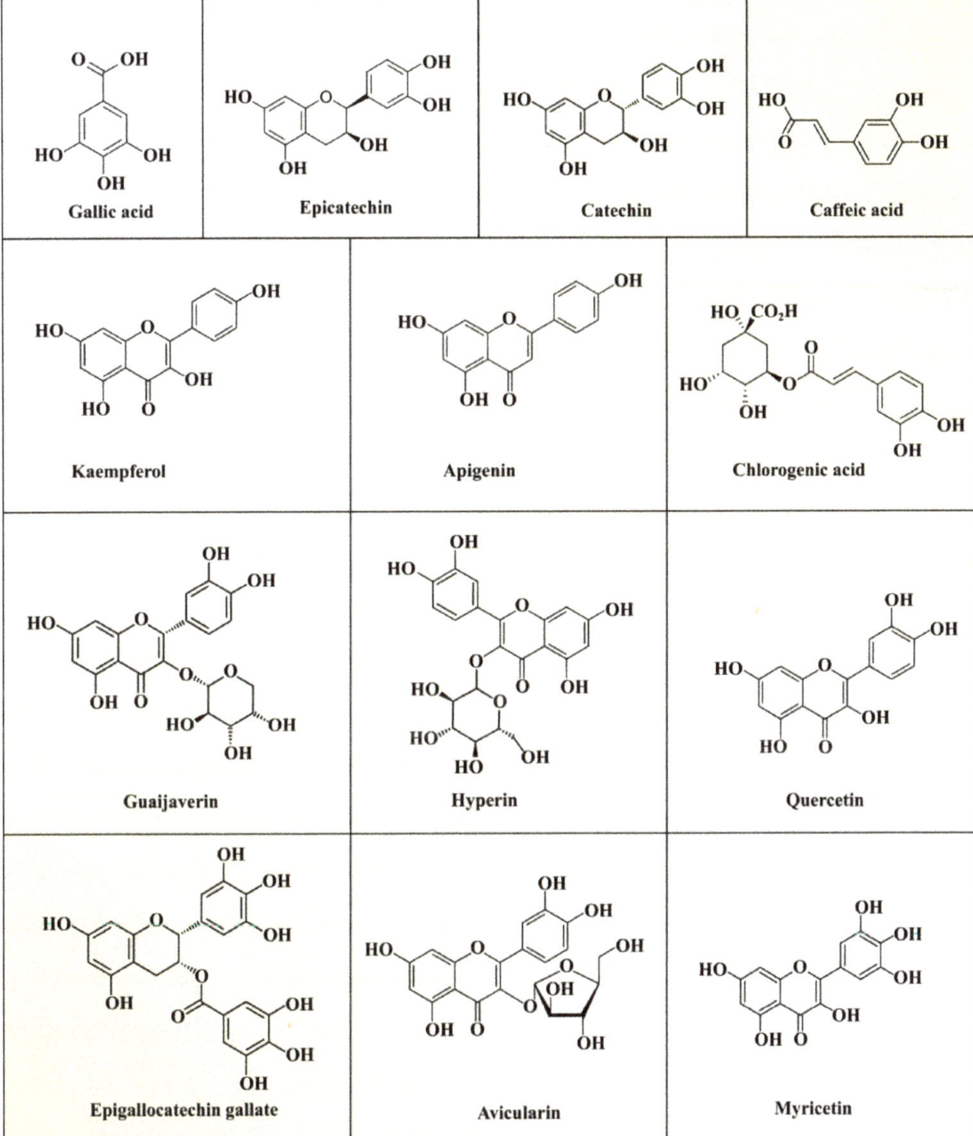

Figure 2. Structures of phenolic compounds present in guava leaf extracts.

3. Biological Activities of Guava Leaf Extracts

The compounds from guava leaf extracts possess multidirectional biological activities, including antioxidant, hypoglycemic, anticancer, and other biological activities. It was also reported that polysaccharide fractions of sulfated GLP possess stronger biological activities, such as antioxidant, antibacterial, and antitumor effects compared to unsulfated ones. The useful bioactivities of GL extract are presented in the following subsections.

3.1. Anticancer/Antitumor Activity

Cancer is a complex health disorder which is identified by the development of cell proliferation or a decrease, causing apoptosis [46]. It can be caused by several exogenous and endogenous factors involved in the excessive production of reactive oxygen species (ROS). This can result in single- or double-strand breaks in DNA or RNA, base mutations, chromosomal breaking and reorganization, DNA cross-linkage, nucleic acid degradation, damage to cell membrane integrity due to lipid peroxidation, and tumor formation [47]. GLs are a good source of triterpenoids, sesquiterpenes, tannins, psiguadials, volatile oils, flavonoids, benzophenone glycosides, and miscellaneous quinones [10]. Psiguadial D and psiguadial C act as inhibitors of human hepatoma cells (HepG2) and protein tyrosine phosphatase 1B (PTP1B). Terpenoids and flavonoids present in GLs exhibit antitumor effects by regulating the immune system, suppression of signal transfer and tumor cell adhesion, and an impediment to tumor angiogenesis and cell proliferation [48]. Studies suggest that these leaves exhibit a potent inhibitory effect against cancer cell lines like MDA-MB-231 and Michigan Cancer Foundation-7 (MCF-7) for breast cancer, Henrietta Lacks (HeLa) for cervical cancer, KB for nasopharyngeal cancer, LNCaP, DU 145, and prostate cancer-3 (PC-3) for prostate cancer, and colorectal 320 double minutes (COLO320DM) for colon cancer [49].

The growth of colorectal tumors chiefly relies on angiogenesis, a process by which new blood vessels develop from pre-existing ones. Prolonged angiogenesis is vital for the progression of tumors towards malignancy since the blood vessels efficiently supply the developing tumor cells with vital metabolites and oxygen and it also functions as an efficient means for cellular waste disposal. A study was conducted to investigate the anticancer and antiangiogenic potential of GL extracts against angiogenesis-dependent colorectal cancer [50]. Guava leaf extracts rich in vitamin E, flavonoids (apigenin), and β-caryophyllene demonstrated strong antiproliferative activity against human colon carcinoma cell lines Caco-2, HT-29, and SW480. The antiangiogenic property of β-caryophyllene is attributable to its interaction with the transcription factor HIF-1α that regulates the biological pathways related to hypoxia, tumor metastasis, and tumor-mediated angiogenesis. HIF-1α also mediates the transcription of vascular endothelial growth factor (VEGF) in the presence of β-caryophyllene, explaining the antiangiogenic and anticolorectal cancer property of guava leaf extract.

A caryophyllene-based meroterpenoid called guajadial from GLs was studied for its antiproliferative and antiestrogenic activities against human breast cancer cell lines MCF-7 BUS and MCF-7 [51]. The authors suggested that guajadial exerts its anticancer activity by acting on estrogenic receptors, induction of apoptosis by blocking DNA synthesis, and inhibition of the cell cycle at the G1 phase [52]. A similar study indicated that three benzophenones, guavinoside B, guavinoside E, and 3,5-dihydroxy-2,4-dimethyl-1-*O*-(6'-*O*-galloyl-β-D-glucopyranosyl)-benzophenone, isolated from guava leaves inhibited the growth of HCT116 human colon cancer cells [53]. These compounds strongly induced cancer cell apoptosis and modulated the expression of key proteins like extracellular signal-related kinases (p-ERK1/2), p53, c-Jun NH_2-terminal kinases (p-JNK), and cleaved caspases 8 and 9, which are involved in apoptotic signaling and cell proliferation. Another study indicated the inhibitor effect of guava leaf extracts on lung cancer genes, primarily involved in signaling pathways like PI3K-Akt [10]. The authors stated that daidzein, ursolic acid, apigenin, genistein, and quercetin in the leaf extract strongly inhibited cyclin-dependent kinase 2,6 (CDK2,6), vitamin D_3 receptor (VDR), hepatocyte growth factor receptor (MET), epidermal growth factor receptor (EGFR), progesterone receptor (PGR), peroxisome proliferator-activated receptor gamma (PPARG) and interleukin-2 (IL-2) proteins and subsequently blocked tumor proliferation and migration, tumor angiogenesis, tumor adhesion, and degradation of the extracellular matrix.

3.2. Antidiabetic Activity

Diabetes is a major chronic disease and about 10% of the world's population suffer from blood glucose metabolic disorder, mainly characterized by a hyperglycemic condition. This situation is either characterized by insufficient secretion of insulin from β-cells of pancreatic islets (type 1 diabetes) or the inability of cells to react in response to the secreted insulin (type 2 diabetes) [12,54]. The International Diabetes Federation (IDF) stated that 451 million people were affected by diabetes mellitus, resulting in 5 million deaths, in 2017 and the global prevalence of diabetes is projected to hit 693 million cases by 2045 [55]. The prolonged condition of hyperglycemia leads to increased production of ROS and dyslipidemia, causing severe cellular damage and complications [56].

GLs have been widely used as ethnomedicine for diabetes management [15]. Flavonoids and polysaccharides of GLs have been reported for their antidiabetic potential in several studies. Guaijaverin and avicularin flavonoids of GL extract were associated with significant improvement in the function of β-cells of pancreatic islets and hepatocyte morphology in diabetic mice [57]. Guaijaverin suppressed the activity of the blood glucose homeostasis enzyme dipeptidyl-peptidase IV [58], while avicularin inhibited intracellular lipid aggregation by impeding glucose uptake through GLUT-4 in vitro and revealed no distinct toxicity for 3T3-L1 adipose cells [59]. Luo et al. [14] extracted GL polysaccharides (GLPs) and further tested the antidiabetic effects on streptozotocin-induced diabetic mice in combination with a high-fat diet. The authors revealed that GLP was associated with a significant reduction in total cholesterol, triglycerides, glycated serum protein, creatinine, fasting blood glucose, and malonaldehyde content, and increased total superoxide dismutase and total antioxidant capacity enzyme activity in vivo. Suboptimal glycemic regulation may lead to elevated postprandial glucose concentrations. Nair et al. [60] suggested that the inhibitors of α-amylase and α-glucosidase enzyme can decline postprandial glucose absorption, and are therefore possible targets for diabetes management. The polysaccharides were isolated from GLs by ultrasound-assisted extraction and the antiglycation activity of extracted polysaccharides was studied [16]. The authors found that GLP showed strong inhibition of α-glucosidase, with a 99.54% inhibition rate at a 100 µg/mL concentration, and less inhibition of α-amylase, with a 14.06% inhibition rate at a 1mg/mL dose concentration. The findings suggests that bioactive compounds from GLs can be effective in reducing the risk of diabetes.

3.3. Antioxidant Activity

Oxygen is an important element for aerobes since it acts as a terminal electron acceptor during the respiration process, which is the key source of energy production. However, free radicals produced during metabolic processes are responsible for numerous ailments in the human body, namely, inflammatory diseases, ischemic diseases, neurological disorders, hemochromatosis, emphysema, acquired immunodeficiency syndrome, and many others [61]. The presence of phenolic compounds, such as gallic acid, pyrocatechol, taxifolin, ellagic acid, ferulic acid, and several others, is responsible for the antioxidant roles of GLs [8,13]. High-performance liquid chromatography analysis of GL extracts revealed the presence of seven major flavonoids: quercetin, hesperetin, kaempferol, quercitrin, rutin, catchin, and apigenin, while other bioactive compounds, such as kaempfertin, isoquinoline, and corilaginoline alkaloids, were also identified [62]. These compounds are the major compounds responsible for the antioxidant properties of GLs.

The significance of antioxidant compounds from GLs in minimizing the harmful effects of free radicals has been shown by numerous studies. Essential oils extracted from GLs were found to function as moderate antioxidants with an IC_{50} value of ~460.37 ± 1.33 µg/mL, as demonstrated by a DPPH assay [27]. The reduction of linoleic acid oxidation and the scavenging effect on peroxyl radicals were revealed by other such analyses on GL extract. The study also showed that there was a linear association between the antioxidant's potency, the ability to scavenge free radicals, and the phenolic content of GL extract [8]. The protective effect of GL polysaccharide was studied in zebrafish. The authors revealed that GL polysaccharides exerted a protective effect against oxidative stress induced by hydrogen peroxide by inhibiting the formation of reactive oxygen species (ROS), reducing lipid peroxidation and cell death [18]. In another study, it was revealed that GL extracts at 4000 ppm or higher can prevent the oxidation of fresh pork sausages, suggesting its application as a functional food ingredient [63]. To release insoluble bound polyphenol components, GLs were co-fermented with yeast and bacterial strains and it was observed that fermentation enhanced the antioxidant ability of soluble guava leaf polyphenols [43]. In an advanced study, silver nanoparticles were synthesized by utilizing crude polysaccharides of GLs, and showed high DPPH radical- and ABTS radical cation-scavenging activity [64]. It is evident from the findings that GL extracts can be a useful antioxidant material in the food preservation and cosmetic industries.

3.4. Antidiarrhea Activity

Currently, diarrhea is one of the prominent root causes of mortality among children in the age group of 0–5 years. Attempts have been made to discover new drugs with minimal side effects on the other organs of the body. In developing nations, attention has been devoted to identifying novel phytochemicals derived from medicinal plants to develop new drugs with minimal side effects [65]. Most pharmaceutical industries are engaged in the innovation of different drugs which have therapeutic potential to combat this disease. A number of therapeutic treatments are available for treating diarrhea in the form of synthetic drugs which cause many side effects in the human body, such as constipation, intestinal obstruction, induction of bronchospasm, and vomiting [66,67]. To combat these side effects, focus should be directed to investigate and isolate potent bioactive compounds from medicinal plants. GLs are considered to possess antidiarrheal properties, as reported by many researchers. Mazumdar et al. [12] reported the antidiarrheal potential of ethanolic isolates of GLs in Wistar rats. The authors reported that a dosage level of extracts at a concentration of 750 and 500 mg/kg had antidiarrheal potential in castor oil-fed rats. Besides this, Ojewole et al. [68] reported similar activity using aqueous extracts of GLs in rodents. They reported that GL extracts at doses of 52–410 mg/kg when administrated orally were found to combat diarrhea, and also resulted in reduced intestinal transit and dilatory removal of unwanted gastric products. Further, loperamide (13 mg/kg, p.o.) reduced the occurrence of defecation, along with the severity of diarrhea in the same animal models. The GL extracts reduced diarrheal symptoms, such as secretion of interstitial fluid and wetness of fecal droppings in a dose-dependent manner. GLs at different concentrations in rabbits showed concentration-dependent pulsing and pendulum retrenchments in the duodenum. These studies suggested that GLs possess excellent potential to combat diarrhea and they have gained the focus of scientific communities regarding their pharmacological prospects in different communities. In another study, Dewi et al. [11] proved the antidiarrheal potency of GL water extracts. The authors used the combination of GL water extract (G) and green tea leaf extract (T). The extracts were used in different combinations, i.e., (G:T) 112.5:110.55, (G:T) 75:221.1, and (G:T) 37.5:331.65 mg/kg body weight. The outcomes proved that all combinations had potent antidiarrheal activities, depicted by enhanced stool weight, stool onset, stool consistency, and diarrhea period. When mice were administered with a water extract of (G:T) 75:221.1 at a time interval of 180–240 min, a significant reduction in diarrhea was observed. Based on the above observations, it can be concluded that GLs demonstrated potent antidiarrheal efficacy. It may interesting to investigate, in future, the associated underlying molecular mechanism/s and also long-term toxicity studies in different animal models, as well as in human patients, to ascertain the therapeutic efficacy and safety.

3.5. Antimicrobial Activity

The evolution of novel disease-causing strains and resistance of microbes to classical antibiotics are currently serious concerns. The incidence of systemic microbial infections such as septicemia, urinary tract infections, meningitis, pneumonia, and gastritis affects the entire human body and contributes significantly to global mortality. Food-borne diseases are mostly caused by pathogens including *Staphylococcus, Shigella, Salmonella, Bacillus, Escherichia coli, Clostridium,* and *Pseudomonas* [69]. Plant-derived bioactive compounds are promising sources of antimicrobials. These compounds act by the inhibition of microbial cell wall development, disruption, and lysis, hampering biofilm formation, repression of DNA replication and transcription, impeding adenosine triphosphate (ATP) production, suppression of bacterial toxins, and the generation of reactive oxygen species (ROS) [70]. GLs, owing to the presence of different organic and inorganic antioxidants and anti-inflammatory compounds, are known to possess antimicrobial properties [71]. GL essential oils display strong antimicrobial properties against *Pseudomonas aeruginosa, Escherichia coli, Streptococcus faecalis, Staphylococcus aureus,* and *Bacillus subtilis* [31]. Studies also indicate their antioxidant and antiproliferative activities.

Qualitative analysis of aqueous and organic extracts of guava leaves revealed the presence of phenolic acids, flavonoids, terpenoids, glycosides, and saponins, in which their presence is positively correlated with antimicrobial activity. HPLC–TOF–ESI/MS analysis of fermented GLs confirmed the presence of gallic acid, chlorogenic acid, rutin, isoquercitrin, avicularin, quercitrin, kaempferol, morin, and quercetin. These compounds have the property of inhibiting ergosterol, which is a fungal cell membrane component, and glucosamine, which is a fungal cell growth indicator. Similarly, water-soluble tannins present in GLs act as bacteriostatic agents, with mechanisms of actions like withholding substratum, hampering oxidative phosphorylation, and extracellular enzyme inhibition. They have been demonstrated to have an inhibitory effect on antibiotic-resistant clinical isolates of *Staphylococcus aureus* [72]. Another study, by Hirudkar et al. [73], identified quercetin as one of the most predominant flavonoids of GLs with the highest pharmacological activity. Additionally, activity against bacterial and fungal pathogens was traced to triterpenoids like betulinic acid and lupeol [74]. A methanolic GL extract demonstrated antibacterial activity against *E. coli* with a minimum inhibitory concentration (MIC) of 0.79 µg/mL, a minimum bactericidal concentration of 51 µg/mL, and a reasonable antifungal activity with a minimum inhibitory concentration of 12.6 µg/mL [75]. A decoction of GLs at various concentrations (1%, 5%, and 10%) exhibited an inhibitory effect on bacterial colonization and binding of bacterial enterotoxins on epithelial cells, thus altering the inflammatory response. Guava extract showed higher activity of the antioxidant enzymes peroxidase, catalase, and polyphenol oxidase [76]. Biogenic production of silver nanoparticles (40 nm in size) using GL extract showed antibacterial activity against *Pseudomonas aeruginosa* owing to high antiradical activity against DPPH radicals and ABTS radical cations [77]. The role of the cytokine interleukin-7 (IL-7) as an immune booster against microbial infections is well studied. It is hypothesized that GL extracts act on the intestinal mucosal cells and aid the upregulation of IL-7 synthesis and help in the development of B and T cells [78]. Ongoing investigations on the antimicrobial activity of plant bioactive compounds encourage the utilization of GL extract in the treatment of microbial infections, oxidative stress-related diseases, and unraveling more prophylactic compounds from GLs. An overall presentation of all the bioactivities demonstrated by GLs is presented in Table 4 and Figure 3.

Table 4. Biological activities of guava leaf (GL) extracts.

Origin of Leaves	Type of Extracts	Bioactive Compounds	Type of Cell Lines, Type of Study	Results	References
Anticancer activity					
Leaves from Chaudhry Wala, Punjab, (Pakistan)	Extracts obtained using methanol, chloroform, and hexane	Phenolics including flavonoids	Human carcinoma cell lines (SCC4, U266, and KBM5)	IC_{50} values of the leaf extracts ranged from 22.73 to 51.65 mg/mL (KBM5); 20.97 to 89.55 mg/mL (U266); 22.82 to 70.25 mg/mL (SCC4). Hexane extract demonstrated strong cytotoxic (IC_{50} value = 32.18 µg/mL) and antitumor (IC_{50} value = 65.02 µg/mL) properties. These extracts also inhibited TNF-α and instigated NF-κB activation in KBM5 cells	[9]
-	Ethanolic extract	Chlorophyll	Glioblastoma cells (U-118 MG), colorectal adenocarcinoma cells (Caco-2), hepatocellular carcinoma cells (HepG2), breast cancer cells (MDA-MB-231 and MCF7)	IC_{50} values of the leaf extracts were >200 µg/mL for Caco-2, HepG2, MDA-MB-231, MCF7 and 133.55 for U-118 MG, demonstrating their potential	[79]
Leaves from Yaoundé (Cameroon)	Ethanolic extract and essential oils	β-Sesquiphellandrene, α-humulene, nerolidol, 1,8-cineole, isodaucene, benzaldehyde, β-bisabolol, β-caryophyllene	Hepatocellular carcinoma cells (HepG2) and healthy human skin fibroblasts (CCD-45-SK)	The IC_{50} values for aqueous and ethanol extracts of guava leaves against CCD-45-SK were >0.1 mg/mL and 0.1 mg/mL for essential oils. The IC_{50} values for aqueous ethanol extracts and essential oils against HepG2 were 0.013, 0.0057, and 0.1, respectively	[80]
Antidiabetic activity					
Leaves from Bangladesh	Ethanolic extract	-	Wistar rats with alloxan-induced diabetes	Administration of guava leaf extract significantly reduced ($p < 0.05$) BGL at doses of 1.00 and 0.50 g/kg, as well as 0.75 g/kg in alloxan-induced diabetic Wistar rats ($p < 0.001$)	[12]
Leaves from Guangdong (China)	Ultrasound-assisted ethanolic extract	Polysaccharides	In vitro	Inhibited α-glucosidase activity and reduced the breakdown of glucose and prevented flatulence by not attenuating α-amylase activity	[16]
-	65% ethanol and ethyl acetate extract	Flavonoids (guaijaverin and avicularin)	Kunming mice with high-fat diet and streptozotocin-induced diabetes	GLF (200 mg/kg/day) not able to prevent loss of body weight, which indicated the inability to remove the damage induced by streptozotocin Hypoglycemic effect, improved glucose tolerance. Decreased TC, TG, LDL-C. Improved the insulin resistance and function of beta cell islets. Reduced liver and kidney index. Reduced liver viscera index by reducing the accumulation of lipids in the liver	[57]
Leaves from Natal Province, (Republic of South Africa)	Lyophilized water extract	-	Sprague Dawley male rats with streptozotocin-induced diabetes	Guava leaf extract (400 mg/kg/d) significantly decreased HSL activity in diabetic rat liver and adipose tissue, which was associated with increased levels of glycogen, decreased total cholesterol, serum triglycerides, LDL-C, and increased HDL-C.	[81]
Leaves from Ambohitantely (Madagascar)		-	Rat hepatoma (H4IIE cells), adipocyte-like cells (3T3-L1), skeletal muscle cells (C2C12)	IC_{50} values of the leaf extract of 1.0 ± 0.3 inhibited α-glucosidase activity and significantly increased the accumulation of triglycerides in 3T3-L1 cells. Results demonstrated the application of guava leaf extract in the treatment of type 2 diabetes	[82]

Table 4. *Cont.*

Origin of Leaves	Type of Extracts	Bioactive Compounds	Type of Cell Lines, Type of Study	Results	References
			Antioxidant activity		
-	Water extract	Low molecular weight polysaccharides (3.64 kDa)	In vitro antioxidant assays	IC_{50} values of 46.49 µg/mL, 175.52 µg/mL, and 102.82 µg/mL for DPPH, OH, and ABTS were recorded, respectively, all higher than that of ascorbic acid or Trolox	[15]
-	Water, methanol, and ethanol	Phenolics including flavonoids	In vitro DPPH assay	IC_{50} value was highest for ethanolic extract and the lowest for methanolic extract	[82]
			Antidiarrheal activity		
-	Water extract	-	In vivo with rats and mice	PGE (50–400 mg/kg p.o.) produced dose-dependent and significant protection of rats and mice against castor oil-induced diarrhea, inhibited intestinal transit, and delayed gastric emptying	[68]
-	Ethanolic extract	-	In vivo Wistar rats	Application of EEPGL at doses of 750 and 500 mg/kg showed antidiarrheal effect in castor oil-induced diarrheal model	[12]
			Antimicrobial activity		
-	Methanolic extract, water extract, extract of flavonoids	Alkaloids, saponins, anthraquinones, tannins, terpenes, flavonoids, coumarins	Antimicrobial activity of leaf extract was studied against *Bacillus subtilis*, *Staphylococcus aureus*, *Escherichia coli*, and *Salmonella typhi*	Methanolic extracts with minimum inhibitory concentration (MIC) of 5.5–11 mg/mL. Aqueous extract found to be least effective with MIC of 2–15 mg/mL. Flavonoid extract was the most effective against bacteria with MIC of 2.5–5 mg/mL	[83]
-	Ethanolic extract	-	Synergistic effect of zinc oxide nanoparticles and guava leaf extract for enhanced antimicrobial activity against enterotoxigenic *Escherichia coli*	Rifampicin (5 µg) zone of inhibition—28 mm Nanoparticle (concentration 128 µg/mL) zone of inhibition—24 mm Nanoparticle (concentration 128 µg/mL) + leaf extract zone of inhibition—20 mm	[84]

Abbreviations: KBM5—human chronic myelogenous leukemia; SCC4—human tongue squamous carcinoma cells; U266—human multiple myeloma cells; IC_{50}—half maximal inhibitory concentration; TNF-α—tumor necrosis factor alpha; NF-κB—nuclear factor kappa light chain enhancer of activated B cells; Caco-2—cancer coli-2; MCF7—Michigan Cancer Foundation-7; GLF—guava leaf flavonoid, HSL—hormone-sensitive lipase; TC—total cholesterol; TG—triglycerides; BGL—blood glucose level; LDL-C—low-density lipoprotein cholesterol; HDL-C—high-density lipoprotein cholesterol.

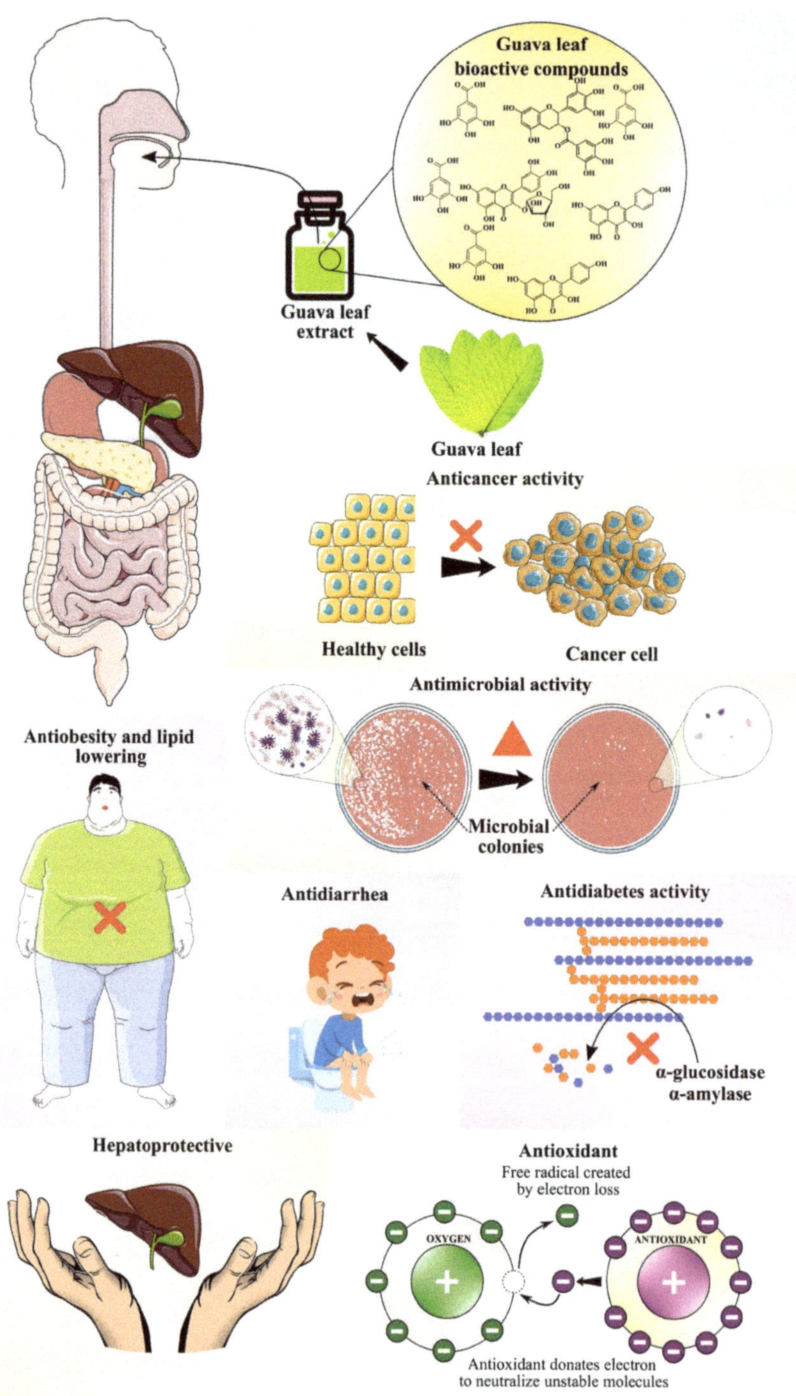

Figure 3. Various bioactivities of guava leaf extracts.

3.6. Hepatoprotective Properties

Liver lipid metabolism requires the activity of adenosine monophosphate-activated protein kinase (AMPK) and PPARα and rats treated with guava leaf extract demonstrated enhanced activity of both parameters. In addition, the guava leaf extracts could ameliorate hepatic insulin resistance. Alanine transaminase (ALT) and aspartate aminotransferase (AST) are associated with the functioning of the liver. Increases in their levels are an indication of fatty liver, which could be restricted with the administration of guava leaf extract [85]. Additionally, it has been found that diabetes has a close association with malfunctioning of the liver, including liver enlargement, steatosis, and fibrosis, as the primary function of the liver is to stabilize blood glucose levels. Any abnormality in the metabolism of glucose, lipid, and insulin is seen as a classic condition in type 2 diabetes mellitus. The bioactive compounds guaijaverin and avicularin present in guava leaves are potent inhibitors of dipeptidyl-peptidase IV and glucose transporter 4 (GLUT4)-mediated glucose uptake, respectively, responsible for raising the blood glucose levels [58,59]. Treatment with guava leaf extract with enhanced flavonoid levels promoted insulin resistance and restricted the rise in glucose as well as lipid levels in type 2 diabetes mellitus rats [57].

3.7. Antiobesity and Lipid-Lowering Activity

GLs are known to produce an antidiabetic effect and, therefore, are used for the treatment of diabetes [86]. Treating diabetic rats with 200 mg/kg body weight (b.w.) GLs caused a reduction in the blood glucose levels and promoted oral glucose tolerance, which is essential to prevent weight loss as a consequence of impaired carbohydrate metabolism. With the improved activity of hexokinase and G6PDH, and reduced activity of gluconeogenic enzymes and glucose-6-phosphatase, the insulin levels stabilized [87]. Similar findings were reported by [82].

Hypercholesterolemia, or high cholesterol levels in the blood, occurs as an effect of faulty dietary habits, genetics, or improper lifestyle. GLs contain many bioactive compounds with antioxidant properties which have health-promoting functions [88]. Wang et al. [40] reported the presence of flavonoids such as quercetin, kaempferol, guaijaverin, avicularin, myricetin, hyperin, and apigenin in guava leaf extract. Additionally, these flavonoids contributed to inhibitory action against α-glucosidase and α-amylase. It could be concluded from the study that the presence of the –OH group on the third position of the flavonoid was responsible for the inhibitory action. The inhibitory actions of myricetin, quercetin, and kaempferol against α-glucosidase and α-amylase were the highest, but a synergistic effect was quite evident. The presence of glycosides is essential to carry out the inhibitory function. Administration of ethanol extracts of guava leaves in the diet of rabbits brought about a significant reduction in the levels of serum triglycerides and low-density lipoprotein, with alleviated high-density lipoprotein levels [89]. Similar findings were reported in rats suffering from chronic diabetes along with hyperlipidemia [90].

4. GLs as a Functional Food Ingredient

Recent articles have shown that plant byproducts, such as fruit or vegetable pomace, seeds, husk/bran/seed coat, peel, and leaves, are important source of bioactive compounds and can be utilized as functional food ingredients [54,91–99]. Numerous reports suggest the beneficial effects of the inclusion of GL extract in food as a functional food ingredient, because of the presence of a myriad of compounds like rutin, naringenin, gallic acid, catechin, epicatechin, kaempferol, isoflavonoids, vitamins, citric acid, and flavonoids such as quercetin and guaijaverin, which are well known for their antimicrobial, antioxidant, and anti-inflammatory actions [100]. A study on the hypoglycemic effects of GL extract, due to the presence of its phenolic compounds, were shown to improve vascular dysfunction in mice with diet-induced obesity [101]. Recently, GL extract has been used in the preparation of jelly with pectin and was subjected to mass spectrometry analysis, which verified the presence of quercetin, gallocatechin, esculin, 3-sinapoylquinic acid, ellagic acid, gallic acid, and citric acid that are responsible for antioxidant and antimicrobial properties.

Additionally, the addition of GL did not cause any change in the texture properties of the jelly [102]. The potentiality of GL as a functional immunostimulant ingredient in fortified foods, owing to the presence of a high level of antioxidant and phenolic compounds, was also studied in detail [7]. Another study on the evaluation of food–drug interactions of guava leaf tea (GLT), which is a functional food and beverage that is commercially available in Japan, showed no possibility of interactions between GLT and medicines, indicating the safety of GLT in terms of food–drug interactions. Borderline diabetics, who are at high risk of the development of diabetes, take GLT to suppress a rapid increase in blood sugar level after meals. GLT consists of carbohydrate and dietary polyphenols which bind to digestive enzymes and are known to contribute to health through poor absorption of dietary sugar or lipids [103]. Furthermore, a recent report that studied herbal tea also stated that guava tea showed no interaction with medicine [104]. Another study on the addition of yellow strawberry GLs with abundant phenolic and flavonoid compounds in the diet of laying hens showed antimicrobial and antioxidant effects which could enhance the quality of eggs through the mechanism of inhibiting the pathways of the enzyme cyclooxygenase (COX), which plays a fundamental role as an inflammatory mediator [105]. In addition, the natural antioxidants present in GLs after fortification in fresh pork sausage were found to be effective in decelerating the process of lipid oxidation in the fresh pork sausage [63]. These examples indicate that GL is an excellent source of active compounds for functional ingredient additives in foods, without altering the rheological and sensory properties.

5. Conclusions and Future Perspectives

GLs are documented as a source of natural compounds that are readily available. GL extracts have been extensively studied for their high levels of antioxidant, anticancer, hypoglycemic, and other biological activities. The rich presence of minerals and proteins, as well as vitamins, in GLs promote their utilization as a direct source of nutrients. The presence of numerous bioactive chemical compounds in GLs have been reported to enhance and stabilize different physiological and metabolic functions in the human body. GL also contains many secondary metabolites, such as flavonoids, triterpenoids, sesquiterpenes, glycosides, alkaloids, saponins, and other phenolic compounds. These compounds play a key role as immune stimulators and modulators of chronic diseases including diabetes, cancer, and gastrointestinal, neurodegenerative, and cardiovascular diseases. GL essential oil also has antioxidant, antimicrobial, and antiproliferative activity. GL extracts that contain high concentrations of vitamin E, flavone (apigenin), or β-caryophyllene show significant antiproliferative activity against colon carcinoma and various forms of human cancer. GL therefore has peculiar characteristics and pharmaceutical and medicinal profiles that promote diverse applications as an essential plant component in medicinal research and a low-cost ingredient in foodstuffs. Furthermore, as an ingredient of plant origin, GL may help in mitigate drug resistance, which is a major problem for the pharmaceutical industry and also can be utilized as a functional food ingredient, which are in high demand. Thus, guava extract with its multiple medicinal properties needs to be further developed for wider applicability. In future studies, the identification and isolation of new chemical components for the development of specific products will be the key research area.

Author Contributions: Conceptualization, M.K., R.A., M.T., and V.S. (Vivek Saurabh); writing—original draft preparation, M.S.N., C.M., M.S., U.P., M.H., S.C., and S.S.; writing—review and editing, R.K.P., M.K.B., and V.S. (Varsha Satankar). All authors have read and agreed to the published version of the manuscript.

Funding: This research received no external funding.

Conflicts of Interest: The authors declare no conflict of interest.

References

1. Newman, D.J.; Cragg, G.M.; Snader, K.M. Natural products as sources of new drugs over the period 1981–2002. *J. Nat. Prod.* **2003**, *66*, 1022–1037. [CrossRef] [PubMed]
2. Kumar, M.; Saurabh, V.; Tomar, M.; Hasan, M.; Changan, S.; Sasi, M.; Maheshwari, C.; Prajapati, U.; Singh, S.; Prajapat, R.K.; et al. Mango (*Mangifera indica* L.) leaves: Nutritional composition, phytochemical profile, and health-promoting bioactivities. *Antioxidants* **2021**, *10*, 299. [CrossRef]
3. Sharma, A.; del Carmen Flores-Vallejo, R.; Cardoso-Taketa, A.; Villarreal, M.L. Antibacterial activities of medicinal plants used in Mexican traditional medicine. *J. Ethnopharmacol.* **2017**, *208*, 264–329. [CrossRef]
4. Amat-ur-Rasool, H.; Symes, F.; Tooth, D.; Schaffert, L.N.; Elmorsy, E.; Ahmed, M.; Hasnain, S.; Carter, W.G. Potential nutraceutical properties of leaves from several commonly cultivated plants. *Biomolecules* **2020**, *10*, 1556. [CrossRef]
5. Mannino, G.; Gentile, C.; Porcu, A.; Agliassa, C.; Caradonna, F.; Bertea, C.M. Chemical profile and biological activity of cherimoya (*Annona cherimola* Mill.) and atemoya (*Annona atemoya*) leaves. *Molecules* **2020**, *25*, 2612. [CrossRef]
6. Mateos-Maces, L.; Chávez-Servia, J.L.; Vera-Guzmán, A.M.; Aquino-Bolaños, E.N.; Alba-Jiménez, J.E.; Villagómez-González, B.B. Edible leafy plants from Mexico as sources of antioxidant compounds, and their nutritional, nutraceutical and antimicrobial potential: A review. *Antioxidants* **2020**, *9*, 541. [CrossRef]
7. Laily, N.; Kusumaningtyas, R.W.; Sukarti, I.; Rini, M.R.D.K. The potency of guava *Psidium guajava* (L.) leaves as a functional immunostimulatory ingredient. *Procedia Chem.* **2015**, *14*, 301–307. [CrossRef]
8. Chen, H.Y.; Yen, G.C. Antioxidant activity and free radical-scavenging capacity of extracts from guava (*Psidium guajava* L.) leaves. *Food Chem.* **2007**, *101*, 686–694. [CrossRef]
9. Ashraf, A.; Sarfraz, R.A.; Rashid, M.A.; Mahmood, A.; Shahid, M.; Noor, N. Chemical composition, antioxidant, antitumor, anticancer and cytotoxic effects of *Psidium guajava* leaf extracts. *Pharm. Biol.* **2016**, *54*, 1971–1981. [CrossRef] [PubMed]
10. Jiang, L.; Lu, J.; Qin, Y.; Jiang, W.; Wang, Y. Antitumor effect of guava leaves on lung cancer: A network pharmacology study. *Arab. J. Chem.* **2020**, *13*, 7773–7797. [CrossRef]
11. Dewi, P.S.; Sutjiatmo, A.B.; Nurdiansyah, A. Antidiarrheal activity of water extracts of guava leaves (*Psidium guajava* L.) and water extracts of green tea leaves (*Camellia sinensis* L.) combination in Swiss Webster mice. *Acta Pharm. Indones.* **2013**, *38*, 67–70.
12. Mazumdar, S.; Akter, R.; Talukder, D. Antidiabetic and antidiarrhoeal effects on ethanolic extract of *Psidium guajava* (L.) Bat. leaves in Wister rats. *Asian Pac. J. Trop. Biomed.* **2015**, *5*, 10–14. [CrossRef]
13. Farag, R.S.; Abdel-Latif, M.S.; Abd El Baky, H.H.; Tawfeek, L.S. Phytochemical screening and antioxidant activity of some medicinal plants' crude juices. *Biotechnol. Rep.* **2020**, *28*, e00536. [CrossRef]
14. Shabbir, H.; Kausar, T.; Noreen, S.; Hussain, A.; Huang, Q.; Gani, A.; Su, S.; Nawaz, A. In vivo screening and antidiabetic potential of polyphenol extracts from guava pulp, seeds and leaves. *Animals* **2020**, *10*, 1714. [CrossRef]
15. Luo, Y.; Peng, B.; Wei, W.; Tian, X.; Wu, Z. Antioxidant and anti-diabetic activities of polysaccharides from guava leaves. *Molecules* **2019**, *24*, 1343. [CrossRef] [PubMed]
16. Luo, Y.; Peng, B.; Liu, Y.; Wu, Y.; Wu, Z. Ultrasound extraction of polysaccharides from guava leaves and their antioxidant and antiglycation activity. *Process. Biochem.* **2018**, *73*, 228–234. [CrossRef]
17. Yu, S.; Feng, Z.; Wu, X.; Kong, F. Optimization of ultrasonic-assisted extraction of antioxidant compounds from Guava (*Psidium guajava* L.) leaves using response surface methodology. *Pharmacogn. Mag.* **2015**, *11*, 463. [CrossRef] [PubMed]
18. Kim, S.Y.; Kim, E.A.; Kim, Y.S.; Yu, S.K.; Choi, C.; Lee, J.S.; Kim, Y.T.; Nah, J.W.; Jeon, Y.J. Protective effects of polysaccharides from *Psidium guajava* leaves against oxidative stresses. *Int. J. Biol. Macromol.* **2016**, *91*, 804–811. [CrossRef] [PubMed]
19. Zhang, Z.; Kong, F.; Ni, H.; Mo, Z.; Wan, J.-B.; Hua, D.; Yan, C. Structural characterization, α-glucosidase inhibitory and DPPH scavenging activities of polysaccharides from guava. *Carbohydr. Polym.* **2016**, *144*, 106–114. [CrossRef] [PubMed]
20. Dutta, P.; Kundu, S.; Bauri, F.K.; Talang, H.; Majumder, D. Effect of bio-fertilizers on physico-chemical qualities and leaf mineral composition of guava grown in alluvial zone of West Bengal. *J. Crop Weed* **2014**, *10*, 268–271.
21. Jassal, K.; Kaushal, S. Phytochemical and antioxidant screening of guava (*Psidium guajava*) leaf essential oil. *Agric. Res. J.* **2019**, *56*, 528. [CrossRef]
22. Rahman, Z.; Siddiqui, M.N.; Khatun, M.A.; Kamruzzaman, M. Effect of guava (*Psidium guajava*) leaf meal on production performances and antimicrobial sensitivity in commercial broiler. *J. Nat. Prod.* **2013**, *6*, 177–187.
23. Thomas, L.A.T.; Anitha, T.; Lasyaja, A.B.; Suganya, M.; Gayathri, P.; Chithra, S. Biochemical and mineral analysis of the undervalued leaves—*Psidium guajava* L. *Int. J. Adv. Sci. Res.* **2017**, *2*, 16–21.
24. Alberts, B.; Johnson, A.; Lewis, J.; Raff, M.; Roberts, K.; Walter, P. Protein Function. In *Molecular Biology of the Cell*, 4th ed.; Garland Science: New York, NY, USA, 2002.
25. Lonnie, M.; Hooker, E.; Brunstrom, J.; Corfe, B.; Green, M.; Watson, A.; Williams, E.; Stevenson, E.; Penson, S.; Johnstone, A. Protein for life: Review of optimal protein intake, sustainable dietary sources and the effect on appetite in ageing adults. *Nutrients* **2018**, *10*, 360. [CrossRef]
26. Adrian, J.A.L.; Arancon, N.Q.; Mathews, B.W.; Carpenter, J.R. Mineral composition and soil-plant relationships for common guava (*Psidium guajava* L.) and yellow strawberry guava (Psidium cattleianum var. Lucidum) tree parts and fruits. *Commun. Soil Sci. Plant Anal.* **2015**, *46*, 1960–1979. [CrossRef]
27. Lee, W.C.; Mahmud, R.; Pillai, S.; Perumal, S.; Ismail, S. Antioxidant activities of essential oil of *Psidium guajava* L. leaves. *APCBEE Procedia* **2012**, *2*, 86–91. [CrossRef]

28. Sacchetti, G.; Maietti, S.; Muzzoli, M.; Scaglianti, M.; Manfredini, S.; Radice, M.; Bruni, R. Comparative evaluation of 11 essential oils of different origin as functional antioxidants, antiradicals and antimicrobials in foods. *Food Chem.* **2005**, *91*, 621–632. [CrossRef]

29. Smith, R.M.; Oliveros-Belardo, L. The composition of leaf essential oils of *Psidium guajava* L. from Manila, Philippines. *Asian J. Pharm.* **1977**, *3*, 5–9.

30. Khadhri, A.; El Mokni, R.; Almeida, C.; Nogueira, J.M.F.; Araújo, M.E.M. Chemical composition of essential oil of *Psidium guajava* L. growing in Tunisia. *Ind. Crop. Prod.* **2014**, *52*, 29–31. [CrossRef]

31. Soliman, F.M.; Fathy, M.M.; Salama, M.M.; Saber, F.R. Comparative study of the volatile oil content and antimicrobial activity of *Psidium guajava* L. and Psidium cattleianum Sabine leaves. *Bull. Fac. Pharm. Cairo Univ.* **2016**, *54*, 219–225. [CrossRef]

32. El-Ahmady, S.H.; Ashour, M.L.; Wink, M. Chemical composition and anti-inflammatory activity of the essential oils of *Psidium guajava* fruits and leaves. *J. Essent. Oil Res.* **2013**, *25*, 475–481. [CrossRef]

33. Díaz-de-Cerio, E.; Gómez-Caravaca, A.M.; Verardo, V.; Fernández-Gutiérrez, A.; Segura-Carretero, A. Determination of guava (*Psidium guajava* L.) leaf phenolic compounds using HPLC-DAD-QTOF-MS. *J. Funct. Foods* **2016**, *22*, 376–388. [CrossRef]

34. Matsuzaki, K.; Ishii, R.; Kobiyama, K.; Kitanaka, S. New benzophenone and quercetin galloyl glycosides from *Psidium guajava* L. *J. Nat. Med.* **2010**, *64*, 252–256. [CrossRef] [PubMed]

35. Shu, J.C.; Chou, G.X.; Wang, Z.T. One new diphenylmethane glycoside from the leaves of *Psidium guajava* L. *Nat. Prod. Res.* **2012**, *26*, 1971–1975. [CrossRef]

36. Shao, M.; Wang, Y.; Liu, Z.; Zhang, D.M.; Cao, H.H.; Jiang, R.W.; Fan, C.L.; Zhang, X.Q.; Chen, H.R.; Yao, X.S.; et al. Psiguadials A and B, two novel meroterpenoids with unusual skeletons from the leaves of *Psidium guajava*. *Org. Lett.* **2010**, *12*, 5040–5043. [CrossRef] [PubMed]

37. Shao, M.; Wang, Y.; Huang, X.J.; Fan, C.L.; Zhang, Q.W.; Zhang, X.Q.; Ye, W.C. Four new triterpenoids from the leaves of *Psidium guajava*. *J. Asian Nat. Prod. Res.* **2012**, *14*, 348–354. [CrossRef] [PubMed]

38. Rasouli, H.; Farzaei, M.H.; Khodarahmi, R. Polyphenols and their benefits: A review. *Int. J. Food Prop.* **2017**, *20*, 1–42. [CrossRef]

39. Luca, S.V.; Macovei, I.; Bujor, A.; Miron, A.; Skalicka-Woźniak, K.; Aprotosoaie, A.C.; Trifan, A. Bioactivity of dietary polyphenols: The role of metabolites. *Crit. Rev. Food Sci. Nutr.* **2020**, *60*, 626–659. [CrossRef]

40. Wang, H.; Du, Y.J.; Song, H.C. α-Glucosidase and α-amylase inhibitory activities of guava leaves. *Food Chem.* **2010**, *123*, 6–13. [CrossRef]

41. Liu, C.W.; Wang, Y.C.; Lu, H.C.; Chiang, W.D. Optimization of ultrasound-assisted extraction conditions for total phenols with anti-hyperglycemic activity from *Psidium guajava* leaves. *Process. Biochem.* **2014**, *49*, 1601–1605. [CrossRef]

42. Wang, L.; Wu, Y.; Bei, Q.; Shi, K.; Wu, Z. Fingerprint profiles of flavonoid compounds from different *Psidium guajava* leaves and their antioxidant activities. *J. Sep. Sci.* **2017**, *40*, 3817–3829. [CrossRef]

43. Wang, L.; Bei, Q.; Wu, Y.; Liao, W.; Wu, Z. Characterization of soluble and insoluble-bound polyphenols from *Psidium guajava* L. leaves co-fermented with *Monascus anka* and *Bacillus sp.* and their bio-activities. *J. Funct. Foods* **2017**, *32*, 149–159. [CrossRef]

44. Díaz-de-Cerio, E.; Pasini, F.; Verardo, V.; Fernández-Gutiérrez, A.; Segura-Carretero, A.; Caboni, M.F. *Psidium guajava* L. leaves as source of proanthocyanidins: Optimization of the extraction method by RSM and study of the degree of polymerization by NP-HPLC-FLD-ESI-MS. *J. Pharm. Biomed. Anal.* **2017**, *133*, 1–7. [CrossRef] [PubMed]

45. Dorenkott, M.R.; Griffin, L.E.; Goodrich, K.M.; Thompson-Witrick, K.A.; Fundaro, G.; Ye, L.; Stevens, J.R.; Ali, M.; O'Keefe, S.F.; Hulver, M.W.; et al. Oligomeric cocoa procyanidins possess enhanced bioactivity compared to monomeric and polymeric cocoa procyanidins for preventing the development of obesity, insulin resistance, and impaired glucose tolerance during high-fat feeding. *J. Agric. Food Chem.* **2014**, *62*, 2216–2227. [CrossRef] [PubMed]

46. Toyokuni, S. Oxidative stress as an iceberg in carcinogenesis and cancer biology. *Arch. Biochem. Biophys.* **2016**, *595*, 46–49. [CrossRef] [PubMed]

47. Gonzalez, H.; Hagerling, C.; Werb, Z. Roles of the immune system in cancer: From tumor initiation to metastatic progression. *Genes Dev.* **2018**, *32*, 1267–1284. [CrossRef]

48. Biswas, S.; Talukdar, P.; Talapatra, D.S.N. Presence of phytochemicals in fruits and leaves of guava (*Psidium guajava* Linn.) for cancer prevention: A mini review. *J. Drug Deliv. Ther.* **2019**, *9*, 726–729. [CrossRef]

49. Correa, M.G.; Couto, J.S.; Teodoro, A.J. Anticancer properties of *Psidium guajava*—A mini-review. *Asian Pac. J. Cancer Prev.* **2016**, *17*, 4199–4204.

50. Lok, B.; Sandai, D.; Baharetha, H.; Nazari, V.; Asif, M.; Tan, C.; Majid, A. Anticancer effect of *Psidium guajava* (guava) leaf extracts against colorectal cancer through inhibition of angiogenesis. *Asian Pac. J. Trop. Biomed.* **2020**, *10*, 293. [CrossRef]

51. Bazioli, J.M.; Costa, J.H.; Shiozawa, L.; Ruiz, A.L.T.G.; Foglio, M.A.; de Carvalho, J.E. Anti-estrogenic activity of guajadial fraction, from guava leaves (*Psidium guajava* L.). *Molecules* **2020**, *25*, 1525. [CrossRef]

52. Huang, B.; Luo, N.; Wu, X.; Xu, Z.; Wang, X.; Pan, X. The modulatory role of low concentrations of bisphenol A on tamoxifen-induced proliferation and apoptosis in breast cancer cells. *Environ. Sci. Pollut. Res.* **2019**, *26*, 2353–2362. [CrossRef] [PubMed]

53. Zhu, X.; Ouyang, W.; Pan, C.; Gao, Z.; Han, Y.; Song, M.; Feng, K.; Xiao, H.; Cao, Y. Identification of a new benzophenone from: *Psidium guajava* L. leaves and its antineoplastic effects on human colon cancer cells. *Food Funct.* **2019**, *10*, 4189–4198. [CrossRef]

54. Punia, S.; Kumar, M. Litchi (*Litchi chinenis*) seed: Nutritional profile, bioactivities, and its industrial applications. *Trends Food Sci. Technol.* **2021**, *108*, 58–70. [CrossRef]

55. Cho, N.H.; Shaw, J.E.; Karuranga, S.; Huang, Y.; da Rocha Fernandes, J.D.; Ohlrogge, A.W.; Malanda, B. IDF Diabetes Atlas: Global estimates of diabetes prevalence for 2017 and projections for 2045. *Diabetes Res. Clin. Pract.* **2018**, *138*, 271–281. [CrossRef] [PubMed]

56. Hu, X.F.; Zhang, Q.; Zhang, P.P.; Sun, L.J.; Liang, J.C.; Morris-Natschke, S.L.; Chen, Y.; Lee, K.H. Evaluation of in vitro/in vivo anti-diabetic effects and identification of compounds from Physalis alkekengi. *Fitoterapia* **2018**, *127*, 129–137. [CrossRef] [PubMed]

57. Zhu, X.; Ouyang, W.; Lan, Y.; Xiao, H.; Tang, L.; Liu, G.; Feng, K.; Zhang, L.; Song, M.; Cao, Y. Anti-hyperglycemic and liver protective effects of flavonoids from *Psidium guajava* L. (guava) leaf in diabetic mice. *Food Biosci.* **2020**, *35*, 100574. [CrossRef]

58. Eidenberger, T.; Selg, M.; Krennhuber, K. Inhibition of dipeptidyl peptidase activity by flavonol glycosides of guava (*Psidium guajava* L.): A key to the beneficial effects of guava in type II diabetes mellitus. *Fitoterapia* **2013**, *89*, 74–79. [CrossRef]

59. Fujimori, K.; Shibano, M. Avicularin, a plant flavonoid, suppresses lipid accumulation through repression of C/EBPα-activated GLUT4-mediated glucose uptake in 3T3-L1 cells. *J. Agric. Food Chem.* **2013**, *61*, 5139–5147. [CrossRef]

60. Nair, S.S.; Kavrekar, V.; Mishra, A. In vitro studies on alpha amylase and alpha glucosidase inhibitory activities of selected plant extracts. *Eur. J. Exp. Biol.* **2013**, *3*, 128–132.

61. Stefanis, L.; Burke, R.E.; Greene, L.A. Apoptosis in neurodegenerative disorders. *Curr. Opin. Neurol.* **1997**, *10*, 299–305. [CrossRef]

62. Taha, T.F.; Elakkad, H.A.; Gendy, A.S.H.; Abdelkader, M.A.I.; Hussein, S.S.E. In vitro bio-medical studies on *Psidium guajava* leaves. *Plant Arch.* **2019**, *19*, 199–207.

63. Tran, T.T.T.; Ton, N.M.N.; Nguyen, T.T.; Le, V.V.M.; Sajeev, D.; Schilling, M.W.; Dinh, T.T.N. Application of natural antioxidant extract from guava leaves (*Psidium guajava* L.) in fresh pork sausage. *Meat Sci.* **2020**, *165*, 108106. [CrossRef] [PubMed]

64. Wang, L.; Xie, J.; Huang, T.; Ma, Y.; Wu, Z. Characterization of silver nanoparticles biosynthesized using crude polysaccharides of *Psidium guajava* L. leaf and their bioactivities. *Mater. Lett.* **2017**, *208*, 126–129. [CrossRef]

65. Kim, H.-S. Do not put too much value on conventional medicines. *J. Ethnopharmacol.* **2005**, *100*, 37–39. [CrossRef] [PubMed]

66. Palombo, E.A. Phytochemicals from traditional medicinal plants used in the treatment of diarrhoea: Modes of action and effects on intestinal function. *Phyther. Res.* **2006**, *20*, 717–724. [CrossRef]

67. Mehra, A.; Sarkar, S.; Basu, D. Lomotil (diphenoxylate) dependence in India. *Indian J. Psychol. Med.* **2013**, *35*, 248–250. [CrossRef] [PubMed]

68. Ojewole, J.A.O.; Awe, E.O.; Chiwororo, W.D.H. Antidiarrhoeal activity of *Psidium guajava* Linn. (Myrtaceae) leaf aqueous extract in rodents. *J. Smooth Muscle Res.* **2008**, *44*, 195–207. [CrossRef] [PubMed]

69. Ullah, F.; Ayaz, M.; Sadiq, A.; Ullah, F.; Hussain, I.; Shahid, M.; Yessimbekov, Z.; Adhikari-Devkota, A.; Devkota, H.P. Potential role of plant extracts and phytochemicals against foodborne pathogens. *Appl. Sci.* **2020**, *10*, 4597. [CrossRef]

70. Mickymaray, S. Efficacy and mechanism of traditional medicinal plants and bioactive compounds against clinically important pathogens. *Antibiotics* **2019**, *8*, 257. [CrossRef]

71. Naseer, S.; Hussain, S.; Naeem, N.; Pervaiz, M.; Rahman, M. The phytochemistry and medicinal value of *Psidium guajava* (guava). *Clin. Phytosci.* **2018**, *4*, 32. [CrossRef]

72. Das, M.; Goswami, S. Antifungal and antibacterial property of guava (*Psidium guajava*) leaf extract: Role of phytochemicals. *Int. J. Health Sci. Res.* **2019**, *9*, 39.

73. Hirudkar, J.R.; Parmar, K.M.; Prasad, R.S.; Sinha, S.K.; Jogi, M.S.; Itankar, P.R.; Prasad, S.K. Quercetin a major biomarker of *Psidium guajava* L. inhibits SepA protease activity of Shigella flexneri in treatment of infectious diarrhoea. *Microb. Pathog.* **2020**, *138*, 103807. [CrossRef]

74. Ghosh, P.; Mandal, A.; Chakraborty, P.; Rasul, M.G.; Chakraborty, M.; Saha, A. Triterpenoids from *Psidium guajava* with biocidal activity. *Indian J. Pharm. Sci.* **2010**, *72*, 504–507. [CrossRef]

75. Dhiman, A.; Nanda, A.; Ahmad, S.; Narasimhan, B. In vitro antimicrobial activity of methanolic leaf extract of *Psidium guajava* L. *J. Pharm. Bioallied Sci.* **2011**, *3*, 226–229. [CrossRef] [PubMed]

76. Almulaiky, Y.; Zeyadi, M.; Saleh, R.; Baothman, O.; Al-shawafi, W.; Al-Talhi, H. Assessment of antioxidant and antibacterial properties in two types of Yemeni guava cultivars. *Biocatal. Agric. Biotechnol.* **2018**, *16*, 90–97. [CrossRef]

77. Bose, D.; Chatterjee, S. Biogenic synthesis of silver nanoparticles using guava (*Psidium guajava*) leaf extract and its antibacterial activity against Pseudomonas aeruginosa. *Appl. Nanosci.* **2016**, *6*, 895–901. [CrossRef]

78. Comber, J.D.; Bamezai, A.K. In vitro derivation of interferon-γ producing, IL-4 and IL-7 responsive memory-like CD4[+] T cells. *Vaccine* **2012**, *30*, 2140–2145. [CrossRef]

79. Wang, Y.T.; Yang, C.H.; Huang, T.Y.; Tai, M.H.; Sie, R.H.; Shaw, J.F. Cytotoxic effects of chlorophyllides in ethanol crude extracts from plant leaves. *Evid.-Based Complement. Altern. Med.* **2019**, *2019*, 9494328. [CrossRef] [PubMed]

80. Kemegne, G.A.; Bettache, N.; Nyegue, M.A.; Etoa, F.-X.; Menut, C. Cytotoxic activities of *Psidium guajava* and Mangifera indica plant extracts on human healthy skin fibroblasts and human hepatocellular carcinoma. *Issues Biol. Sci. Pharm. Res.* **2020**, *8*, 58–64. [CrossRef]

81. Tella, T.; Masola, B.; Mukaratirwa, S. The effect of *Psidium guajava* aqueous leaf extract on liver glycogen enzymes, hormone sensitive lipase and serum lipid profile in diabetic rats. *Biomed. Pharmacother.* **2019**, *109*, 2441–2446. [CrossRef] [PubMed]

82. Beidokhti, M.N.; Eid, H.M.; Villavicencio, M.L.S.; Jäger, A.K.; Lobbens, E.S.; Rasoanaivo, P.R.; McNair, L.M.; Haddad, P.S.; Staerk, D. Evaluation of the antidiabetic potential of *Psidium guajava* L. (Myrtaceae) using assays for α-glucosidase, α-amylase, muscle glucose uptake, liver glucose production, and triglyceride accumulation in adipocytes. *J. Ethnopharmacol.* **2020**, *257*, 112877. [CrossRef] [PubMed]

83. Seddiek, A.S.; Hamad, G.M.; Zeitoun, A.A.; Zeitoun, M.A.M.; Ali, S. Antimicrobial and antioxidant activity of some plant extracts against different food spoilage and pathogenic microbes. *Eur. J. Nutr. Food Saf.* **2020**, *12*, 1–12. [CrossRef]
84. Naili, M.; Errayes, A.; Alghazeer, R. Evaluation of antimicrobial and antioxidant activities of *Psidium guajava* L. growing in Libya. *Int. J. Adv. Biol. Biomed. Res.* **2020**, *8*, 419–428. [CrossRef]
85. Yoshitomi, H.; Guo, X.; Liu, T.; Gao, M. Guava leaf extracts alleviate fatty liver via expression of adiponectin receptors in SHRSP.Z-Leprfa/Izm rats. *Nutr. Metab.* **2012**, *9*, 13. [CrossRef] [PubMed]
86. Schmid, R.; Van Wyk, B.-E.; Van Oudtshoorn, B.; Gericke, N. Medicinal Plants of South Africa. *TAXON* **1998**, *47*, 787. [CrossRef]
87. Vinayagam, R.; Jayachandran, M.; Chung, S.S.M.; Xu, B. Guava leaf inhibits hepatic gluconeogenesis and increases glycogen synthesis via AMPK/ACC signaling pathways in streptozotocin-induced diabetic rats. *Biomed. Pharmacother.* **2018**, *103*, 1012–1017. [CrossRef] [PubMed]
88. Jiménez-Escrig, A.; Rincón, M.; Pulido, R.; Saura-Calixto, F. Guava fruit (*Psidium guajava* L.) as a new source of antioxidant dietary fiber. *J. Agric. Food Chem.* **2001**, *49*, 5489–5493. [CrossRef] [PubMed]
89. Olaniyan, M.F. Cholesterol lowering effect of guava leaves (*Psidium guajava*) extract on egg yolk induced hypercholesterolaemic rabbits. *J. Biol. Nat.* **2017**, *7*, 24–28.
90. Bahrani, A.H.M.; Zaheri, H.; Soltani, N.; Kharazmi, F.; Keshavarz, M.; Kamalinajad, M. Effect of the administration of Psidium guava leaves on blood glucose, lipid profiles and sensitivity of the vascular mesenteric bed to Phenylephrine in streptozotocin-induced diabetic rats. *J. Diabetes Mellit.* **2012**, *2*, 138–145. [CrossRef]
91. Kumar, M.; Dahuja, A.; Sachdev, A.; Kaur, C.; Varghese, E.; Saha, S.; Sairam, K.V.S.S. Valorisation of black carrot pomace: Microwave assisted extraction of bioactive phytoceuticals and antioxidant activity using Box–Behnken design. *J. Food Sci. Technol.* **2019**, *56*, 995–1007. [CrossRef]
92. Kumar, M.; Dahuja, A.; Sachdev, A.; Kaur, C.; Varghese, E.; Saha, S.; Sairam, K.V.S.S. Evaluation of enzyme and microwave-assisted conditions on extraction of anthocyanins and total phenolics from black soybean (*Glycine max* L.) seed coat. *Int. J. Biol. Macromol.* **2019**, *135*, 1070–1081. [CrossRef]
93. Kumar, M.; Dahuja, A.; Sachdev, A.; Kaur, C.; Varghese, E.; Saha, S.; Sairam, K.V.S.S. Black carrot (*Daucus carota* ssp.) and black soybean (*Glycine max* (L.) Merr.) anthocyanin extract: A remedy to enhance stability and functionality of fruit juices by copigmentation. *Waste Biomass Valoriz.* **2020**, *11*, 99–108. [CrossRef]
94. Kumar, M.; Tomar, M.; Punia, S.; Amarowicz, R.; Kaur, C. Evaluation of cellulolytic enzyme-assisted microwave extraction of Punica granatum peel phenolics and antioxidant activity. *Plant Foods Hum. Nutr.* **2020**, *75*, 614–620. [CrossRef]
95. Nishad, J.; Dutta, A.; Saha, S.; Rudra, S.G.; Varghese, E.; Sharma, R.R.; Tomar, M.; Kumar, M.; Kaur, C. Ultrasound-assisted development of stable grapefruit peel polyphenolic nano-emulsion: Optimization and application in improving oxidative stability of mustard oil. *Food Chem.* **2021**, *334*, 127561. [CrossRef]
96. Kumar, M.; Potkule, J.; Patil, S.; Saxena, S.; Patil, P.G.; Mageshwaran, V.; Punia, S.; Varghese, E.; Mahapatra, A.; Ashtaputre, N.; et al. Extraction of ultra-low gossypol protein from cottonseed: Characterization based on antioxidant activity, structural morphology and functional group analysis. *LWT* **2021**, *140*, 110692. [CrossRef]
97. Kumar, M.; Potkule, J.; Tomar, M.; Punia, S.; Singh, S.; Patil, S.; Singh, S.; Ilakiya, T.; Kaur, C.; Kennedy, J.F. Jackfruit seed slimy sheath, a novel source of pectin: Studies on antioxidant activity, functional group, and structural morphology. *Carbohydr. Polym. Technol. Appl.* **2021**, *2*, 100054. [CrossRef]
98. Kumar, M.; Dahuja, A.; Tiwari, S.; Punia, S.; Tak, Y.; Amarowicz, R.; Bhoite, A.G.; Singh, S.; Joshi, S.; Panesar, P.S.; et al. Recent trends in extraction of plant bioactives using green technologies: A review. *Food Chem.* **2021**, *353*, 129431. [CrossRef] [PubMed]
99. Punia, S.; Sandhu, K.S.; Grasso, S.; Purewal, S.S.; Kaur, M.; Siroha, A.K.; Kumar, K.; Kumar, V.; Kumar, M. Aspergillus oryzae fermented rice bran: A byproduct with enhanced bioactive compounds and antioxidant potential. *Foods* **2020**, *10*, 70. [CrossRef]
100. Shaheena, S.; Chintagunta, A.D.; Dirisala, V.R.; Kumar, N.S.S. Extraction of bioactive compounds from *Psidium guajava* and their application in dentistry. *AMB Express* **2019**, *9*, 1–9. [CrossRef]
101. Díaz-de-Cerio, E.; Rodríguez-Nogales, A.; Algieri, F.; Romero, M.; Verardo, V.; Segura-Carretero, A.; Duarte, J.; Galvez, J. The hypoglycemic effects of guava leaf (*Psidium guajava* L.) extract are associated with improving endothelial dysfunction in mice with diet-induced obesity. *Food Res. Int.* **2017**, *96*, 64–71. [CrossRef] [PubMed]
102. Kumar, N.S.S.; Sarbon, N.M.; Rana, S.S.; Chintagunta, A.D.; Prathibha, S.; Ingilala, S.K.; Kumar, S.P.J.; Anvesh, B.S.; Dirisala, V.R. Extraction of bioactive compounds from *Psidium guajava* leaves and its utilization in preparation of jellies. *AMB Express* **2021**, *11*, 36. [CrossRef] [PubMed]
103. Kaneko, K.; Suzuki, K.; Iwadate-Iwata, E.; Kato, I.; Uchida, K.; Onoue, M. Evaluation of food-drug interaction of guava leaf tea. *Phyther. Res.* **2013**, *27*, 299–305. [CrossRef] [PubMed]
104. Matsuda, K.; Nishimura, Y.; Kurata, N.; Iwase, M.; Yasuhara, H. Effects of continuous ingestion of herbal teas on intestinal CYP3A in the rat. *J. Pharmacol. Sci.* **2007**, *103*, 214–221. [CrossRef] [PubMed]
105. Dos Santos, A.F.A.; Da Silva, A.S.; Galli, G.M.; Paglia, E.B.; Dacoreggio, M.V.; Kempka, A.P.; Souza, C.F.; Baldissera, M.D.; da Rosa, G.; Boiago, M.M.; et al. Addition of yellow strawberry guava leaf extract in the diet of laying hens had antimicrobial and antioxidant effect capable of improving egg quality. *Biocatal. Agric. Biotechnol.* **2020**, *29*, 101788. [CrossRef]

 *foods*

Review

Autoclaved and Extruded Legumes as a Source of Bioactive Phytochemicals: A Review

Mercedes M. Pedrosa [1,*], **Eva Guillamón** [2] and **Claudia Arribas** [1]

[1] Food Technology Department, SGIT-INIA, Ctra de La Coruña, Km 7.5, 28040 Madrid, Spain; arribas.claudia@inia.es

[2] Centre for the Food Quality, INIA, C/Universidad s/n, 42004 Soria, Spain; guillamon.eva@inia.es

* Correspondence: mmartin@inia.es

Abstract: Legumes have been consumed since ancient times all over the world due to their easy cultivation and availability as a low-cost food. Nowadays, it is well known that pulses are also a good source of bioactive phytochemicals that play an important role in the health and well-being of humans. Pulses are mainly consumed after processing to soften cotyledons and to improve their nutritive and sensorial characteristics. However, processing affects not only their nutritive constituents, but also their bioactive compounds. The final content of phytochemicals depends on the pulse type and variety, the processing method and their parameters (mainly temperature and time), the food matrix structure and the chemical nature of each phytochemical. This review focuses on the changes produced in the bioactive-compound content of pulses processed by a traditional processing method like cooking (with or without pressure) or by an industrial processing technique like extrusion, which is widely used in the food industry to develop new food products with pulse flours as ingredients. In particular, the effect of processing methods on inositol phosphates, galactosides, protease inhibitors and phenolic-compound content is highlighted in order to ascertain their content in processed pulses or pulse-based products as a source of healthy phytochemicals.

Keywords: pulses; autoclaving; extrusion; galactosides; phytates; protease inhibitors; phenols

Citation: Pedrosa, M.M.; Guillamón, E.; Arribas, C. Autoclaved and Extruded Legumes as a Source of Bioactive Phytochemicals: A Review. *Foods* **2021**, *10*, 379. https://doi.org/10.3390/foods10020379

Academic Editor: Juan E. Andrade

Received: 17 December 2020

Accepted: 4 February 2021

Published: 9 February 2021

Publisher's Note: MDPI stays neutral with regard to jurisdictional claims in published maps and institutional affiliations.

1. Introduction

Pulses, which are the dry seeds, separated from their pod, of the Leguminosae family, have been cultivated and consumed for centuries as a staple food for humans. The most common pulses consumed by humans around the world are peas, lentils, beans, soybeans, faba beans, lupins, cowpeas and mung beans [1–3]. From a nutritional point of view, they are highly valuable foods since they are high in protein (17–50%), slow-digestion carbohydrates (0.4–55%) and dietary fiber (3–15%), and low in fat (0.8–6.6%) (except the oil seeds: soybeans, peanuts and some lupins) [1–3]. Their proteins complement those of cereals well and are gluten-free, being a food of choice for celiacs; furthermore, the demand for pulses has increased recently since more people such as vegetarians, vegans or flexitarians are looking for alternative sources of animal proteins [4,5]. Pulses are cheap, easy to prepare, versatile and in many cases, non-perishable [6]. The growing interest in pulses as both nutritious and healthy foods, led the WHO/FAO to declare 2016 as the "International Year of Pulses" [7], with its main objective to raise awareness of the multiple varieties and benefits of pulses for food security, nutrition, health and environment, and to encourage consumption. In addition, the dietary guidelines of many countries [8] recommend the consumption of 2.5–3.5 cups of pulses per week as part of a healthy diet.

In addition to nutritive compounds, pulses contain numerous phytochemicals, traditionally considered as anti-nutritional factors, and in recent decades as non-nutritional components; although most of them are non-toxic for humans, they can produce some discomfort (e.g., flatulence) and interfere with protein digestibility and the bioavailability of some nutrients such as minerals [3,6,9]. However, nowadays these phytochemicals are

25

well recognized as bioactive compounds able to exert a beneficial/healthy effect when ingested on a regular basis, reducing the incidence of several chronic diseases such as type 2 diabetes, cardiovascular diseases or some types of cancer; thus, pulses are recognized as functional foods [6,9]. These health effects may be associated with more than one bioactive compound, and synergistic combinations may exist [3,6,9,10]. These bioactive substances are not equally present in all legumes (seeds and varieties); for example, the common bean shows the highest levels of lectins, while soybean is rich in trypsin inhibitors, and peas contain higher amounts of α-galactosides [9,11]. According to different authors [3,9], depending on the compound, its concentration in the food, the time at which it is consumed and its interaction with other food matrix components, these compounds can act as anti-nutrients or as bioactive compounds Therefore, depending on the compound, it may be desirable to reduce or increase its content but not remove it completely from food products [3,9,11]. It is noteworthy that processing of pulses or pulse-based mixtures can also increase, reduce/inactivate or produce minor changes in the content of other non-nutritive components such as protease inhibitors, galactosides, lectins, phenols or phytates [1,5,12–16]. Although the beneficial effects of these components mainly depend on their bioavailability in the gut, and since there currently is not a recommended daily intake of bioactive compounds, it would be of great interest to know the level of each bioactive compound to be consumed in the processed foods, because processing can disrupt the food matrix, making phytochemicals more or less bioaccessible [17].

There are a number of processing techniques available, such as soaking, dehulling, germination, malting, fermentation, cooking, autoclaving, microwaving, roasting or extrusion, that make it possible to achieve the suitable nutritional and organoleptic characteristics of pulses, as well as to improve the content of bioactive compounds in comparison with the raw products. Different processing methods can affect, to a different extent, the content of a specific compound. Soaking is very effective in reducing water-soluble compounds such as oligosaccharides and some phenolic compounds. Germination reduces phytic acid effectively, while cooking is more effective in reducing bioactive compounds that are heat-labile, such as protease inhibitors and lectins [9]. In relation to the phytochemical content of the processed pulses, there is a great variability in the literature on the data for the same processing method [5,9,12,18–20]. For example, for a specific compound, some authors report contradictory findings. In general, the final effect of a processing method depends on the pulse type and variety, the processing parameters (mainly amount of water, temperature and time), the food matrix structure, the chemical nature of each phytochemical and the presence of additional compounds that may protect each other during processing [3,9,11,21].

Although different review papers can be found in the literature on the content of several bioactive compounds in pulses [9–11,22,23], reviews focused on the bioactive-compound content in processed pulses or in pulse-based foods are scarce, and their content is more limited than the present review, since some of them are concerned only with one type of treatment (for example, boiling), or in one type of bioactive compound (mainly phenolic compounds), and to the best of our knowledge, there are no reviews about the cold extrusion process [11,17,21,24].

This review is focused on the changes produced in the bioactive phytochemical content of pulses processed either by a traditional processing method such as cooking with pressure either domestically or industrially, or by industrial processing techniques like extrusion/cooking or cold extrusion, which are widely used in the food industry to develop new food products with pulse flours as ingredients. In particular, the effect of the processing method on some of the bioactive phytochemicals present in pulses (inositol phosphates, galactosides, protease inhibitors and phenolic compounds) is highlighted in order to ascertain the content of these bioactive compounds in processed pulses or pulse-based products as a source of healthy phytochemicals.

2. Methodology

The methodology followed in the elaboration of this systematic review first included first the selection of the topics addressed (Figure S1). For this, domestic and/or industrial processing techniques were taken in consideration. Second, the most common bioactive compounds (α-galactosides, myo-inositol phosphates, protease inhibitors and phenolic compounds) present in the legumes usually consumed and related to healthy roles were chosen. With the information obtained about these specific issues, the structure of this review is as follows: (i) content of some bioactive compounds in raw pulses and their health effect in humans; (ii) effect of autoclaving on the bioactive compounds of pulses; (iii) effect of extrusion/cooking on the bioactive compounds of pulses; and (iv) effect of cold extrusion on the bioactive compounds of pulses.

After establishing the subject of study (autoclaved, extruded legumes and bioactive compounds), a search was conducted on different scientific databases: SciELO, Science Citation Index, Science Direct, Google Scholar, Medline and Mendeley, and thanks to cross-references, some papers were found and reviewed. Although no restriction was applied for the publication dates, almost all the records identified (85%) corresponded to papers in the last 20 years, and only 15% of the records corresponded to papers published from 1980 to 2000. For each topic, appropriate keywords were used to search for relevant papers. Terms such as phytochemicals, bioactive compounds, the specific name of the phytochemical (α-galactosides, myo-inositol phosphates, protease inhibitors and phenolic compounds), the names of the specific processing method (autoclaving, extrusion/cooking, cold extrusion), canning, snack, pasta and the names of the most common legumes consumed were used as keywords in all possible combinations during the paper-identification step. The results of this search showed that the effect of processing on bioactive compounds of legumes should be organized by taking into account three main factors: the processing procedure, the type of bioactive compound and type of legume. After the search step, some papers were excluded based on the following criteria: (i) record duplicates, (ii) age (before 2000, except those papers that include relevant information that was not reported in more recent papers), (iii) missing information about the processing conditions, and (iv) no clear information about the effect of the processing, making the comparison to the results of other papers difficult. The quality assessment of data was evaluated, taking into account the accessibility of the publication, studies relevant in the area and cited by many other authors, papers that studied more than one phytochemical and papers that evaluated the processing method effect in interaction with more than one other factor (temperature, time, cultivars or the presence of other materials/ingredients). In the case of papers considering more than one treatment or more than one legume, data were considered separately. Finally, 164 research papers (Figure S1) about the phytochemical content on legumes and comparing the effect of autoclaving, extrusion/cooking and cold extrusion on the main bioactive compounds present in legumes (galactosides, myo-inositol phosphates, protease inhibitors and phenolic compounds) were selected. Among these, 143 papers corresponded to studies published in the last 20 years. Data were summarized in tables reporting, for each processing method and legume or their mixtures, the content of each bioactive compound and the effect of the processing treatment on each compound and the source of the data.

3. Content of Some Bioactive Compounds in Raw Pulses and Their Health Effects in Humans

3.1. α-Galactosides

The most common oligosaccharides in pulse seeds are α-galactosides or the raffinose family of oligosaccharides (raffinose, stachyose, verbascose and ajugose), raffinose and stachyose being the most ubiquitous sugars. Some authors [3,5,6,9,10,23] have reviewed the content of α-galactosides in different seeds and varieties, ranging from 0.4 to 16.1% (dry matter—d.m). Singh et al. [11] reported that lentils presented the lowest amount of total galactosides (37.5 mg/g), followed by faba beans (52.0 mg/g), beans (60.9 mg/g) and peas (66.3 mg/g). Muzquiz et al. [9] reported concentrations of raffinose, ciceritol, stachyose and verbascose in different varieties of Spanish beans, peas, chickpeas, faba beans, soybeans and lupins; raffinose ranged from 1.0 mg/g in *Phaseolus vulgaris* var. Palmeña to 33.15 mg/g in *Lupinus mariae-josephi*; stachyose varied from 9.22 mg/g in *Vicia faba* var. Alameda to 59.08 mg/g in *Lupinus albus* var Multolupa. Verbascose was not detected in the reported chickpeas, the highest amount being found in *Pisum sativum* var. Luna (50.25 mg/g). Another α-galactoside is ciceritol, an α-D-digalactoside of pinitol that does not belong to the raffinose family of oligosaccharides. It is not present in all pulses, and its amount ranged from 1.61 mg/g to 29.65 mg/g in pea var. Iceberg and chickpea var. Duraton, respectively.

The α-galactosides are not digested or hydrolyzed by humans. However, they are fermented by colonic bacteria with the production of hydrogen, carbon dioxide, methane and short-chain fatty acids (SCFA), mainly butyric and propionic. While the gases produced are responsible for flatulence, bloating and diarrhea, the SCFA are mainly related to the prebiotic effect associated with the α-galactosides, promoting the growth of beneficial gut microflora (bifidobacterias and lactobacilli) and reducing the enterobacteria population. α-Galactosides have also been shown to help in normalizing bowel function, reducing potentially carcinogenic compounds (such as N-nitroso compounds), and enhancing the immune system and increasing resistance to infection [3,9,11,25]. In addition, it has also been reported that propionic acid reduces serum cholesterol, helping to reduce the risk of cardiovascular diseases, and butyric acid induces apoptosis and stops the growth and differentiation in colon cancer cells [3,9,11]. Despite there not being a recommended dietary intake for α-galactosides, Martinez-Villaluenga et al. [25] documented that a dose of 3 g/day of α-galactosides produces an increase in the intestinal bacteroides, bifidobacterias and eubacteria without any flatulence discomfort.

3.2. Myo-Inositol Phosphates

Phytate (IP6) or myo-inositol hexakisphosphate is the main form of phosphorous storage in pulse seeds (up to 75% of total seed phosphorous), and is stored as salts (phytate–mineral complex) or bound with proteins or starch [3,9]. According to the literature, the total content of inositol phosphates ranges from 0.2 to 2.34% [3,9–11,23], and as described above for α-galactosides, the amount of inositol phosphates varies between species and varieties, as well as with the soil phosphorous [26]. Sparvoli et al. [3] reported that varieties of peas (3.1–7.1 mg/g), chickpeas (2.8–13.6 mg/g), lentils (2.5–12.2 mg/g) and mung beans (1.8–5.8 mg/g) contain relatively lower amounts of IP6 than those of common beans (3.4–28.7 mg/g), faba beans (5.9–15.0 mg/g) and soybeans (4.8–20.1 mg/g). Muzquiz et al. [9] reported the total inositol phosphate content in different varieties of some Spanish legumes, the average content being 0.4% in beans, 0.75% in faba beans, 0.6% in chickpeas and beans, 0.7% in lupins and 1.2% in soybean varieties.

In general, phytate, myo-inositol hexaphosphate or IP6 has been considered as an antinutrient that interferes with nutrient (mineral, protein and starch) digestibility and bioavailability [27]. However, this mechanism of action also produces health benefits. As IP6 binds starch and the calcium necessary for α-amylase activity, the starch digestibility is reduced, improving the glycaemic response of pulses and the management of diabetes type 2 [3,23]. Due to its mineral-binding capacity, IP6 has been linked with other beneficial health effects, such as the prevention of kidney stone formation, the prevention of cavities and plaque in teeth and protection from demineralization. It has been also reported that a diet with 1% sodium phytate added can control hypercholesterolaemia and atherosclerosis, and can reduce the risk of colon cancer and improve irritable bowel syndrome [22,28]. Further, it has been reported that the lower phosphorylated forms (IP5–IP3) can promote the absorption of minerals and show strong antioxidant and anti-inflammatory activities, inducing apoptosis and normalizing abnormal cell proliferation [3,9,22,23]. In addition, some authors [29,30] have reported that iron absorption can be improved when IP6 is below 10 mg/g protein in one serving dose.

3.3. Protease Inhibitors

Protease inhibitors in legumes belong to the Kunitz and the Bowman–Birk families, and both are capable of inhibiting trypsin and chymotrypsin enzymes. There is a high number of isoforms of both inhibitors that vary with the legume species and variety [31]. The trypsin inhibitor content ranges from 5.75 to 15 trypsin inhibitor units (TIU)/mg in peas, from 5 to 10 TIU/mg in faba beans, from 12.60 to 19 TIU/mg in chickpeas, from 3 to 8 TIU/mg in lentils and from 8.57 to 83.70 TIU/mg in soybeans [9,19]; while the chymotrypsin inhibitor content varies from 2.19 chymotrypsin inhibitor units (CIU)/mg in *Vicia narbonensis* to 17.30 CIU/mg in bean var. Riñón, but is not detected in soybean var. Ostrumi [9,10]. Protease inhibitors have a negative effect on animal growth due to the inhibition of gut protein digestion. However, in a Western diet, there have not been any reported toxic problems related to the intake of these compounds, mainly because pulses are cooked prior to consumption and protease inhibitors are thermal-labile compounds. Over the past two decades, different studies [3,6,9,22,23,32,33] have shown that protease inhibitors are effective in preventing or reducing colon, lung, liver, prostate and breast cancer progression. Sánchez-Chino et al. [23] reported that some possible mechanisms of action are: (i) the reduction of protein digestibility reduces the availability of essential amino acids for the cancer cells; (ii) the protease inhibitors act as insoluble dietary fiber able to absorb carcinogens (such as free radicals) in the gut; and (iii) the inhibiting of proteases produced by cancer cells. Even though there is not a recommended amount of protease inhibitor consumption, it is important to note that the traditional Japanese diet contains about 420 protease inhibitor units/day; further, it has been reported that the consumption of the purified protease inhibitor at 25–800 CIU per day during a period of 12 weeks exerted a protective effect against cancer development, and doses of up to 2000 CIU/day did not cause health problems in humans [9–11,22,32–36].

3.4. Phenolic Compounds

Pulses are rich in phenolic compounds and include different subcategories such as tannins, flavonoids, isoflavones, phenolic acids (such as caffeic, ferulic, sinapic and p-coumaric acids) and anthocyanins. Many of these compounds are located in the seed coat and are responsible for seed color, and are related to the taste and flavor of seeds. In general, the darkest legume varieties tend to have higher amounts of phenolic compounds than the light seeds/varieties [3,11,37,38]. Even though a great variability in the phenolic content can be found due to the different methods used in their extraction and quantification (spectrophotometrically or by HPLC (high-performance chromatography)), in general, a high variability can be found among legumes and varieties [5,24,38–42]. For example, among beans, there can be found values from 0.3 mg/g of phenolic compounds in white varieties to 12.6 mg/g in black varieties; peas show values from 0.6 to 2.7 mg/g and chickpeas from 0.6 to 2.7 mg/g [19]. Pedrosa et al. [5] reported that the raw Curruquilla bean, a cream-colored variety, shows a higher content of total phenols and anthocyanins (2.70 mg/g and 40.10 µg/g) than the Almonga bean (2.38 mg/g and 38.47 µg/g), a white variety. Some authors [24,41] reviewed the phenolic content of various legume seeds and documented contents from 11.2 to 48.3 mg/g in dry beans, from 117.8 to 157.6 mg/g in different faba bean genotypes, from 4.9 to 68 mg/g in lentils and from 0.98 to 183 mg/g in chickpeas. Condensed tannins are associated with an astringent taste, as well as with some anti-nutritional effects due to their ability to bind and precipitate proteins, reducing their digestibility. However, from a health point of view, different studies found in the literature report phenolic compounds as bioactive molecules with antioxidant, antimicrobial, anticarcinogenic, immunomodulating, cardio-protective, anti-hypertensive and anti-inflammatory properties, lowering the risk of colon cancer and osteoporosis [6,10,22,38,41]. Most of these actions are linked to the antioxidant and antiradical activities of the different phenolic compounds. In general, antioxidant activity is positively correlated with total phenolic content [41]. Cardador-Martinez et al. [43] reported that white beans possess low antioxidative activity in comparison to black, brown and red bean varieties. As documented in some reviews [11,41], different authors related this high antioxidant activity to pro-anthocyanidins or condensed tannins, as well as to flavonoid content, and concluded that pulses can be a useful natural source of antioxidants. Zhang et al. [44] reported that lentil phenols also inhibit glucosidase and lipase, and would therefore contribute to controlling blood glucose levels and obesity in humans. In spite of these health benefits, there is not in fact a recommended daily intake of phenolic compounds, mainly due to the differences in their total intake for the overall population; however, some reports recommend a minimum daily dose of 300 mg of total phenolic compounds to benefit from their health properties [45].

4. Effect of Autoclaving on the Bioactive Compounds of Pulses

Forty-eight papers about the effect of the autoclaving on legume phytochemicals have been consulted for this review. Nine papers evaluated the effect on α-galactosides; 15 papers were about the effect on inositol phosphates content; 18 papers were about the effect on trypsin inhibitor activity and 25 were the effect on phenolics compounds and their antioxidant activity.

4.1. Autoclaving Process

Autoclaving (cooking with pressure) is a thermal process usually undertaken at home in a pressure cooker and by the food industry in an autoclave to prepare pulses for human consumption. It is an alternative to boiled legumes, which require cooking in water for very long time. A processing technique such as autoclaving reduces the cooking time and promotes pulse consumption by providing products that are quickly prepared and easy to use.

Autoclaving is a high-pressure cooking technique that uses high temperature and pressure over a short period of time. It is a high-intensity thermal method used in industry to produce tinned legumes commercially. The autoclave is a pressure chamber that operates at 1.8 to 2.0 bar and works by subjecting the item to pressurized saturated steam at 121 °C.

The autoclaving parameters of pressure, temperature and time vary from 15 psi, 116 °C to 127 °C and from 10 to 60 min, respectively. This processing affects the nutritional and antinutritional composition of legumes in different ways. Khatoon et al. [46] reported that pressure cooking of eight whole legumes did not affect the nutrient composition. The mean of in vitro protein digestibility of these pressure-cooked legumes was 79.8%, a higher percentage than for raw materials or even when other cooking techniques were applied, such as microwaving. However, Pedrosa et al. [5] observed significant differences in the nutritional composition of two industrial tinned bean varieties (*P. vulgaris* var. Almonga and var. Curruquilla): increasing protein (>7%) and dietary fiber (>5%); and decreasing fat (>16%), carbohydrate (>15%) and, in general, mineral content (P, Mg, Ca, Fe and Zn) after treatment at 116 °C for 42 min. In autoclaved (121 °C) chickpea for 35 min [47], and faba bean [48] for 30 min, similar effects on nutritional composition were observed, except on protein content, which did not change, nor their amino acid composition. However, in vitro protein digestibility and quality increased with autoclaving [47]. The increment of protein digestibility by autoclaving may be due to the heat-denaturation of protein, and reduction or even abolition of bioactive compounds/anti-nutritional factors such as a trypsin inhibitor, tannins and phytic acid. In addition, autoclaving reduced the B-vitamin content of legumes [46–48]. Finally, it should be noted that autoclaving was the most effective in reducing or even abolishing allergenicity in lentil, chickpea and lupin [49,50]. Consumption of tinned beans reduced metabolic risk factors associated with obesity [51], and may be used in a renal patient diet [52]. Autoclaving can also affect the content of the bioactive compounds/anti-nutritional factors present in raw materials, increasing or decreasing their contents, mainly depending on the raw material and the extrusion parameters. Moreover, as has been stated previously in this review, in many cases the elimination of some phytochemicals is not desirable, because small amounts are able to produce beneficial health effects against different chronic diseases.

Foods **2021**, 10, 379

The effect of autoclaving on the content of the different bioactive phytochemicals reviewed is shown in Table 1.

Table 1. Content of different bioactive compounds (α-galactosides, inositol phosphates, protease inhibitors, phenolic compounds and antioxidant activity) in autoclaved legumes.

Legumes	Autoclaving Conditions	α-Galactosides	Inositol Phosphates	Protease Inhibitors	Phenolic Compounds	Antioxidant Activity	References
Adzuki bean	121 °C; 15 min			0.10 TIU/mg ↑			[53]
Pre-soaked bean (two vars.)	116 °C; 42 min	30.2–24.25 mg/g ↓	10.59–10.53 mg/g ↓	0.47–0,43 TIU/mg ↓	Total phenols: 9.65–11.80 µg/g ↓	ORAC: 16.03–16.03 µmol TE/g ↓	[5]
Pre-soaked pinto bean	116 °C;10–30 min	46.9–57.6 g/100 g ↑					[54]
Unsoaked kidney bean (three vars.)	121 °C; 30 min	0.55–0.65 g/100 g ↓	6.94–8.43 mg/g ↓	n.d. * ↓	Tannins: 1.51-14.37 mg/g ↓		[55]
Pre-soaked kidney bean (three vars.)	121 °C; 30 min	0.36–0.49 g/100 g ↓	6.59–8.90 mg/g ↓	n.d. ↓	Tannins: 1.34–12.91 mg/g ↓		[55]
Wild and cultivated Mexican bean	20 min	60.4 mg/g (mean) ↓	8.7 mg/g (mean) ↓				[56]
Brazilian bean (eight vars.)	121 °C; 15 min			0.48–1.00 TIU/mg ↓			[57]
Pre-soaked Brazilian bean (five vars.)	121 °C; 15 min			n.d. ↓			[57]
Pre-soaked kidney, pinto, black and borlotti bean	115 °C; 20 min				Total phenols: 0.38–0.94 mg GAE/100 g =	DPPH: 2.29–5.20 µmolTE/g = ABTS: 0.41–1.56 µmol TE/g = FRAP: 4.24–7.91 µmol TE/g ↓	[58]
Bean (two vars.)	121 °C; 7–12 min				Total phenols: 77.1–78.6 mg GAE/100 g ↓		[59]
Pre-soaked chickpeas	121 °C; 35 min	2.27 g/100 g ↓	0.71 g/100 g ↓	0.44 TIU/mg protein ↓	Tannins: 2.42 mg/g ↓		[47]
Whole chickpea (five cvs.)	120 °C; 60 min		112–335 mg/100 g ↓				[60]
Dehulled chickpea (five cvs.)	120 °C; 60 min		105–241 mg/100 g ↓				[60]
Chickpea	121 °C; 15 min			0.10 TIU/mg ↑			[53]
Dolichos lablab (Vulgaris var)	121 °C; 15–45 min		452–482 mg/100 g ↓	0.85–2.36 g/100 g ↓	Tannins: 0.05–0.13 g/100 g ↓ Total free phenols: 0.22–0.98 g/100 g ↓		[61]
Pre-soaked faba bean	120 °C; 30 min	1.43 g/100 g ↓	0.23 g/100 g ↓	0.35 TIU/mg ↓	Tannins: 0.58 g/100 mg ↓		[48]
Unsoaked faba bean (white and green)	121 °C; 15 min		8.90–9.27 mg/g ↑	0.46–1.23 mg/g ↓	Tannins: 4.72–6.51 mg/g ↑ in green; ↓ in white		[62]
Pre-soaked faba bean (white and green)	121 °C; 15 min		5.12–9.21 mg/g ↓ in green; ↑ in white	0.74–1.86 mg/g ↓	Tannins: 2.21–4.02 mg/g ↓		[62]
Pre-soaked faba bean (three genotypes)	115 °C; 20 min				Total phenols: 0.7–1.9 mg GAE/100 g ↓	DPPH: 4.6–9.9 µmol TE/g ↑ TEAC: 8.35–11.25 µmol TE/g ↓ ORAC: 20.54–33.25µmol TE/g ↓	[63]
Hyacinth bean	121 °C; 15 min			0.23 TIU/mg ↑			[53]
Pre-soaked lentil	121 °C; 35 min	1.42 g/100 g ↓	2.4 g/ g ↓	0.19 TIU/mg ↓	Tannins: 0.82 g/100 g ↓		[64]

Table 1. *Cont.*

Legumes	Autoclaving Conditions	α-Galactosides	Inositol Phosphates	Protease Inhibitors	Phenolic Compounds	Antioxidant Activity	References
Lentil (two vars.)	121 °C; 7 min				Total phenols: 318.3–533.4 mg GAE/100 g ↓		[65]
Lima bean	121 °C; 10–20 min		8.89 mg/ g ↓	n.d. ↓	Tannins: 0.23–0.32 g/100 g ↓		[66]
Black-eyed pea	121 °C; 15 min			0.22 TIU/mg ↑			[53]
Pea (two vars.)	121 °C; 12 min				Total phenols: 80.0–162.1 mg GAE/100 g ↓		[59]
Pigeon pea	121 °C; 15 min			0.41 TIU/mg ↑			[53]
Mung bean	121 °C; 15 min			0.11 TIU/mg ↑			[53]
Soybean	121 °C; 15 min			0.09 TIU/mg ↑			[53]
Pre-soaked soybean	115 °C; 20 min				Total phenols: 0.71 mg GAE/100 g ↓	DPPH: 0.91 μmol TE/g ↓ ABTS: 0.27 μmol TE/g ↓ FRAP: 2.18 μmol TE/g ↓	[58]

* n.d. not detected; TIU: trypsin inhibitor units; GAE: gallic acid equivalents; ORAC: oxygen radical absorbance capacity assay; DPPH: 2,2-diphenyl-1- picrylhydrazyl assay; TEAC: Trolox equivalent antioxidant capacity assay; FRAP: ferric reducing antioxidant power assay; ABTS: free radical scavenging ability assay; TE: Trolox equivalents; ↑: increase; ↓: decrease; =: no change.

4.2. Autoclaving Effect on the α-Galactosides of Pulses

Autoclaving has been reported to be a more effective heat process in reducing α-galactosides, particularly when processed broths are discarded by leaching them through soaking and also into the canning solutions [54]. The pressure applied during this process forces water into the seeds, increasing the extraction of α-galactosides during cooking [67].

In soaked pinto beans (*P. vulgaris*), autoclaving at 115 °C for 30 min resulted in a more effective heat process in reducing α-galactosides, even with cooking for 90 min. Autoclaving for a shorter time (10 or 20 min) also significantly reduces oligosaccharides (46.9% and 49.4%, respectively), particularly stachyose (50.1% and 51.8%, respectively) [54]. Autoclaving (116 °C, 42 min) of beans var. Almonga and Curruquilla was less effective in reducing α-galactosides (25.11% and 39.42%, respectively), with stachyose again reducing more than raffinose, particularly in Curruquilla (50.26%) [5]. Shimelis and Rakshit [55] found that a combination of soaking (in water or NaHCO₃) and autoclaving caused a significant reduction in stachyose (75.81–77.72%) and raffinose (64.71–72.41%) contents of var. Roba, var. Awash and var. Beshdesh kidney beans, and the combined effect was significantly higher than autoclaving (lower than 62.5% reduction in raffinose and 71.2% in stachyose) or soaking alone. However, it was necessary to combine sprouting for more than 48 h and autoclaving to abolish or completely eliminate α-galactosides in these varieties of bean seeds. Similar results were found when lentils, previously soaked, were autoclaved at 121 °C for 35 min; raffinose, stachyose and verbascose were significantly reduced, but reductions were mainly observed in raffinose (75%) and verbascose (54%) [64]. The same conditions of processing in chickpea provoked similar reductions in the total α-galactoside content (45.95%), with a 44.14% reduction in raffinose and 42.97% in stachyose, while verbascose was completely eliminated after autoclaving treatment [47]. A more effective heat treatment in the reduction of raffinose (67.44%), stachyose (67.26%) and verbascose (87.40%) in *Dolichos lablab* var. Vulgaris was also found to be autoclaving (121 °C) for 45 min; shorter durations (15 and 30 min) provoked fewer reductions (20.93–63.77%) [61]. Stachyose was significantly decreased in faba beans (21%) through autoclaving for 30 min [48].

In the reviewed studies, without exceptions, autoclaving reduced the α-galactosides content in legumes between 14 and 77% with respect to those of the raw seeds, corresponding to the lowest reduction to autoclaved beans and faba beans and the highest reduction to pinto beans. The content of α-galactosides in the autoclaved legumes ranged from 3.6 mg/g to 30.2 mg/g in different beans, becoming higher when cooking broth was not discarded (60.4 mg/g), 27.7 mg/g in chickpeas, from 23.6 mg/g to 8.50 mg/g in *D. lablab* (depending on processing time: 15–45 min), and around 14 mg/g in faba beans and lentils. The interest in ready-to-eat products such as canned legumes is growing, so, taking into account the remaining α-galactosides content in legumes after this treatment, one serving dose of autoclaved legume (100 g) could supply 0.36–6.04 g for beans (with and without broth), 2,27 g for chickpeas, around 1.4 g for faba bans and lentils and 0.85–2.36 g for *D. lablab*. Therefore, autoclaved legumes will maintain the prebiotic activity due to α-galactosides, according to Villanueva et al. [25].

4.3. Autoclaving Effect on the Myo-Inositol Phosphates of Pluses

Autoclaving of legumes affects phytic acid by reducing its content [68,69]. The partial elimination of this phytochemical can contribute to improving digestibility of proteins and starch in legumes, which is mainly observed after this cooking treatment [70]. Autoclaving for 30 min caused significant losses, from 33.72% to 41.61% in phytic acid contents of lentils [64,71] faba beans [48] and chickpeas [47,60,71], and higher losses (59.98–65.93%) have been reported in several bean varieties [5,55,71]. Even autoclaving lima beans for 20 min caused the complete removal of phytic acid content [66].

In addition, the phytate phosphorous contents of the lima bean and five different chickpea cultivars were reduced after autoclaving [60,66]. The increase in total phosphorus after autoclaving (120 °C) of chickpea for 60 min demonstrated that phytic acid was hydrolyzed or decomposed during this processing [60]. Phytic acid (IP6) was the main form that was detected in raw and industrial tinned beans var. Almonga and var. Curruquilla, with about 61% reductions after autoclaving at 116 °C for 42 min, and the contents of less phosphorylated forms (IP5 and IP4) and IP3 were detected only in tinned beans [5]. Similar reductions in phytic acid content were found in other varieties of kidney beans after cooking at 121 °C for 30 min (var. Roba, var. Awash and var. Deshbesh) by Shimelis et al. [55]. Soaking of these beans provoked a significant decrease in their phytic acid content (>17%); however, no differences between the reducing effect of autoclaving without soaking and with previous soaking (in water or NaHCO$_3$) were observed. Soaking could contribute to reducing total IP content by leaching during processing and by activation of endogenous phytases; additionally, cooking also contributes by formation of insoluble complexes between phytate and other components [72]. In fact, when the autoclaved bean seeds were analyzed with their corresponding broth, the percentage losses of phytic acid were lower (24.6%) than when the broths were discarded [56]. In contrast, Wang et al. [73] reported that soaking (in water) of six pea (*P. sativum*) varieties did not cause significant reductions in phytic acid content (1.2–1.6%). Autoclaving was more effective in reducing phytic acid content than cooking, but it was necessary to combine autoclaving and sprouting to eliminate phytic acid content. The breakdown of phytate during germination is attributed to the increased activity of the endogenous phytase [55]. Myo-inositol phosphate content in legumes was reduced by autoclaving (2–100%) depending on the legume type and the processing time, except in unsoaked faba beans and pre-soaked green faba beans [62]. Data evaluated in 15 reviewed papers showed a reduction in myo-inositol phosphate content from 53% to 100% for lima beans, from 24% to 66% for beans, from 33% to 46% for chickpeas, from 24% to 46% for lentils, and from 23% to 41% for faba beans. The lowest reductions were found for *D. lablab* (2–8%). Just one reviewed paper showed an increase of myo-inositol phosphate content, from 4 to 10% in unsoaked green and white faba beans, and from 7% to 12% in pre-soaked white bean. Taking into account this effect, one serving dose of autoclaved legumes (100 g) could be beneficial for human health, since they can supply from 8.89 to 1059 mg inositol phosphates/100 g) [3,9,22,23]. Considering a protein content around 20%, the autoclaved legumes can supply from 0.44 to 53 mg inositol phosphates per gram of protein.

4.4. Autoclaving Effect on the Protease Inhibitors of Pulses

Trypsin inhibitor activity in legumes was reduced, partially or even completely, by heat treatments, particularly by autoclaving [68,69,74]. The thermostability of the protease inhibitor in legumes depends on legume sources, and also processing conditions such as pH, humidity, time, temperature and pressure [74]. Reactions involving deamidation splitting of covalent bonds, such as hydrolysis of peptide bonds in aspartic acid residues, and interchange or destruction of disulfide bonds, might be involved in the thermal inactivation [75]. Several authors have reported this reduction in trypsin inhibitor activity provoked by autoclaving (with pre-soaking) in chickpeas (83.87%), beans (94–98%) [4,41], lentils (80.87%) [64], faba beans (85%) [48]; and even total inactivation of the trypsin inhibitor in the seeds of beans [55,57,61], lima beans [66] and faba beans [62,76,77]. Different effects on the trypsin inhibitor activity by soaking of legumes can be found in the literature, from significant reduction in soaked kidney beans (6–17%) [55] to even an increase (3.20–19.30%) in soaked peas [73]. Independent of this fact, the elimination effect of autoclaving on the trypsin inhibitor activity did not depend on a previous soaking treatment of kidney beans [55]. Luo and Xie [62] studied the effect of autoclaving combined with soaking and/or dehulling on green and white faba beans. Autoclaving at 121 °C for 20 min caused a reduction in trypsin inhibitor activity of both faba beans (50.40 and 84.03%), but combining with dehulling and/or soaking caused an increase of the trypsin inhibitor activity vs. autoclaved seeds. The increase by dehulling demonstrated that trypsin inhibitors are mainly located in the seed cotyledons. In contrast, Choi et al. [53] observed an increase in trypsin inhibitor activity levels by autoclaving at 121 °C for 15 min in the adzuki bean (*Phaseolus angularis*), chickpea (*Cicer arietinum* L.), hyacinth bean (*Lablab purpureus* L.), black-eyed pea (*Vigna unguiculata* L.), pigeon pea (*Cajanus cajan* L. Millsp.), bambara groundnut (*Vigna subterranea* L. Verdc) and soybean (*Glycine max* L.). The exception was mung bean (*Vigna radiata* L.), which showed a 35% reduction.

Autoclaving of legumes resulted in an efficient thermal process in a reduction or even total inactivation of the trypsin inhibitors (22–100%) except for some underutilized legumes that, after this thermal treatment, exhibited an increase in their inhibitor activity. It was necessary to autoclave at 121 °C for 30 min to the complete inactivation of trypsin inhibitors in beans, while it was not possible in faba beans treated under the same autoclaving conditions (22–85%). These differences could be due to the different isoforms present in each legumes that can differ even between varieties [31]. Therefore, the efficiency of this process in the reduction of trypsin inhibitor activity depends on the legume type and processing time (15–45 min). The low remaining trypsin inhibitor activity exhibited after autoclaving in legumes included in this review would have a positive effect in human health (1 to 1.5 TIU/mg) [74].

4.5. Autoclaving Effect on the Phenolic Compounds of Pulses and Their Antioxidant Activity

The phenolic compounds of legumes were significantly affected by autoclaving with respect to raw materials. However, several authors have reported both an increment [58,78–81] and a decrease [5,47,48,55,61,64,68,69,78,82–84], depending on the type of legume, conditions and the detection methods used to determine their amount [5]. It is known that high temperatures provoke polymerization and/or decomposition in the structure of aromatic rings of polyphenols, which makes their quantification difficult [85,86]. Moreover, contact with water at high temperatures could increase the solubility of polyphenols, thereby increasing their release into the cooking water [87]. As more water volume was used for cooking beans, higher loses of polyphenols were observed [88].

The phenolic content in the lablab bean was reduced under pressure cooking [61,86]. Such reduction can be explained by a lixiviation phenomenon that drives phenols into cooking water and by their being bound to other compounds, forming insoluble complexes [89]. The percentages of total free phenolic reduction are extended by the duration of processing, varying from 31% for 15 min to 85% for 45 min, and from 28 to 72% for tannins [61]. Vijayakumari et al. [84] reported that autoclaving seemed to be the most efficient processing for reducing polyphenol and tannin contents of *Vigna sinensis* and *Vigna aconitifolia* seeds compared to soaking and cooking. This loss of phenolic contents by autoclaving may also be related to the interaction of polyphenols with other components of seeds, such as insoluble tannin-protein complexes [82]. Pedrosa et al. [5] observed that different effects after industrial autoclaving (116 °C, 42 min) depend on the methods (spectrophotometric or HPLC) used to determine the content of total phenols in two varieties of *P. vulgaris* (var. Almonga and var. Curriquilla). A drastic reduction (70%) in phenolic content was observed by HPLC in both tinned beans, while with spectrophotometry, tinned Almonga showed a significant increase (12%), and tinned Curruquilla a significant decrease (9%). In addition, according to the HPLC results, industrial processing induces qualitative changes among varieties and processed vs. raw beans. A lower percentage of reduction in the total phenolic content of beans (var. Raba and Warta) was observed under shorter processes (7–12 min) and without previous soaking (26% and 18%, respectively) [59]. The combination of soaking and autoclaving caused a significant reduction in tannin contents of kidney bean, and the combined effect was significantly higher than cooking or soaking alone [55]. Siah et al. [63] found light losses (31– 48%) of phenolic compounds by soaking *V. faba*. A decrease also was observed in autoclaved *Lens culinaris* seeds var. Anita and Tina (32 and 30%, respectively), and even lower in pea var. Milwa (14%) [59]. A 50% reduction in tannins was observed in autoclaved lentils and chickpeas autoclaved for 35 min [47,64], a 60% loss in autoclaved faba beans autoclaved for 30 min [48] and a 35.94% loss in lentils autoclaved for 35 min [64]. Autoclaving processes (115 °C for 20 min, including pre-soaking for 12 h) of several faba beans (*V. faba* var. Rossa, Nura and TF (Ic*As)*483/13 line) reduced total phenolic (35–63%) and flavonoid (33–42%) contents and their antioxidant activity, as evaluated by different methodologies (ORAC: oxygen radical absorbance capacity assay; TEAC: Trolox equivalent antioxidant capacity assay; and FRAP: ferric reducing antioxidant power assay). When these contents were evaluated on the autoclaved seeds separated from their respective broths, some differences were observed. These contents and their antioxidant activities in all autoclaved faba beans were lower than their respective broths. Compared to the cooking process, high pressure (autoclaving) caused more disruption to seed membranes, which allowed the release of the greater amounts of some components into the cooking broth. In addition, autoclaving faba beans caused the disappearance of the less-polar phenolic compounds. Autoclaved beans were also darker in color compared to cooked beans, suggesting the formation of new compounds as a result of Maillard reactions [63]. The phenolic content reduction by cooking could be due to either their destruction or the formation of new insoluble components with other organic compounds [90]. The leaching of phenols could be influenced by the physical properties of the seeds, such as size, coat thickness, color, shape and hardness [91]. From a health benefit point of view, it was recommended that the legume seed be consumed together with the autoclaving broth. In contrast, several researchers reported that autoclaving increases the total phenolic content and antioxidant activity in *P. vulgaris* and *G. max* [58,80] and in other legumes (*C. arietinum*, *Lathyrus sativum* and Brazilian beans) [79,81]. In different varieties of *P. vulgaris* L. (kidney bean, pinto bean, black bean and borlotti bean), pressure cooking (115 °C, 20 min, pre-soaking) increased the total phenolic contents, flavonoid and ortho-diphenol content and antioxidant activity [58]. In soybean, the same pressure cooking modestly increased the total phenolic content and antioxidant capacity, and substantially increased the genistein and daidzein content. Discarding the cooking water significantly decreased the content of total phenols and their antioxidant activity in beans and soybeans. This suggested that heat could cause destruction of the seed structure, enabling the release of phenols into the cooking broths, and thereby increasing the extraction efficiency of the used solvent [58]. In addition, vegetable phenolic compounds are generally bound covalently to amine functional groups, and therefore heat treatment can hydrolyze them, increasing their extractability [78]. Osman [83] also reported a significant increase in tannins in heat-treated *D. lablab* bean. Amarowicz and Pegg [24] reviewed the profile and content of phenolic compounds in different legumes and the influence of soaking, boiling, fermentation and germination on their content and their antioxidant activity. They documented both increased and decreased total phenol content in boiled or autoclaved beans, lentils or peas. The DPPH activity was decreased, while the ORAC values of boiled legumes decreased, but increased in autoclaved legumes. Yeo and Shahidi [42,92] reported the effect of boiling on total, soluble and insoluble phenolic compounds and on their antioxidant activity in four lentil varieties. The phenolic content in the boiled samples ranged from 5.50 mg/g to 6.56 mg/g; the highest reduction corresponded to the insoluble phenols, and on average, the total phenolic compounds were reduced by 13%. The antioxidant activity was measured by means of DPPH and ORAC activities. Boiling reduced both parameters from 3% to 39% and from 75 to 17%, respectively.

Autoclaved legumes can be consider a sources of phenolic compounds and tannins, particularly beans with a tannin content up to 14.37 mg/g, followed by lentils (up to 8.2 mg/g), chickpeas and faba beans (up to 6.10 and 6.51 mg/g, respectively). In the majority of cited studies, autoclaving reduced the phenolic compounds and tannin content in legumes (14–84%) with respect to raw materials, except for some types of beans when they were autoclaved under less drastic conditions (around 115 °C). However antioxidant activity was reduced in some legumes by autoclaving but, on the contrary, in other cases they increased or were not affected. These different effects may be due to different methods used to quantify them by the authors cited, other than processing conditions. Despite these reductions, the inclusion of a serving dose of autoclaved legumes (100 g) in our diet will provide a certain amount of phenolic compounds to contribute to the recommended minimum intake of 300 mg/g, and consequently, it would maintain a beneficial effect on human health related to this kind of compound [38].

5. Effect of Extrusion/Cooking on the Bioactive Compounds of Pulses

Forty-three papers we selected were concerned with the effect of extrusion/cooking on the bioactive compounds; of these, 13 papers examined the effect on the content of a-galactosides, 16 studied the effect on myo-inositol phosphates, 10 reported the effect on protease inhibitors (mainly trypsin inhibitors), 13 examined the effect on the phenolic compounds (mainly total phenols and tannins) and 10 papers examined their antioxidant activity.

5.1. Extrusion/Cooking Process

There are different processes that allow the inclusion of pulse flours in new food products with high nutritional values and health-giving attributes. One of these processing methods is extrusion/cooking, since it is a versatile technology that allows the development of different food products such as snacks, breakfast cereals, instant soups, sports foods, baby foods, meat analogues, etc. Extruded products show good nutritional, functional and sensorial properties able to satisfy the consumer's demands for nutritious, convenient and healthy foods adapted to the modern lifestyle [93].

The extrusion parameters used, mainly temperature (<100 °C) and moisture (<10%), shear forces, screw speed (<50 rpm) and pressure generated in the extruder (up to 200 MPa), help to knead, compress, plasticize and cook the starchy and/or proteinaceous raw materials [94]. When the modified material exits from the die of the extruder, there is an instant pressure drop that produces a puffing effect; therefore, extrusion/cooking can modify both the composition and the texture of the raw materials, producing more appealing foods [19,93]. Depending on the parameters selected, the products obtained can be very different: directly expanded, indirectly expanded, textured or co-extruded, which include food products such as ready-to-eat cereals, snacks, infant formulas, textured meat-like products, etc. [95]. In general, these kinds of products are developed with corn or rice starches. However, these products are considered to be low-nutrition and high-energy-density foods that may promote obesity, cardiovascular events or metabolic syndromes [96]. The use of pulses as ingredients in the formulation of these kinds of products is an economic way to improve their nutritional quality [13]. The high protein and dietary fiber contents of pulses make it necessary to optimize the extrusion parameters for each pulse and mixture used as raw materials. Recently, Pasqualone et al. [19] reviewed the optimal processing conditions for different legumes to obtain nutritious and well-expanded legume-based foods or extruded pulse ingredients. From a nutritional point of view, extrusion/cooking has positive effects, since it causes starch gelatinization, increases the soluble dietary fiber, reduces lipid oxidation and improves the retention of nutritive components, flavors and colors of extrudates [96,97]. Extrusion/cooking also affects the content of the bioactive compounds/anti-nutritional factors present in the raw materials. In the literature, there are studies that reported either their increase or their reduction, mainly depending on the raw material and the extrusion parameters; moreover, not all the bioactive compounds were affected to the same extent under the same extrusion conditions [13,18,37,65,93,98–104]. As has been stated previously in this review, in many cases, the elimination of some phytochemicals is not desirable, since small amounts are able to produce beneficial health effects against different chronic diseases [3,9,21,105].

The effect of extrusion/cooking on the content of the different bioactive phytochemicals reviewed is shown in (Tables 2 and 3).

Table 2. Content of different bioactive compounds (α-galactosides, inositol phosphates, protease inhibitors, phenolic compounds and antioxidant activity) in extruded legumes.

Legumes	Extrusion Conditions	α-Galactosides	Inositol Phosphates	Protease Inhibitors	Phenolic Compounds	Antioxidant Activity	References
Kidney bean (unsoaked)	140 °C and 180 °C; 18% and 22% moisture		9.64–10.90 mg/g ↓	n.d. *	Tannins: 196–223 mg/100 g ↓ Total phenols: 539–621 mg/100 g ↓		[104]
Kidney bean (pre-soaked)	140 °C and 180 °C; 18% and 22% moisture		9.53–10.41 mg/g ↑	n.d.	Tannins: 171–190 mg/100 g ↓ Total phenols: 413–494 mg/100 g ↓		[104]
Bean (four varieties)	120–140 °C; 25–30% moisture	27.75–36.20 mg/g ↑					[65]
Kidney bean var. Pinto	150–155 °C; 20% moisture	37.7 mg/g ↓	4.71 mg/g ↓		Tannins:2.75 g eq cat/kg ↓		[106]
Pinto bean	110 °C to 163 °C. 18.8 to 28.95% moisture	33.3–47.8 mg/g ↑					[107]
Navy and pinto bean (pre-soaked)	85 °C 36% moisture	30.45–33.65 mg/g ↓					[108]
Bean (three cultivars)	120 °C-180 °C. 14%-20% moisture				Total Phenols: 24.12–102.68 mg/100 g	TEAC: 69.26–77.88 µM Trolox/g ↓	[109]
Canavalia ensiformis	155 °C; 20% moisture			1.40 TIU/mg			[110]
Chickpea (unsoaked)	140 °C and 180 °C; 18% and 22% moisture		7.33–8.16 mg/g ↓	n.d.	Tannins: 190–245 mg/100 g ↓ Total phenols: 190–245 mg/100 g ↓		[104]
Chickpea (pre-soaked)	140 °C and 180 °C; 18% and 22% moisture		7.35–8.00 mg/g ↓ =	n.d.	Tannins: 195–214 mg/100 g ↑ Total phenols: 270–380 mg/100 g ↓		[104]
Chickpea	160 °C; 17% moisture	26.39 mg/g ↓					[111]
Chickpea (deffated)	130 °C; 14% moisture		12.6 µmol/g ↓		Total Phenols: 48.7 mg/100 mg ↑		[112]

Table 2. *Cont.*

Legumes	Extrusion Conditions	α-Galactosides	Inositol Phosphates	Protease Inhibitors	Phenolic Compounds	Antioxidant Activity	References
Chickpea (germinated and dehulled)	180 °C; 16% moisture		1.33 mg/g ↓		Total phenols: 7.35 mg GAE/g ↑	DPPH: 45.46% ↑	[113]
Faba bean (unsoaked)	140 °C and 180 °C; 18% and 22% moisture		6.05–6.86 mg/g ↑	n.d.	Tannins: 397–438 mg/100 g ↓ Total phenols: 635–750 mg/100 g ↓		[104]
Faba bean (pre-soaked)	140 °C and 180 °C; 18% and 22% moisture		4.80–5.00 mg/g ↑	n.d.	Tannins: 362–426 mg/100 g Total phenols: 618–644 mg/100 g ↓		[104]
Lentil	160 °C 17% moisture	14.02 mg/g ↓					[111]
Split lentils	140 °C-160 °C 100 °C; 14%-18%-22% moisture		0.08–0.57 mg/g ↓	0.013–0.049 TIU/mg ↓	Tannins: 0.011–0.065 mg CE/100 g ↓ Total phenols 2.4–5.1 mg GAE/g ↓		[114]
Pea cv Ballet	145 °C-25% moisture	46.9 mg/g ↑	11.23 mg/g ↓	0.34 TIU/g ↓	Tannins: 0.02 g CE/kg ↓ Total Phenols: 0.23 g/kg ↓		[106]
Pea (unsoaked)	140 °C and 180 °C; 18% and 22% moisture		7.90–8.34 mg/g ↓	n.d.	Tannins: 236–278 mg/100 g ↓ Total phenols: 392–430 mg/100 g ↓		[104]
Pea (pre-soaked)	140 °C and 180 °C; 18% and 22% moisture		7.14–7.60 mg/g↓	n.d.	Tannins. 200–233 mg/100 g ↓ Total phenols: 343–379 mg/100 g ↓		[96]
Pea	160 °C 17% moisture	23.45 mg/g ↓					[111]
Pea	150–155 °C; 20% moisture	46.9 mg/g =	4.10 mg/g ↓		Tannins: 0.02 g CE/kg ↓		[106]

* n.d. not detected; TIU: trypsin inhibitor units; CIU: chymotrypsin inhibitor units; CE: caechin equivalents; GAE: gallic acid equivalents; DPPH: 2,2-diphenyl-1- picrylhydrazyl assay; TEAC: Trolox equivalent antioxidant capacity assay; ↑: increase; ↓: decrease; =: no change.

Table 3. Bioactive compounds (α-galactosides, inositol phosphates, protease inhibitors, phenolic compounds and antioxidant activity) content in extruded legume mixtures with other materials.

Legume Mixtures	Extrusion Conditions	α-Galactosides	Inositol Phosphates	Protease Inhibitors	Phenolic Compounds	Antioxidant Activity	References
Bean/rice (30/70)	80 °C; 14% moisture		16.59 mg/g		Tannins: 7.57 mg/g		[14]
Red common bean (15, 30 and 45%)/corn starch	160 °C; 22% moisture		0.54–2.33 mg/g ↓	n.d. *	40.94–94.82 mg FAE/100 g ↓	DPPH: 213.93–642.33 μmol TE/100 g ↓ ORAC: 388.69–1527.27 μmol TE/100 g ↓	[37]
Navy common bean (15. 30 and 45%)/corn starch	160 °C; 22% moisture		0.56–2.19 mg/g ↓	n.d.	28.27–45.96 mg FAE/100 g ↓	DPPH: 41.42–126.23 μmol TE/100 g ↓ ORAC: 254.36–584.46 μmol TE/100 g ↓	[37]
Bean/carob/rice (different formulas: 20–40%/0–10%/50–80%)	125 °C; 20% moisture	19.73–34.30 mg/g ↑	3.65–6.11 mg/.g ↓	n.d.	Total phenols: 0.92–3.25 mg CE/g ↓	ORAC: 8.92–11.89 μmol Trolox/g ↑	[100]
Kidney bean/corn starch (80/20)	150 °C. 20% moisture	45.63 mg/g ↓	7.08 mg/g ↓	0.25 TIU/mg ↓			[115]

Table 3. *Cont.*

Legume Mixtures	Extrusion Conditions	α-Galactosides	Inositol Phosphates	Protease Inhibitors	Phenolic Compounds	Antioxidant Activity	References
Black bean with (0–4%) sodium bicarbonate	160 °C; 20% moisture	15.2–22.5 mg/g ↑					[116]
Bean/corn (60/40)	from 150 °C to 190 °C; from 14.5% to 18.0% moisture				Phenols: 6.46–17.40 mg GAE/g	CUPRAC: 9.55–37.02 μM Trolox eq/g; β-carotene 8.94–34.20%	[117]
Chickpea mixed with starch and fiber	150 °C. 20% moisture	14.38 mg/g ↓					[111]
Chickpea/corn starch (80/20)	150 °C. 20% moisture	44.21 mg/g ↓	5.62 mg/g ↑	2.52 TIU/mg ↓			[115]
Chickpea (germinated and dehulled) (10–30%)/corn/tomato pomace (5%)	180 °C; 16% moisture		1.20–1.00 mg/g ↓	n.d.	9.37–11.11 mg GAE/g ↑	DPPH: 53.82–56.33% ↑	[113]
Fermented chick-pea/yogurt/locus bean gum	140 °C and 150 °C; 17% moisture		6.12–10.38 mg/g↓	2.02–8.82 TIU/mg ↓	Tannins: 0.37–1.03 mg CE/g ↓		[118]
Faba bean/corn starch (80/20)	150 °C; 20% moisture	31.66 mg/g ↓	6.25 mg/g ↓	0.44 TIU/mg ↓			[115]
Lentil mixed with nutritional yeast	160 °C; 17% moisture	33.35–49.55 mg/g ↑	2.72–3.94 mg/g ↓	0.20–0.28 TIU/mg ↓			[97]
Lentil mixed with starch and fiber	150 °C. 20% moisture	12.82 mg/g ↓					[111]
Red lentil (dehulled)/fiber (wheat, apple, nutriose®) (4 different formulations with at least 68% lentil)	160 °C; 17% moisture	24.60–42.52 mg/g ↑	1.38–4.62 mg/g ↓	n.d	4.51–9.38 mg GAE/g ↑	DPPH 6.63–63.56 EC_{50} mg/mL ↓ β-carotene assay 2.66–9.75 EC50 mg/mL ↑↓ TBARS: 1.52–4.59 EC_{50} mg/mL ↓	[119,120]
Lupin/corn starch (80/20)	150 °C; 20% moisture	83.76 mg/g ↓	5.99 mg/g ↑	0.29 TIU/mg ↓			[115]
Pea/carob/rice (different formulas: 20–40%/0–10%/50–80%)	125 °C; 20% moisture	24.80–50.21 mg/g ↑	3.14–4.60 mg/g ↓	n.d*.	Total phenols: 2.19–5.55 mg CE/g ↓	ORAC:9.81–12.00 μmol Trolox/g ↑	[18]
Pea mixed with starch and fiber	150 °C. 20% moisture	38.7 mg/g ↑					[111]
Pea/corn starch (80/20)	150 °C. 20% moisture	74.63 mg/g ↑	7.86 mg/g ↑	0.24 TIU/mg ↓			[115]

* n.d. not detected; TIU: trypsin inhibitor units; GAE: gallic acid equivalents; eq cat: catechin equivalents; DPPH: 2,2-diphenyl-1-picrylhydrazyl assay; ORAC: oxygen radical absorbance capacity assay; CUPRAC: cupric reducing antioxidant capacity *assay*; TBARS: thiobarbituric acid reactive substance assay; TE: Trolox equivalents; ↑: increase; ↓: decrease.

5.2. Extrusion/Cooking Effect on the α-Galactosides of Pulses

Regarding the extrusion effect on the content of α-galactosides, the results found in the literature are controversial, since contradictions are reported for pulse extrudates. Moreover, the different oligosaccharides may be affected in a different way [75]. The extent of these modifications depends not only on the extrusion parameters, but also on the raw materials submitted to the processing. Varela et al. [115] studied the changes in total galactoside content in different pulse/corn starch mixtures extruded at 150 °C and 20% moisture. They reported an increase in the total galactoside content (around 13%) in extruded pea blends, but a decrease in kidney bean (3%) and faba bean (47%). Verbascose showed a reduction from 26% (lupin) to 90% (faba bean) after extrusion, while this oligosaccharide increased by 15% in extruded pea. After extrusion, the stachyose content increased from 11% to 35% (pea and faba beans, respectively), but decreased 15%, 18% and 66% in lupin, chickpea and common bean, respectively. On other hand, extruded pea, common bean and chickpea showed an increase in their raffinose content (23%, 97% and 2%, respectively), but a reduction was reported for faba bean (61%) and lupin (20%). In general, the observed increase has been related to the release of carbohydrates from the food matrix due to the cell damage that occurs during extrusion, which improved the oligosaccharide extraction. A similar trend was reported by Alonso et al. [106] in comparison to the raw seeds, in which the content of raffinose, stachyose and verbascose were reduced in extruded kidney bean at 150 °C and 20% moisture; however, in extruded peas (under the same conditions) only stachyose showed a drop (20%) in its content, and raffinose and verbascose increased their content (4% and 34%, respectively). These authors reported that some sugars can interact with proteins (Maillard reaction), resulting in reduced extractability of these sugars, which may explain the reduction observed after extrusion; whereas hydrolysis reactions that can occur during extrusion can lead to an increased content of some sugars [99]. Berrios and Pan [121] reported a reduction of the α-galactosides in extruded black bean, although the extent of this reduction was different according to the extrusion conditions used; however, Berrios et al. [116] reported that the extrusion of black bean with different percentages of sodium bicarbonate (160 °C, 20% moisture) did not significantly modify the content of α-galactosides. Berrios et al. [111] found both an increase and a decrease of raffinose and stachyose in extruded peas, chickpeas and lentils at 170 °C and 17% moisture. In extruded peas, raffinose and stachyose showed a reduction (52% and 75%, respectively) compared to the raw seeds; the opposite trend was reported in extruded chickpeas, where the raffinose and stachyose content increased after extrusion (25% and 33%, respectively). Ai et al. [65] studied the extrusion effect at 120 °C and 25% moisture, and 140 °C and 30% moisture, on four bean varieties. All the bean varieties showed the same trend, with extrudates containing lower amounts of raffinose but higher amounts of stachyose than the raw samples. Borejszo and Khan [107] reported a reduction of raffinose and stachyose from 44% to 60% in extruded pinto bean, with a higher reduction at a higher extrusion temperature (110–163 °C). Extrusion of pre-soaked pinto bean seeds at a low temperature (85 °C) and 36% moisture [108] resulted in a 20% decrease in raffinose and a 10% decrease in stachyose content compared to the raw seeds. The different reduction trends observed may be due to differences in raffinose and stachyose solubility. Some authors [18,97,100,103,120] have developed pulse-based snacks fortified with different fiber sources with good sensorial acceptability. Morales et al. [120] reported that the snacks formulated with lentils presented a higher α-galactoside amount than those of the unprocessed mixes, from 2% to 31% depending on the formula ingredients (wheat bran or corn and apple fibers). This effect could be due to a more effective extraction of sugars in the extruded food matrix. Verbascose decreased in almost all formulations (from 1% to 21%), while stachyose and raffinose increased from 4% to 45% and from 8% to 55%, respectively. This could be due to the partial hydrolysis of the longer oligosaccharide under the high temperature (160 °C) and pressure of the extrusion process, which produced fewer raffinose family oligosaccharides. Ciudad-Mulero et al. [97] also developed lentil-based snacks, but enriched with different percentages of a nutritional yeast (*Saccharomyces cerevisiae*) extract using two extrusion temperatures (140 °C and 160 °C) at 17% moisture. After extrusion, with the exception of the sample formulated with 12% yeast, the samples showed a significant increase in total α-galactosides, with this increase being higher in the samples extruded at 160 °C. While raffinose and verbascose increased in all the samples, the stachyose content did not show a consistent tendency. Arribas et al. [18,100] developed snacks formulated with different percentages of rice, and pea or bean, fortified with whole carob bean flour and extruded at 125 °C and 20% moisture. All the extrudates showed higher content of α-galactosides than the non-extruded samples. Rice/bean and pea/rice snacks contain up to twice as many α-galactosides than their corresponding controls, and the highest pulse content produced the highest sugar increase. The sugar pattern was verbascose > stachyose > raffinose in the rice/pea extrudates and stachyose > raffinose in the rice/bean extrudates. In the pea-based snacks, on average, extrusion produced an increase of verbascose and stachyose of around 52% and around 100% of raffinose; whereas in bean-based snacks, the increase was around 48% and 200% for stachyose and raffinose, respectively. The authors related these increases to the higher extractability of sugars from the extruded matrix and to a hydrolysis of the sugars during the extrusion. The data evaluated show that the content in α-galactosides in the extruded products from different legume seeds ranged from 15.2 mg/g to 47.8 mg/g in beans, from 14.38 mg/g to 44.21 mg/g in chickpeas and from 12.82 mg/g to 14.02 mg/g in lentils. Only one paper studied the a-galactosides content in extruded faba bean and lupin seeds, with their content after extrusion/cooking being 31.66 mg/g and 83.76 mg/g, respectively. Despite many authors reporting an influence of the seed variety on the results, only five papers detailed the variety used in their studies. We can conclude that, in general, the α-galactosides are reduced after extrusion cooking up to 68%, depending on the legume seed and the extrusion conditions. In different types of common beans, the reduction reported in eight papers was between 3% and 33%, but five papers reported an increase in the α-galactosides content, from 2% to 97%. The reduction documented for chickpeas was from 17% to 68%; regarding the extrusion effect on lentils, both a decrease from 60% to 63% and an increase from 2% to 31% has been reported. In extruded peas, the a-galactosides content was reduced from 1.2% to 15%, although two papers reported an increase from 13% to 35%. Faba bean and lupin were reduced by 47% and 23%, respectively. Many of the extruded legumes are of great interest in the development of snacks or breakfast cereals; therefore, considering one serving of these types of extruded product (40 g) can supply 483–560 mg of α-galactosides for lentils, 938–1876 mg for peas, 5575–1768 mg for chickpeas and 608–1912 mg for beans. All the extruded legumes are a good and adequate source of galactosides able to exert a healthy influence [25].

5.3. Extrusion/Cooking Effect on the Myo-Inositol Phosphates of Pulses

The inositol phosphates are also affected by extrusion. A thermal hydrolysis of the phytic acid (IP6) into less phosphorylated forms has been reported, with the phytic acid reduction being higher as the extrusion temperature is increased [27,30,114]. As phytate can bind other macromolecules, the formation of insoluble complexes may occur during extrusion, which reduces their extractability, thus explaining its reduction [122]. This reduction could improve the mineral bioavailability and allow the health roles associated with the inositol phosphates. Most of the literature reported a decrease in the total inositol phosphates and the phytic acid (IP6), although the extent of this reduction varied with the raw materials and the extrusion conditions. El-Hady and Habiba [104] observed a reduction in IP6 content (2–9%) in extruded faba bean, pea, chickpea and kidney bean, with the higher temperature (180 °C) and moisture (22%) used being more effective in decreasing phytic content than the lower ones (140 °C and 18% moisture). A similar trend was reported by Anton et al. [37] in an extruded navy or small red bean/corn starch blend. They observed an overall reduction in total inositol phosphates of 44%. A higher reduction of phytic acid has been reported by Rathod and Annapure [114]. These authors found that the increase in extrusion temperature (140 °C, 160 °C and 180 °C) and moisture (14%, 18% and 22%) produced a decrease in phytic acid up to 99%. In contrast, Lombardi-Boccia et al. [123] observed that the legume phytate content in mottle and white beans, faba beans, chickpeas and lentils was not affected by extrusion at 130 °C and 28% moisture. In extruded chickpeas at 130 °C and 14.5% moisture, Poltronieri et al. [112] observed a reduction in neither total inositol phosphates nor phytic acid; however, they observed a significant reduction in IP5. Alonso et al. [99] also reported that the phytate content did not change significantly in extruded breakfast meals formulated with 30% bean flour and broken rice [14], or in extruded peas at 145 °C and 25% moisture. Morales et al. [119] reported a general reduction in total inositol phosphates from 14% to 50% in snacks produced at 160 °C and based on lentils and different fibers, depending on the formulation. The addition of different types of fibers produced different matrixes with a visco-rheological behavior during extrusion; thus, the extrudate formulas containing corn and apple fiber showed a non-significant reduction. IP6 showed the highest reductions, while IP5 and IP4 forms increased their content, on average, by 13% and 26%, respectively. Similarly, Alonso et al. [106] reported an increase in IP5 in extruded kidney beans (5.2-fold) and peas (1.5-fold) compared to raw seeds. However, Ciudad-Mulero et al. [97] reported not only an IP6 decrease ranging from 35% to 48% for lentil/yeast samples extruded at 140 °C and 160 °C, but also an IP5 and IP4 reduction from 19% to 51% and 17% to 43%, respectively. Hegazy et al. [113] produced snacks based on germinated and dehulled chickpeas (up to 30%) and corn. The extrusion at 180 °C and 16% moisture decreased the phytic acid content by 41–46% compared to extruded corn as a control. After extrusion at 150 °C and 20% moisture of different pulse/corn starch blends (80/20), Varela et al. [115] observed both an increase and a reduction of the total inositol phosphate content, depending on the pulse type. While extruded bean, chickpea and faba bean showed a significant reduction (7%, 8% and 36%, respectively), pea and lupin showed a significant increase in the total inositol phosphate content (3% and 24%, respectively). Arribas et al. [18,100] documented a reduction of phytic acid by 10% in extruded snacks (125 °C, 20% moisture) formulated with different proportions of rice/bean or pea/carob fruit; moreover, an increase of 16–70% of the less phosphorylated forms (IP4–IP5) was observed. These authors also reported that the higher legume content caused the higher phytate reduction. Thus, from a nutritional and health point of view, extrusion would be a good processing method to reduce phytic acid content and obtain the benefits of the less phosphorylated forms. Yağci et al. [118] reported the effect of extrusion and locus bean gum content on the phytic acid content of fermented chickpea-yogurt blends. Extrusion was developed at 140 °C and 150 °C with 17% moisture, and the gum content varied from 2.5% to 4%. The phytic content in the fermented blends was reduced by 31.3–41.3% after the extrusion treatment. The highest reduction was obtained in the blends containing 4% of locus bean gum extruded at 150 °C, and the lowest one corresponded to the blends with 1% of gum extruded at 150 °C.

In addition, by comparing the results reported in the literature reviewed, the content of myo-inositol phosphates ranged from 0.54 mg/g to 16.59 mg/g in beans, from 1.00 mg/g to 12.6 mg/g in chickpeas, from 6.05 mg/g to 6.86 mg/g in faba beans, from 0.08 mg/g to 0.57 mg/g in lentils and from 3.14 mg/g to 11.23 mg/g in peas. Regarding the effect of extrusion/cooking on the myo-inositol phosphates content, in general a reduction of its content was observed, with a high variability in the reduced percent (from 1% to 99%), although in five reviewed papers, an increase was reported. The inositol phosphates content decreased from 1% to 21% in beans, from 1% to 53% in chickpeas, from 95% to 99% in lentils and from 6% to 21% in peas; however, an increase was documented in the myo-inositol phosphates content of 3% for chickpeas, lupin and peas, and of 22% for beans. Assuming a protein content for these extrudates of around 20%, one serving (40 g) of the extruded products, similar to a snack than can supply less than 10 mg/g to improve the mineral absorption [30], were the extruded lentils (0.4–2.85 mg/g protein), some of the extruded chickpeas (5 mg/g protein), beans (0.27 mg/g protein) and faba bean (5.99 mg/g protein).

5.4. Extrusion/Cooking Effect on the Protease Inhibitors of Pulses

Protease inhibitors are proteinaceous heat-labile compounds, so their content is significantly reduced by extrusion cooking; depending on the seed, the formulation of the raw flours and the extrusion parameters used, the extent of the reduction will vary, even to zero [122]. Extrusion at 150 °C and 20% moisture of some pulse/starch mixtures produced a different percentage of reduction in the trypsin inhibitory activity. The lowest reduction corresponded to lupin blends (77%), followed by chickpea (89%), faba bean (93%), pea (95%) and bean (99%). These differences could be related to the presence of different inhibitor isoforms, some of which are more resistant to thermal treatment. This fact was confirmed by zymogram gels. Other authors [18,37,100,104] reported that the trypsin inhibitors were abolished in corn starch-based snacks with navy or red bean produced at 160 °C and 22% moisture, as well as in pre-soaked and extruded (140 °C–180 °C, 18–22% moisture) pea, chickpea, faba bean and kidney bean, and in extrudates (125 °C, 20% moisture) based on rice, pea or bean and carob bean mixtures. Zamora [110] reported that the extrusion of *Canavalia ensiformis* (155 °C, 20% moisture) affects the chymotrypsin and trypsin inhibitor activity by 99% and 95%, respectively. In contrast, other authors [99,124] reported that chymotrypsin inhibitor activity was lowered less extensively than trypsin inhibitor activity in pea cv. Ballet when extruded at 145 °C, 25% moisture (65% vs. 95%), and in faba bean var. Aguadulce when extruded at 140 °C and 40% moisture (42% vs. 92%). Rathod and Annapure [114] reported a drastic reduction in trypsin inhibitors (up to 99.5%) in extruded split lentils; as the extrusion temperature (from 140 °C to 180 °C) and moisture (from 14% to 22%) increased, a higher increase in the removal of trypsin inhibitors was observed. Reductions in the trypsin inhibitor activity from 93% to 95% have been reported in extruded lentil enriched with nutritional yeast at 140 °C and 160 °C. The highest reduction corresponded to the samples with the highest lentil content, while the higher yeast inclusion produced the lower trypsin inhibitor activity reduction [103]. Morales et al. [120] observed different trypsin inhibitor content in extruded lentil-based blends enriched with different types of dietary fiber. After extrusion (160 °C, 17% moisture), all samples presented a reduction in the trypsin inhibitor content, although the highest decrease (97%) corresponded to extruded lentils plus apple fiber and/or corn flour, and the lowest one (93%) corresponded to blends of lentils with wheat bran. Extrusion cooking of pigeon peas and African yam beans at 100 °C and 140 °C and 16% moisture significantly reduced the protease inhibitor activity (up to 98%); however, the trypsin activity of bambara groundnut extruded at 100 °C was not modified [125]. The extrusion of fermented chickpea-yogurt and locus bean gum blends reduced the trypsin inhibitor activity in the range of 77–79%, with the higher reduction corresponding to the extrudates obtained at 150 °C with a 4% of gum content [118].

The protease inhibitors are thermo-labile compounds; therefore, they were reduced from 77% in lupin to 99% in beans. These differences are related to the different thermo-stability of the different isoforms present in each legume [31]. However, in general, protease inhibitors were not detected after extrusion. The higher activity was determined in chickpea (2.52 TIU/mg), and the lowest one in lentil (0.013 TIU/mg). Bean, faba bean, pea and lupin showed protease activity of around 0.25 TIU/mg. One serving of 40 g can supply 0.52–9.6 TIU, and since these amounts are below the concentration to impair protein digestibility, they are considered as safe products.

5.5. Extrusion/Cooking Effect on the Phenolic Compounds of Pulses and Their Antioxidant Activity

Considering the effects of extrusion on the content of phenols, both increases and decreases in their content have been reported, mainly depending on the type of phenolic compounds studied. In general, decreases have been related to decarboxylation of phenolic acids and/or their polymerization during extrusion, which reduce their extractability; whereas increases have been related to the phenols released from the cell-wall matrix; consequently, the antioxidant activity associated with the phenols is also affected by the extrusion. Korus et al. [109,126] studied the effect that extrusion parameters (two temperatures and two moistures) had on the phenols and their antioxidant activity in five cultivars of dry common bean. They observed that extrusion at the lower temperature (120 °C) and the higher moisture (20%) retained the highest proportion of phenols. In comparison to the raw samples, a reduction in the total phenols and the antioxidant activity was observed, and these changes were cultivar-dependent. On average for the two temperatures used, the Rawela cv. (dark-red bean) increased its total phenol content (14%) after extrusion, while Toffi (black-brown bean) and Tip-top (cream bean) varieties decreased their content (by 19% and 21%, respectively). The different phenolic-compound contents after extrusion also are affected by the cultivar. Other authors [102,104,114] have also reported a significant decrease in the phenolic-compound content and the antioxidant activity of whole peas, faba beans, chickpeas and kidney beans, and the extent of this reduction was linked to the extrusion parameters used. Changes in the phenolic composition of extrudates, based on cereal/pulses or vegetable/pulse blends, have also been studied, since the food matrix highly conditions phenol bioaccessibility. Delgado-Lincon et al. [117] examined the effect of five extrusion temperatures (150 °C–190 °C) and four moisture levels (14.5–18%) on the phenol content and the antioxidant activity of a bean/corn blend (60/40). They observed a significant decrease in total phenols, flavonoids and antioxidant capacity, and they concluded that extrusion at 142 °C and 16.3% moisture retained the highest amount of total phenols and flavonoids, and showed the highest antioxidant capacity. Anton et al. [37] reported a significant decrease in the total phenol content and antioxidant activity of corn-starch snacks containing two different bean flours. The red bean/corn snack showed a higher reduction in total phenol content and antioxidant activity (around 70% and 65%, respectively) than the navy bean/corn snacks (around 10% and 22%, respectively). Lentil and nutritional yeast extruded at 140 °C and 160 °C showed decreases in total phenolic compounds, individual phenolic compounds and antioxidant activity compared to the raw blends. The observed decrease was correlated to the extrusion temperature. Also, a reduction in the tannin content was reported by Carvalho et al. [14] in extruded rice/bean flour blends, by Alonso et al. [91,104] in pea and kidney bean meals and by Rathod and Annapure [114] in split lentils, reaching a reduction of up to 50%, 92% and 99%, respectively. Conversely, other authors reported increases in both total phenol and antioxidant activity; for example, in extruded lentil flours enriched in fibers, Morales et al. [119] found an increase in total phenolics, as well as in most polyphenolic fractions (except in flavonols), with an increase in antioxidant activity. Arribas et al. [100] reported an increase in tartaric esters (on average 11%), anthocyanins (on average 24%) and total phenol content (on average 36%), while flavonoids did not vary significantly in different extruded rice/bean/carob flour formulations compared to their raw counterparts. A slight increase in the antioxidant activity (on average 5.4%) was observed, and the antioxidant activity of the extrudates was positively correlated with their phenolic content. Díaz-Batalla et al. [127] reported that extrusion at 150 °C of *Prosopis laevigata* (legume tree) seeds reduced the content of total phenolic compounds by 3%, but increased the DPPH radical scavenging capacity by 2.30%. They did not observe an increase the content of Maillard reaction product compared to the raw samples. Hegazy et al. [113] reported an increase in the total phenolic content (1.92–7.94%) and antioxidant activity (1.07–5.55%) in corn snacks enriched with different proportions of germinated chickpea and tomato pomace after extrusion. The selection of the raw materials, their blends with other food ingredients and the selection of the extrusion parameters (mainly temperature and moisture) are very important in order to obtain end-products containing adequate amounts of all types of phytochemicals with health-giving benefits.

In summary, the phenolic compound and tannin content was mainly reduced after extrusion by up to 99%, although a high variability can be observed depending on the seed and the extrusion conditions, as well as the method of analysis used. The antioxidant activity was only determined in nine papers; however, it is difficult to compare the results due to the different methods used (DPPH, ORAC, TBARS, etc.). The phenolic-compound content in extruded beans was reduced up to 74% and 70% for total phenols and tannins, respectively; however, one paper reported an increase in the total phenol content in extruded beans from 30% to 38%. Four papers reported an increase in the antioxidant activity, and five papers described its reduction. These differences are mainly related to the different methods of analysis. One serving (40 g) of the extruded product can supply more than the minimum daily dose (300 mg) of total phenolic compounds to obtain health benefits.

6. Effect of Cold Extrusion on the Bioactive Compounds of Pulses

Thirty-seven papers were reviewed in this section. From them, 16 studied the elaboration of pasta by cold extrusion, and the effect of the process and/or its subsequent cooking necessary for its consumption. Two papers evaluated the effect of the cooking process on galactosides, while three of them studied the modification in the inositol phosphate content, the inhibitor protease activity and/or the phenols content after the cooking process. Furthermore, four papers showed the effect of the cooking process on the antioxidant capacity.

6.1. Cold Extrusion Process

Cold extrusion is the standard process used for pasta production. It is a process similar to extrusion/cooking, with a single screw and high pressures, below 50 °C (in comparison to the temperatures in extrusion/cooking (Section 5), which are above 100 °C) [128]. In this type of extrusion, the product is made without cooking, so that the raw materials are subjected to minimal modifications both by friction and temperature [129].

The conditions used in extrusion/cooking might not be ideal, since they reduce the content of thermolabile compounds such as vitamins, functional proteins and flavors that can be incorporated into the formulation. However, cold extrusion allows these to be kept in the final product [130]. Pasta consumption is very high due to its palatability, long shelf life and simplicity of cooking, in addition to its nutritional characteristics [131]. The process of making pasta consists of transforming the flours into pasta in a homogeneous way; to achieve the required texture, the flour is hydrated by mixing, and then it is extruded. There are four different stages in the preparation of pasta. First, the kneading, which consists of adding water to the mixture of flours used for the preparation of the pasta. The final dough obtained contains an average moisture between 30% and 32% [132,133]. At this stage, the mixing is carried out for approximately 10 min, avoiding the incorporation of air into the mass. The presence of small air bubbles can weaken the structure of the pasta and activate enzymes [134]. Second, the cold extrusion process, during which the moisture content of the dough is approximately 28%. The extrusion process occurs at a low temperature and with continuous pressure along the worm screw, ensuring that a temperature of 50 °C is not reached to avoid any deterioration of the proteins (as this would reduce the quality of the pasta during subsequent cooking). The dough comes out in a tube (the exit hole through which the dough comes out has the desired shape of the pasta: spaghetti, fettuccine, macaroni, etc.), where a minimum expansion takes place. The third step is drying, with the aim of producing a resistant and stable pasta. The moisture on the surface of the pasta is reduced by hot air currents, creating a moisture gradient within the pasta [135]. The drying cycles, during which the pasta acquires its final color, are finished by reaching a moisture content of less than 12% [136]. Cold extrusion is an ideal process to produce pasta fortified with legumes, although this fortification is a new process. The principal studies in the literature showed the fortification of wheat flour with legumes such as soya, pea, lentil, chickpea, lupin and bean [137–154]. These authors revealed the modifications observed in the pastas fortified in comparison to the control pasta (in general, durum wheat semolina). However, only some studies showed the differences between the bioactive compounds in the uncooked and cooked fortified pasta.

The effect of cold extrusion on the content of the different bioactive phytochemicals reviewed is shown in (Tables 4 and 5).

Table 4. Content of different bioactive compounds (α-galactosides, inositol phosphates, protease inhibitors, phenolic compounds and antioxidant activity) in uncooked legume-based pasta.

Legumes	α-Galactosides	Inositol Phosphates	Protease Inhibitors	Phenolic Compounds	Antioxidant Activity	References
100% legume (faba, lentil or black-gram flours)	44–57 mg/g	14.5–18 IP6 mg/g	7.8–11.3 TIU/mg			[155]
Commercial samples manufactured with lentil, bean and pea				Total phenols: 632–1743 mg GAE./100 g		[156]
Bean and carob fruit/rice	10–45 mg/g	3–9 mg/g IP total 2–5 mg/g IP6	2–13 TIU/mg 2–17 CIU/mg	Total phenols: 2.9–7.3 mg CE/g	ORAC: 5.3–13.7 μmol Trolox/g	[157]
Hard-to-cook bean protein hydrolysate/wheat semolina					ABTS: 15–31 mM/mg	[146]
Lima bean and cowpea/wheat					TAC: 26.09–31.84 mg Trolox eq./g	[158]
Faba bean/wheat				Total phenols: 185 mg GAE/100 g	ORAC: 1017 mg Trolox eq./100 g	[159]
Germinated and fermented cowpea flour/wheat semolina		677 mg IP6/g			TAC:31.9 μmol Trolox eq./g	[160]
Lentil				Total phenols: 1.42–2.14 mg GAE/g		[131]
Pea			0.45–1.09 TIU/mg			[151]
Pea flour, red lentil flour or 60% grass pea and 40% chickpea flour				Total phenols: 87–176 mg GAE/100 g	ORAC: 1851–3789 mg Trolox eq./100 g	[161]

CE: catechin equivalents; GAE: gallic acid equivalents; TAC: total antioxidant capacity; ORAC: oxygen radical absorbance capacity assay; ABTS: free radical scavenging ability assay.

Table 5. Content of different bioactive compounds (α-galactosides, inositol phosphates, protease inhibitors, phenolic compounds and antioxidant activity) in cooked legume-based pasta.

Legumes	α-Galactosides	Inositol Phosphates	Protease Inhibitors	Phenolic Compounds	Antioxidant Activity	References
100% legume (faba, lentil or black-gram flours)	12–34 mg/g ↓	12.8–15 IP6 mg/g ↓	1.52–2.5 TIU/mg ↓			[155]
Commercial samples manufactured with lentil, bean and pea				Total phenols: 323–814 mg GAE/100 g ↓		[156]
Bean and carob fruit/rice	4–27 mg/g ↓	3–8 mg/g IP total = 2–5 mg/g IP6 =	2–17 TIU/mg ↓ 3–17 CIU/mg ↑	Total phenols: 2.3–5.8 mg CE/g ↓	ORAC: 5.1–10.3 μmol Trolox/g ↓	[157]
Bean/semolina				Total phenols: 6.45–9.68 mg CE/g		[162]
Lima bean and cowpea/wheat					TAC: 16.63–21.39 mg Trolox eq./g ↓	[158]
Hard-to-cook bean protein hydrolysate/wheat semolina					ABTS: 6.6.-7.9 mM/mg ↓	[146]
Carob fiber/wheat semolina				Total phenols: 3.21–4.8 mg GAE/g		[163]

Table 5. *Cont.*

Legumes	α-Galactosides	Inositol Phosphates	Protease Inhibitors	Phenolic Compounds	Antioxidant Activity	References
Carob pod/wheat semolina				Total phenols: 5.3–12.2 mg GAE/g	ABTS: 0.25–1.35 mg Trolox eq./g FRAP: 2.05–6.9 mg Trolox eq./g	[164]
Chickpea/soya/different cereals		1050–9640 mg IP6/Kg				[138]
Germinated and fermented cowpea flour/wheat semolina		574 mg IP6/g ↓			TAC: 27.5 μmol Trolox eq./g	[154]
Lentil				Total phenols: 0.93–1.42 mg GAE/g ↓		[131]
Pea			0.23–0.29 TIU/mg ↓			[151]
Fermented pigeon pea/wheat semolina	n.d. *	0.24 g/100 g	n.d.	Tannins: 0.28 g GAE/100 g	TAC: 3.35 μmol Trolox eq./g	[153]
Germinated pigeon pea/wheat semolina		0.27 g/100 g	1.57 TIU/mg	Tannins: 0.19 g GAE/100 g	TAC: 5.8 μmol Trolox eq./g	[160]

* n.d.: not detected; GAE: gallic acid equivalents; CE: catechin equivalents; TAC: total antioxidant capacity; ORAC: oxygen radical absorbance capacity assay; FRAP: ferric reducing antioxidant power assay; ABTS: free radical scavenging ability assay; ↑: increase; ↓: decrease; =: no change.

6.2. Cold Extrusion Effect on the α-Galactosides of Pasta Fortified with Legumes

Torres et al. [146] produced pasta products based on whole durum wheat semolina fortified with proportions of 5%, 10% and 12% of fermented pigeon pea (*C. cajan*). Fermentation brought about an 82% reduction in α-galactosides in the seeds used, since the oligosaccharides were hydrolyzed by α–galactosidase and invertase either as an enzyme or from microorganisms present. Taking into account the previous fact, this study showed that the pasta fortified with 10% fermented pigeon flour did not report α-galactoside content (raffinose, stachyose and verbascose), as with the control pasta analyzed. Nevertheless, Laleg et al. [155] showed that the content of total α-galactosides ranged from 44 to 57 mg/g (d.m.) in the uncooked pasta, higher in the 100% lentil pasta than the 100% faba pasta and the 100% black-gram (*Vigna mungo*) pasta. The stachyose and verbascose were present in the oligosaccharide profiles in the uncooked pasta, and raffinose was the minor oligosaccharide in the pasta studied. On the other hand, after the cooking process, the α-galactoside content was reduced from 41–73% because of the heat hydrolysis or leaching in the cooking water. The optimal cooking time was a parameter that determined the concentration of soluble compounds such as α-galactosides in the pasta ready to eat. This effect was observed in the cooking of fettuccine based on rice and fortified with two different legumes (bean (*P. vulgaris* L.) and carob fruit (*Ceratonia siliqua* L.)) [148]. The cooked pasta showed around 4–27 mg/g content of α-galactosides, while the uncooked pasta showed around 10–57 mg/g. The difference observed (about 40–70%) could be related to the water solubility in the cooking process. These compounds migrated to the cooking water and were discarded. Taking into account the reviewed data, we can conclude that the cooking process after the preparation of the pasta by cold extrusion modified the content of α-galactosides, reducing it between 21 and 82%, in pasta elaborated with bean/carob fruit and 100% legume, respectively. The content of α-galactosides ranged in these samples between 4 and 57 mg/g. In addition, it is important to note that the processing reduced but did not eliminate these compounds, so their consumption would be beneficial to human health. The consumption of one serving of cooked pasta (70 g) will provide around 280–2380 mg of α-galactosides. This range allows the consumption of a quantity close to the one that has been observed to convey benefits on the intestinal flora [25].

6.3. Cold Extrusion Effects on the Myo-Inositol Phosphate Content of Pasta Fortified with Legumes

Torres et al. [153] reported that pasta products fortified with pigeon pea had a higher content of phytic acid than the control produced with 100% semolina. On the other hand, in the pasta fortified with germinated *C. cajan* in proportions of 5%, 8% and 10%, the phytic acid presented twice the amount than the control pasta [149]. Herken et al. [154] analyzed the phytic acid content using the colorimetric method in unprocessed, germinated, fermented and cooked pasta made from cowpeas. These authors showed that the germination, fermentation and thermal processes reduced the phytic acid content from 13.3% to 15.2%, with respect to the unprocessed samples. The 100% legume pasta revealed phytic acid contents from 14.5 to 18 mg/g in the uncooked pasta. A slight reduction in this content was observed in the cooked pasta (from 12.8 to 15 mg/g) related to the corresponding flours. A low content in phytic acid in the cooked pasta was associated with high iron bioavailability and a general nutritional quality [155]. On the other hand, Arribas et al. [157] showed a slight increase ($p > 0.05$) in the total inositol phosphate (the different IP forms included phytic acid or IP6), which can be explained by losses during cooking; these compounds migrated to the cooking solution, which caused changes in the percentage of the specific components on a dry-matter basis. Bilgicli et al. [138] studied the phytic acid content in pasta based on wheat, maize and rice flour fortified with soya and chickpea grains. They showed ranges of phytic acid from 7 to 9 times higher than the control noodle made from wheat flour.

In this case, the reduction in inositol phosphate content, regardless of the seed used in the production of the pasta, was around 15%. In the cooked pasta reviewed, the phytic acid content ranged from 3 to 574 mg IP6/g in the rice pasta fortified with bean and carob fruit and in germinated and fermented cowpea flours, respectively. Therefore, the consumption of one serving (70 g) of pasta would be associated with the health benefits of these compounds [3,9,22,23].

6.4. Cold Extrusion Effects on the Protease Inhibitors of Pasta Fortified with Legumes

Several authors have studied the inhibitor protease content in pasta fortified with different proportions of legumes. Torres et al. [146,149] showed, in pasta based on semolina and fortified with fermented and germinated *C. cajan* seed in proportions of 5% or 8% and 10% or 12%, respectively, that the trypsin inhibitor activity in the fermentation of pigeon pea seeds affected trypsin inhibitor activity and caused a reduction of 39%. This was not detected in the pasta produced with 8–10% fermented *C. cajan*. However, in the pasta with 10% germinated *C. cajan*, the trypsin inhibitor activity determined was 1.57 TIU/mg, which is an insignificant amount when compared to the trypsin inhibitor activity value found in the germinated flour (15.9 TIU/mg). The difference between the two processes, fermented and germinated, revealed that the fermentation of pigeon pea produces a significant reduction in the trypsin inhibitor activity, and the inclusion of this flour in small quantities produced pasta without trypsin inhibitor activity. That said, however, the use of germinated pigeon pea flours could be associated with the positive effect of these bioactive compounds.

Frias et al. [151] produced macaroni pea (*P. sativum* L.) products, and the trypsin inhibitor activity analyzed revealed that as the cooking time of the pasta increased, the reduction of this activity was 48–59%. Taking this into account, the cooked pasta had lower reported contents than the uncooked pasta, since heat inactivated the trypsin inhibitor activity, although it increased the legume protein digestibility. This effect was observed in pasta made with 100% legume [147]. The analyzed samples ranged from 7.8 to 11.3 mg/g, but the cooked pasta revealed five times lower trypsin inhibitor activity, and the cooking process in legume pasta was reduced by 38% to 96% of this activity. Thermal processing of legumes can be applied to totally or partially reduce the protease inhibitor activities because they are thermo-labile compounds. However, Arribas et al. [148] showed, in the fettuccine made with rice, bean and carob, a slight increase ($p > 0.05$) in trypsin and chymotrypsin inhibitor activities after the cooking process, in comparison to the uncooked samples. This could possibly be due to the short cooking time applied (on average 3 min) and to the different heat-sensitivity of both types of protease inhibitors.

The cooking process of the pasta showed in general a reduction in the protease inhibitor content (around 50–80%). However, in the pasta formulated with bean and carob fruit included in the review, a small increase in this content was observed. The protease inhibitor activity after the thermal process ranged between 2–17 TIU/mg and 3–17 CIU/mg. The cooking process (100 °C, 3–10 min) was not so severe to eliminate these compounds, unlike in other products. One serving (70 g) of the pasta would supply from 140 to 1190 TIUs and from 210 to 1190 CIU. These amounts are below the content reported to impair protein digestibility (2000 CIU per day), and most of the reviewed pasta contained amounts in the range of 25–800 CIU/day, which is able to exert the health benefits associated with these bioactive compounds (Section 3.3).

6.5. Cold Extrusion Effects on the Phenols and Their Antioxidant Activity of Pasta Fortified with legumes

Phenols and their associated antioxidant activity are some of the most commonly analyzed features of products made by cold extrusion. Pasta fortified with 1, 2, 3, 4 and 5% carob fruit (*C. siliqua* L.) revealed higher contents of phenols using the Folin–Ciocalteau reagent (3.21–4.8 mg gallic acid equivalent (GAE)/g d.m.) and antioxidant activity analyzed by chelating power (CHEL), ABTS radical scavenging assay, ability to inhibit lipoxygenase and ferric-reducing antioxidant power (FRAP), more than the control sample made with common wheat flour [163]. Seczyk et al. [164] studied the fortification from 1%–5% of carob pod in pasta based on durum wheat semolina. The results obtained demonstrated that the carob pod flour increased the phenolic content determined (5.3–12.2 mg GAE/g d.m.) and the antioxidant activity analyzed (0.25–1.35 mg Trolox equivalent/g in the ABTS assay and 2.05–6.9 mg Trolox equivalent/g in the FRAP assay) showed the highest values and activity in the formulation with 5% of carob, such as in previous studies [163]. In comparison to the control, the pasta fortified with 5% carob pods revealed double the amount of phenolic content, and a three- to 18-fold increase in antioxidant activity. In the gluten-free pasta made with rice and fortified with different proportions of legumes (carob fruit 10%), the results obtained showed a total phenol content in the uncooked samples of around 2.9–7.3 mg catechin equivalent (CE)/g d.m., but after the cooking process, the total phenol content was reduced to around 17–48% (2.3–5.8 mg CE/g dw). The pasta fortified with 10% carob fruit presented a 2.5–3.2-fold higher amount of total phenols than the commercial sample made with rice and used as a control. The antioxidant activity by ORAC assay in the uncooked fettuccine revealed 5.2 to 13.7 μmol Trolox/g d.m. After the cooking process, the ORAC assay exposed 5.1–10.3 μmol Trolox/g d.m. in the fortified fettuccine. The cooking process showed a slight and non-significant reduction in the antioxidant activity [157]. These studies showed the potentialities of this legume, carob, to fortify products made with cold extrusion, since it is a good source of polyphenols and therefore of antioxidant activity, improving the health characteristics of traditional pasta based on wheat. In Turco et al. [161], pasta made of 100% pea flour, 100% red lentil flour, and 60% grass pea and 40% chickpea flour revealed that total polyphenols and flavonoids determined by the Folin–Ciocalteu method and antioxidant properties determined by ORAC were higher in the legume pasta than in the control made with durum wheat semolina. The pasta made with 60% grass pea and 40% chickpea was the formulation with the highest total polyphenol content, followed by red lentil pasta and pea pasta. All the experimental formulations revealed higher content than the pasta made with durum wheat semolina (control). Turco et al. [159], in pasta based on semolina and fortified with 35% faba bean, showed a total phenol content around two times higher compared to the control. Also, the higher content of flavonoids and antioxidant activity found in the ORAC assay was lower in the control pasta. Torres et al. [153,160] showed that the tannin content in the pasta fortified with 10% germinated pigeon pea flour (0.19 g galic acid/100 g d.m.) and fortified with 10% fermented pigeon pea flour (0.28 g galic acid/100 g d.m.) was lower than the control elaborated with 100% semolina. The fermentation process could improve the phenolic content in relation to germination. The antioxidant activity was analyzed in the fermented samples, and the results revealed that the fortification increased the total antioxidant activity by around 80% with respect to the control sample. Herken et al. [154] showed that in samples fortified with different proportions of cowpea flour (10%, 15% and 20%), the inclusion of 20% legume, fermented or germinated, increased the total antioxidant activity (31.16–34.51 µmol Trolox equivalent/g, respectively) in comparison to the control made with 100% semolina. However, it was observed that the antioxidant activity detected in the macaroni with 20% fermented flour showed lower activity than the macaroni with unprocessed flour ($p < 0.05$), and lower than the germinated ones, in contrast to other authors [153,160]. However, cooking had a negative effect on the antioxidant activity determined.

The inclusion of bean in pasta has been studied by different authors; Gallegos-Infante et al. [162] reported that the highest phenolic compound content detected was in the samples fortified with common bean flour (15–30%). Arribas et al. [157] showed the same tendency in the samples based on rice and fortified with two different legumes. The inclusion of bean in these samples increased the total phenolic content, and the antioxidant activity detected in the uncooked samples was 5.3–13.7 µmol Trolox/g, with higher legume content revealing higher antioxidant activity. After the thermal process, the same tendency was observed; however, the antioxidant activity was reduced in all formulations. Segura-Campos [146] showed in pasta made with durum wheat semolina and hard-to-cook bean protein hydrolysate (0%, 5%, and 10%) that the antioxidant activity determined by the ABTS decolorization assay increased antioxidant values. The highest antioxidant value determined was in the pasta fortified with 10% of hard-to-cook bean protein hydrolysate (31.4 mM/mg); however, the cooking process revealed lower antioxidant values (around 6.6–7.9 mM/mg sample in the pasta fortified with 5% and 10% hard-to-cook bean protein hydrolysate, respectively). In all these cases, we can conclude that the thermal process can degrade the bioactive compounds (phenols, peptides, etc.) and decrease the antioxidant activity analyzed. Spaghettis fortified with 5–10% legume hydrolysates showed that the pasta made with cowpea (*V. unguiculata*) hydrolysate revealed higher antioxidant activity (294 µmol Trolox/g), determined by the ABTS assay, than the lima bean (*P. lunatus*) hydrolysate (245 µmol Trolox/g) and the control analyzed. In general, a higher amount of legumes in the formulation increased the antioxidant activity, so the inclusion of 10% hydrolysate of *P. lunatus* in the pasta showed a twofold increase in antioxidant activity than the control, which could be due to the presence of low molecular weight peptides in the hydrolysate [158]. In commercial samples manufactured with lentils, beans and peas, these types of pasta revealed higher total phenolic compounds than cereal-based pastas, and the black lentil pasta revealed the highest total phenol content (1743.3 mg GAE/100 g d.m.). Black bean pasta, pea pasta and red lentil pasta also presented higher total phenol content than the cereal and pseudo-cereal-based pasta studied (781, 657.1 and 631.6 mg GAE/100 g d.m., respectively). After cooking, the free polyphenol fraction decreased by around 50% in all the legume pastas (323.3 to 813.5 mg GAE/100 g d.m.) in comparison to the original content in the samples, due to the solubilization in the cooking water and the thermal sensibility of these compounds [156]. Di Stefano et al. [131], in pasta based on semolina and fortified with 40% of lentil, showed around a 28% reduction of the total phenolic content after the cooking process.

The content of the total phenolic compounds in the cooked pasta was 0.93–1.42 mg GAE/g. All the data analyzed in this section showed that the cooking process reduced the content of phenolic compounds from 20–51%. The range obtained in the papers included in this review was between 0.93 and 814 mg GAE/g. in pasta fortified with 40% lentil and commercial pasta elaborated with legumes, respectively. One serving (70 g) of almost cooked pasta fortified with the legumes studied would provide an average of 0.56 g GAE/g, which is enough to obtain a healthy intake of phenolic compounds [45].

7. Concluding Remarks

It is well known that pulses are an inexpensive and sustainable source of nutrients, and they represent functional foods that are rich in different bioactive compounds, providing many health benefits when consumed regularly in a balanced diet. Pulses are processed prior to their consumption; therefore, even though a pulse seed contains a high amount of bioactive compounds, it does not mean that the processed pulse has a high content of the bioactive components, because their processing modifies their content. The extent of these modifications depends on the seed, the variety, the food matrix, the formulations and the technological processing. Galactosides were reduced up to 77%, 68% and 82% in autoclaved, extrusion/cooked and cold-extruded cooking samples, respectively, although some papers reported an increase in some of the extrusion/cooking processed samples. Myo-inositol phosphates were more affected by autoclaving and extrusion/cooking, reaching reduction of 2–100% and 1–99%, respectively. Cooked pasta reduced their content up to 15%. All the processes reviewed drastically reduced the protease inhibitor content (80–100%); therefore, we concluded that these samples would show an improved protein digestibility. The majority of the revised papers reported a reduction in phenolic compounds of up to 80%, 99% and 51% for autoclaved, extrusion/cooking and cold-extruded/cooked products. One serving of autoclaved legumes, extruded/cooked and cold-extruded/cooked products can supply on average 3.2, 1.8 and 1.3 g of galactosides, respectively; even though there is no recommended daily intake, according to the literature, it would be enough to obtain prebiotic benefits on the gut. Considering one serving of each product and depending on the legume seed and the processing conditions, some of the processed legumes could supply less than 10 mg/g protein, and thus improve mineral absorption. The low remaining trypsin inhibitor activity (<1 to 1.5 TIU/mg) in the processed legumes included in this review would have a positive effect on human health. One serving of a majority of the reviewed products can supply more than the minimum daily dose (300 mg) of total phenolic compounds to obtain health benefits. Therefore, from the research documented in this review, it can be concluded that processed pulses and pulse-based foods can supply not only nutritive compounds, but also significant amounts of bioactive compounds, such as galactosides, phytates, protease inhibitors and phenolic compounds, with the potential to contribute to human health and wellbeing. Notably, the health benefits reported for legumes include anticarcinogenic and antihypertensive effects, and improvements in cardiovascular disease, diabetes type 2 and obesity could be related to a synergistic combination of the bioactive components present in legume seeds. With the increasing innovations in food-processing technology, the use of pulses in the development of new food products can meet consumers' requirements for more nutritious and healthier products. Finally, it is interesting not only to know the phytochemical content of the processed legumes, but it also is necessary to know the amount of these compounds available in the gut once they are absorbed to exert their beneficial effect. In the near future, it will be necessary to carry out in vitro digestibility studies to determine the amount that reach the gut and are available to exert their healthy effect.

Supplementary Materials: The following are available online at https://www.mdpi.com/2304-8158/10/2/379/s1, Figure S1. A PRISMA flow chart for this manuscript.

Author Contributions: Conceptualization, M.M.P.; methodology, M.M.P., E.G., C.A.; resources, M.M.P., E.G., C.A.; data curation, M.M.P., E.G., C.A.; writing—original draft preparation, M.M.P., E.G., C.A.; writing—review and editing, M.M.P., E.G., C.A. All authors have read and agreed to the published version of the manuscript.

Funding: This research received no external funding.

Institutional Review Board Statement: Not applicable.

Informed Consent Statement: Not applicable.

Conflicts of Interest: The authors declare no conflict of interest.

References

1. Boye, J.; Zare, F.; Pletch, A. Pulse proteins: Processing, characterization, functional properties and applications in food and feed. *Food Res. Int.* **2010**, *43*, 414–431. [CrossRef]
2. Duranti, M. Grain legume proteins and nutraceutical properties. *Fitoterapia* **2006**, *77*, 67–82. [CrossRef] [PubMed]
3. Sparvoli, F.; Bollini, R.; Cominelli, E. Nutritional Value. In *Grain Legumes*; Ron, A.M.D., Ed.; Springer: New Work, NY, USA, 2015; pp. 291–326. [CrossRef]
4. Dogan, H.; Gueven, A.; Hicsasmaz, Z. Extrusion Cooking of Lentil Flour (*Lens Culinaris*–Red)–Corn Starch–Corn Oil Mixtures. *Int. J. Food Prop.* **2013**, *16*, 341–358. [CrossRef]
5. Pedrosa, M.M.; Cuadrado, C.; Burbano, C.; Muzquiz, M.; Cabellos, B.; Olmedilla-Alonso, B.; Asensio-Vegas, C. Effects of industrial canning on the proximate composition, bioactive compounds contents and nutritional profile of two Spanish common dry beans (*Phaseolus vulgaris* L.). *Food Chem.* **2015**, *166*, 68–75. [CrossRef] [PubMed]
6. Maphosa, Y.; Jideani, V.A. The Role of Legumes in Human Nutrition. In *Functional Food—Improve Health through Adequate Food*; Hueda, M.C., Ed.; IntechOpen: London, UK, 2017. [CrossRef]
7. FAO. Resolution 68/231. International Year of Pulses. 2016. Available online: http://www.fao.org/ (accessed on 6 May 2020).
8. *US Dietary Guidelines for Americans*; U.S. Department of Health and Human Services: Washington, DC, USA, 2005; Volume 2011.
9. Muzquiz, M.; Varela, A.; Burbano, C.; Cuadrado, C.; Guillamón, E.; Pedrosa, M.M. Bioactive compounds in legumes: Pronutritive and antinutritive actions. Implications for nutrition and health. *Phyt. Rev.* **2012**, *11*, 227–244. [CrossRef]
10. Carbonaro, M. 14-Role of pulses in nutraceuticals. In *Pulse Foods*; Tiwari, B.K., Gowen, A., McKenna, B., Eds.; Academic Press: San Diego, CA, USA, 2011; pp. 385–418. [CrossRef]
11. Singh, B.; Singh, J.P.; Shevkani, K.; Singh, N.; Haur, A. Bioactive constituents in pulses and their health benefits. *J. Food Sci. Technol.* **2017**, *54*, 858–870. [CrossRef] [PubMed]
12. Aguilera, Y.; Diaz, M.F.; Jimenez, T.; Benitez, V.; Herrera, T.; Cuadrado, C.; Martin-Pedrosa, M.; Martin-Cabrejas, M.A. Changes in nonnutritional factors and antioxidant activity during germination of non-conventional legumes. *J. Agric. Food Chem.* **2013**. [CrossRef]
13. Berrios, J. Extrusion Cooking: Legume Pulses. In *Encyclopedia of Agricultural, Food, and Biological Engineering*, 2nd ed.; CRC Press: Boca Raton, FL, USA, 2010; pp. 453–464. [CrossRef]
14. Carvalho, A.; de Andrade Mattietto, R.; Bassinello, P.; Nakamoto Koakuzu, S.; Rios, A.; Maciel, R.; Nunes Carvalho, R. Processing and characterization of extruded breakfast meal formulated with broken rice and bean flour. *Ciência e Tecnologia de Alimentos* **2012**, *32*, 515–524. [CrossRef]
15. Dueñas, M.; Sarmento, T.; Aguilera, Y.; Benitez, V.; Mollá, E.; Esteban, R.M.; Martín-Cabrejas, M.A. Impact of cooking and germination on phenolic composition and dietary fibre fractions in dark beans (*Phaseolus vulgaris* L.) and lentils (*Lens culinaris* L.). *LWT-Food Sci. Technol.* **2016**, *66*, 72–78. [CrossRef]
16. El Siddig, O.; El Tinay, A.; Abd Alia, A.; Elkhalifa, A. Proximate composition, minerals, tannins, in vitro protein digestibility and effect of cooking on protein fractions of hyacinth bean (*Dolichos lablab*). *J. Food Sci. Technol.* **2002**, *39*, 111–115.
17. Kamau, E.H.; Nkhata, S.G.; Ayua, E.O. Extrusion and nixtamalization conditions influence the magnitude of change in the nutrients and bioactive components of cereals and legumes. *Food Sci. Nutr.* **2020**, *8*, 1753–1765. [CrossRef]
18. Arribas, C.; Cabellos, B.; Cuadrado, C.; Guillamón, E.; Pedrosa, M. The effect of extrusion on the bioactive compounds and antioxidant capacity of novel gluten-free expanded products based on carob fruit, pea and rice blends. *Innov. Food Sci. Emer. Technol.* **2019**, *52*, 100–107. [CrossRef]
19. Pasqualone, A.; Costantini, M.; Coldea, T.E.; Summo, C. Use of legumes in extrusion cooking: A review. *Foods* **2020**, *9*, 958. [CrossRef]
20. van Boekel, M.; Fogliano, V.; Pellegrini, N.; Stanton, C.; Scholz, G.; Lalljie, S.; Somoza, V.; Knorr, D.; Jasti, P.R.; Eisenbrand, G. A review on the beneficial aspects of food processing. *Mol. Nutr. Food Res.* **2010**, *54*, 1215–1247. [CrossRef]
21. Singh, S.; Gamlath, S.; Wakeling, L. Nutritional aspects of food extrusion: A review. *Int. J. Food Sci. Technol.* **2007**, *42*, 916–929. [CrossRef]
22. Barman, A.; Marak, C.M.; Barman, R.M.; Sangma, C.S. Nutraceutical Properties of Legume Seeds and Their Impact on Human Health, Legume Seed Nutraceutical Research. In *Legume Seed Nutraceutical Research*; Jimenez-Lopez, J.C., Clemente, A., Eds.; IntechOpen: London, UK, 2018. [CrossRef]
23. Sanchez-Chino, X.; Jimenez-Martinez, C.; Davila-Ortiz, G.; Alvarez-Gonzalez, I.; Madrigal-Bujaidar, E. Nutrient and nonnutrient components of legumes, and its chemopreventive activity: A review. *Nutr. Cancer* **2015**, *67*, 401–410. [CrossRef] [PubMed]
24. Amarowicz, R.; Pegg, R.B. Legumes as a source of natural antioxidants. *Eur. J. Lipid Sci. Technol.* **2008**, *110*, 865–878. [CrossRef]
25. Martínez-Villaluenga, C.; Frias, J.; Vidal-Valverde, C. Alpha-galactosides: Antinutritional factors or functional ingredients? *Crit. Rev. Food Sci. Nutr.* **2008**, *48*, 301–316. [CrossRef] [PubMed]
26. Burbano, C.; Muzquiz, M.; Ayet, G.; Cuadrado, C.; Pedrosa, M.M. Evaluation of antinutritional factors of selected varieties of *Phaseolus vulgaris*. *J. Sci. Food Agric.* **1999**, *79*, 1468–1472. [CrossRef]
27. Greiner, R.; Konietzny, U. Improving enzymatic reduction of myo-inositol phosphates with inhibitory effects on mineral absorption in black beans (*Phaseolus vulgaris* var. Preto). *J. Food Process. Preserv.* **1999**, *23*, 249–261. [CrossRef]
28. Schlemmer, U.; Frølich, W.; Prieto, R.M.; Grases, F. Phytate in foods and significance for humans: Food sources, intake, processing, bioavailability, protective role and analysis. *Mol. Nutr. Food Res.* **2009**, *53*, S330–S375. [CrossRef]

29. Fredrikson, M.; Biot, P.; Alminger, M.; Carlsson, N.; Sandberg, A. Production Process for High-Quality Pea-Protein Isolate with Low Content of Oligosaccharides and Phytate. *J. Agric. Food Chem.* **2001**, *49*, 1208–1212. [CrossRef]

30. Hurrell, R.F.; Juillerat, M.A.; Reddy, M.B.; Lynch, S.R.; Dassenko, S.A.; Cook, J.D. Soy protein, phytate, and iron absorption in humans. *Am. J. Clin. Nutr.* **1992**, *56*, 573–578. [CrossRef] [PubMed]

31. Guillamón, E.; Pedrosa, M.M.; Burbano, C.; Cuadrado, C.; Sánchez, M.C.; Muzquiz, M. The trypsin inhibitors present in seed of different grain legume species and cultivar. *Food Chem.* **2008**, *107*, 68–74. [CrossRef]

32. Clemente, A.; Mackenzie, D.; Johnson, I.T.; Domoney, C. Investigation of legume seed protease inhibitors as potential anticarcinogenic proteins. In *Recent Advances of Research in Antinutritional Factors in Legume Seeds and Oilseeds*; Muzquiz, M.C., Cuadrado, M.M.P., Burbano, C., Hill, G.D., Eds.; EAAP Publication: Toledo, Spain, 2004; pp. 137–142.

33. Kennedy, A.R. Chemopreventive agents: Protease inhibitors. *Pharm. Ther.* **1998**, *78*, 167–209. [CrossRef]

34. Clemente, A.; Sonnante, G.; Domoney, C. Bowman-Birk Inhibitors from Legumes and Human Gastrointestinal Health: Current Status and Perspectives. *Curr. Protein Pept. Sci.* **2011**, *12*, 358–373. [CrossRef]

35. Kennedy, A.R. The Bowman-Birk inhibitor from soybeans as an anticarcinogenic agent. *Am. J. Clin. Nutr.* **1998**, *68*, 1406s–1412s. [CrossRef]

36. Roy, F.; Boye, J.I.; Simpson, B.K. Bioactive proteins and peptides in pulse crops: Pea, chickpea and lentil. *Food Res. Int.* **2010**, *43*, 432–442. [CrossRef]

37. Anton, A.; Fulcher, R.; Arntfield, S. Physical and nutritional impact of fortification of corn starch based extruded snacks with common bean (*Phaseolus vulgaris* L.) flour: Effects of bean addition and extrusion cooking. *Food Chem.* **2009**, *133*, 989–996. [CrossRef]

38. Oomah, B.D.; Cardadro-Martínez, A.; Loarca-Piña, G. Phenolics and antioxidative activities in common beans (*Phaseolus vulgaris* L.). *J. Sci. Food Agric.* **2005**, *85*, 935–942. [CrossRef]

39. Amarowicz, R.; Estrella, I.; Hernández, T.; Robredo, S.; Troszyńska, A.; Kosińska, A.; Pegg, R.B. Free radical-scavenging capacity, antioxidant activity, and phenolic composition of green lentil (*Lens culinaris*). *Food Chem.* **2010**, *121*, 705–711. [CrossRef]

40. Amarowicz, R.; Shahidi, F. Antioxidant activity of broad bean seed extract and its phenolic composition. *J. Func. Foods* **2017**, *38*, 656–662. [CrossRef]

41. Singh, B.; Singh, J.P.; Kaur, A.; Singh, N. Phenolic composition and antioxidant potential of grain legume seeds: A review. *Food Res. Int.* **2017**, *101*, 1–16. [CrossRef] [PubMed]

42. Yeo, J.; Shahidi, F. Identification and quantification of soluble and insoluble-bound phenolics in lentil hulls using HPLC-ESI-MS/MS and their antioxidant potential. *Food Chem.* **2020**, *315*, 126202. [CrossRef]

43. Cardador-Martinez, A.; Castano-Tostado, E.; Loarca-Pina, G. Antimutagenic activity of natural phenolic compounds present in the common bean (*Phaseolus vulgaris*) against aflatoxin B1. *Food Addit. Contam.* **2002**, *19*, 62–69. [CrossRef]

44. Zhang, J.; Li, L.; Kim, S.H.; Hagerman, A.E.; Lü, J. Anti-cancer, anti-diabetic and other pharmacologic and biological activities of penta-galloyl-glucose. *Pharm. Res.* **2009**, *26*, 2066–2080. [CrossRef]

45. Del Bo', C.; Bernardi, S.; Marino, M.; Porrini, M.; Tucci, M.; Guglielmetti, S.; Cherubini, A.; Carrieri, B.; Kirkup, B.; Kroon, P.; et al. Systematic Review on Polyphenol Intake and Health Outcomes: Is there Sufficient Evidence to Define a Health-Promoting Polyphenol-Rich Dietary Pattern? *Nutrients* **2019**, *11*, 1355. [CrossRef]

46. Khatoon, N.; Prakash, J. Nutritional quality of microwave-cooked and pressure-cooked legumes. *Int. J. Food Sci. Nutr.* **2004**, *55*, 441–448. [CrossRef]

47. Alajaji, S.A.; El-Adawy, T.A. Nutritional composition of chickpea (*Cicer arietinum* L.) as affected by microwave cooking and other traditional cooking methods. *J. Food Compos. Anal.* **2006**, *19*, 806–812. [CrossRef]

48. Khalil, A.H.; Mansour, E.H. The effect of cooking, autoclaving and germination on the nutritional quality of faba beans. *Food Chem.* **1995**, *54*, 177–182. [CrossRef]

49. Alvarez-Alvarez, J.; Guillamón, E.; Crespo, J.F.; Cuadrado, C.; Burbano, C.; Rodríguez, J.; Fernández, C.; Muzquiz, M. Effects of extrusion, boiling, autoclaving, and microwave heating on lupine allergenicity. *J. Agric. Food Chem.* **2005**, *53*, 1294–1298. [CrossRef]

50. Cuadrado, C.; Cabanillas, B.; Pedrosa, M.M.; Varela, A.; Guillamón, E.; Muzquiz, M.; Crespo, J.F.; Rodriguez, J.; Burbano, C. Influence of thermal processing on IgE reactivity to lentil and chickpea proteins. *Mol. Nutr. Food Res.* **2009**, *53*, 1462–1468. [CrossRef]

51. Luhovyy, B.L.; Mollard, R.C.; Panahi, S.; Nunez, M.F.; Cho, F.; Anderson, G.H. Canned Navy Bean Consumption Reduces Metabolic Risk Factors Associated with Obesity. *Can. J. Diet. Pract. Res.* **2015**, *76*, 33–37. [CrossRef]

52. Martínez-Pineda, M.; Yagüe-Ruiz, C.; Caverni-Muñoz, A.; Vercet-Tormo, A. Cooking Legumes: A Way for Their Inclusion in the Renal Patient Diet. *J. Ren. Nutr.* **2019**, *29*, 118–125. [CrossRef] [PubMed]

53. Choi, W.C.; Parr, T.; Lim, Y.S. The impact of four processing methods on trypsin-, chymotrypsin- and alpha-amylase inhibitors present in underutilised legumes. *J. Food Sci. Technol.* **2019**, *56*, 281–289. [CrossRef]

54. Song, D.; Chang, S.K. Enzymatic degradation of oligosaccharides in pinto bean flour. *J. Agric. Food Chem.* **2006**, *54*, 1296–1301. [CrossRef]

55. Shimelis, E.A.; Rakshit, S.K. Effect of processing on antinutrients and in vitro protein digestibility of kidney bean (*Phaseolus vulgaris* L.) varieties grown in East Africa. *Food Chem.* **2007**, *103*, 161–172. [CrossRef]

56. Díaz-Batalla, L.; Widholm, J.M.; Fahey, G.C., Jr.; Castaño-Tostado, E.; Paredes-López, O. Chemical components with health implications in wild and cultivated Mexican common bean seeds (*Phaseolus vulgaris* L.). *J. Agric. Food Chem.* **2006**, *54*, 2045–2052. [CrossRef] [PubMed]

57. Genovese, M.I.; Lajolo, F.M. Inibidores de tripsina do feijaõ (*Phaseolus vulgaris*): Isoformas e estabilidade térmica. In Proceedings of the Anais XVI Congresso Brasileiro de Ciencia e Tecnologia de Alimentos, Rio de Janeiro, Brazil, 15–17 July 1998; pp. 1335–1338.

58. Teixeira-Guedesa, C.I.; Oppolzer, D.; Barros, A.I.; Pereira-Wilso, C. Impact of cooking method on phenolic composition and antioxidant potential of four varieties of *Phaseolus vulgaris* L. and *Glycine max* L. *LWT-Food Sci. Technol.* **2019**, *103*, 238–246. [CrossRef]

59. Piecyk, M.; Wołosiak, R.; Drużynska, B.; Worobiej, E. Chemical composition and starch digestibility in flours from Polish processed legume seeds. *Food Chem.* **2012**, *135*, 1057–1064. [CrossRef]

60. Hussain, B.; Khan, S.; Ismail, M.; Sattar, A. Effect of roasting and autoclaving on phytic acid content of chickpea. *Nahrung* **1989**, *33*, 345–348. [CrossRef]

61. Vijayakumari, K.; Siddhuraju, P.; Janardhanan, K. Effects of various water or hydrothermal treatments on certain antinutritional compounds in the seeds of the tribal pulse, *Dolichos lablab* var. vulgaris L. *Plant Foods Hum. Nutr.* **1995**, *48*, 17–29. [CrossRef] [PubMed]

62. Luo, Y.-W.; Xie, W.-H. Effect of different processing methods on certain antinutritional factors and protein digestibility in green and white faba bean (*Vicia faba* L.). *CyTA J. Food Sci.* **2013**, *11*, 43–49. [CrossRef]

63. Olah, O.; Wood, J.A.; Agboola, S.; Koniczak, I.; Blanchard, C.L. Effects of soaking, boiling and autoclaving on the phenolic contents and antioxidant activities of faba beans (*Vicia faba* L.) differing in seed coat colours. *Food Chem.* **2014**, *142*, 461–468. [CrossRef] [PubMed]

64. Hefnawy, T.H. Effect of processing methods on nutritional composition and anti-nutritional factors in lentils (*Lens culinaris*). *Ann. Agric. Sci.* **2011**, *56*, 57–61. [CrossRef]

65. Ai, Y.; Cichy, K.A.; Harte, J.B.; Kelly, J.D.; Ng, P.K.W. Effects of extrusion cooking on the chemical composition and functional properties of dry common bean powders. *Food Chem.* **2016**, *211*, 538–545. [CrossRef]

66. Adeparusi, E.O. Effect of processing on the nutrients and anti-nutrients of lima bean (*Phaseolus lunatus* L.) flour. *Nahrung* **2001**, *45*, 94–96. [CrossRef]

67. Malki, E.; Waisel, Y. Effects of pressure on germination of seeds of wheat (*Triticum aestivum* cv. Barqai) in saline and in non-saline media. *Physiol. Plant.* **1987**, *70*, 73–77. [CrossRef]

68. Abbas, Y.; Ahmad, A. Impact of processing on nutritional and antinutritional factors of legumes: A review. *Ann. Food Sci. Technol.* **2018**, *19*, 199–215.

69. Fabbri, A.D.T.; Crosby, G.A. A review of the impact of preparation and cooking on the nutritional quality of vegetables and legumes. *Int. J. Gastr. Food Sci.* **2016**, *3*, 2–11. [CrossRef]

70. Rehman, Z.U.; Shah, W.H. Thermal heat processing effects on antinutrients, protein and starch digestibility of food legumes. *Food Chem.* **2005**, *91*, 327–331. [CrossRef]

71. Margier, M.; Georgé, S.; Hafnaoui, N.; Remond, D.; Nowicki, M.; Du Chaffaut, L.; Amiot, M.J.; Reboul, E. Nutritional Composition and Bioactive Content of Legumes: Characterization of Pulses Frequently Consumed in France and Effect of the Cooking Method. *Nutrients* **2018**, *10*, 1668. [CrossRef]

72. Kumar, K.G.; Venkataraman, L.V.; Java, T.V.; Krishnamurthy, K.S. Cooking characteristics of some germinated legumes: Changes in phytins, Ca++, Mg++ and pectins. *J. Food Sci.* **1978**, *43*, 85–88. [CrossRef]

73. Wang, N.; Hatcher, D.W.; Gawalko, E.J. Effect of variety and processing on nutrients and certain anti-nutrients in field peas (*Pisum sativum*). *Food Chem.* **2008**, *2008*, 132–138. [CrossRef]

74. Avilés-Gaxiola, S.; Chuck-Hernández, C.; Serna-Saldívar, S.O. Inactivation Methods of Trypsin Inhibitor in Legumes: A Review. *J. Food Sci.* **2018**, *83*, 17–29. [CrossRef] [PubMed]

75. Alonso, R.; Orue, E.; Marzo, F. Effects of extrusion and conventional processing methods on protein and antinutritional factor contents in pea seeds. *Food Chem.* **1998**, *63*, 505–512. [CrossRef]

76. Kozlowska, H.; Borowska, J.; Scheneider, C. Preparation of faba bean (*Vicia faba*, L. minor) products. V. Effect of hydrothermal treatment of faba bean on the quality of flour. *Acta Aliment. Pol.* **1990**, *15*, 171–177.

77. Jourdan, G.A.; Noreña, C.P.Z.; Brandelli, A. Inactivation of Trypsin Inhibitor Activity from Brazilian Varieties of Beans (*Phaseolus vulgaris* L.). *Food Sci. Technol. Int.* **2007**, *13*, 195–198. [CrossRef]

78. Jiratanan, T.; Liu, R.H. Antioxidant activity of processed table beets (*Beta vulgaris* var, conditiva) and green beans (*Phaseolus vulgaris* L.). *J. Agric. Food Chem.* **2004**, *52*, 2659–2670. [CrossRef]

79. Ranilla, L.; Genovese, M.I.; Lajolo, F.M. Effect of different cooking conditions on phenolic compounds and antioxidant capacity of some selected brazilian bean (*Phaseolus vulgaris* L.) cultivars. *J. Agric. Food Chem.* **2009**, *57*, 5734–5742. [CrossRef]

80. Rocha-Guzmán, N.E.; González-Laredo, R.F.; Ibarra-Pérez, F.J.; Nava-Berúmen, C.A.; Gallegos-Infante, J.A. Effect of pressure cooking on the antioxidant activity of extracts from three common bean (*Phaseolus vulgaris* L.) cultivars. *Food Chem.* **2007**, *100*, 31–35. [CrossRef]

81. Sarmento, A.; Barros, L.; Fernandes, Â.; Carvalho, A.M.; Ferreira, I.C. Valorization of traditional foods: Nutritional and bioactive properties of *Cicer arietinum* L. and *Lathyrus sativus* L. pulses. *J. Sci. Food Agric.* **2015**, *95*, 179–185. [CrossRef]

82. Babar, V.S.; Chavan, J.K.; Kadam, S.S. Effects of heat treatments and germination on trypsin inhibitor activity and polyphenols in jack bean (*Canavalia ensiformis* L. DC). *Plant Foods Hum. Nutr.* **1988**, *38*, 319–324. [CrossRef] [PubMed]

83. Osman, M.A. Effect of different processing methods, on nutrient composition, antinutritional factors and in vitro protein digestibility of Dolichos Lablab bean (*Lablab purpureus* (L) sweet). *Pak. J. Nutr.* **2007**, *6*, 299–303. [CrossRef]

84. Vijayakumari, K.; Siddhuraju, P.; Pugalenthi, M.; Janardhanan, K. Effect of soaking and heat processing on the levels of antinutrients and digestible proteins in seeds of *Vigna aconitifolia* and *Vigna sinensis*. *Food Chem.* **1998**, *63*, 259–264. [CrossRef]

85. Granito, M.; Brito, Y.; Torres, A. Chemical composition, antioxidant capacity and functionality of raw and processed *Phaseolus lunatus*. *J. Sci. Food Agric.* **2007**, *87*, 2801–2809. [CrossRef]

86. Granito, M.; Torres, A.; Frias, J.; Guerra, M.; Vidal-valverde, C. Influence of fermentation on the nutritional value of two varieties of *Vigna sinensis*. *J. Eur. Food Res. Technol.* **2005**, *220*, 176–181. [CrossRef]

87. Turkmen, N.; Sari, F.; Velioglu, Y. The effect of cooking methods on total phenolics and antioxidant activity of selected green vegetables. *J. Food Chem.* **2005**, *93*, 713–718. [CrossRef]

88. Andlauer, W.; Stumpf, C.; Hubert, M.; Rings, A.; Fürst, P. Influence of Cooking Process on Phenolic Marker Compounds of Vegetables. *Int. J. Vitam. Nutr. Res.* **2003**, *73*, 152–159. [CrossRef]

89. Maheshu, V.; Priyadarsini, D.T.; Sasikumar, J.M. Effects of processing conditions on the stability of polyphenolic contents and antioxidant capacity of *Dolichos lablab* L. *J. Food Sci. Technol.* **2013**, *50*, 731–738. [CrossRef] [PubMed]

90. Satwadhar, P.N.; Kadam, S.S.; Salunkhe, D.K. Effects of germination and cooking on polyphenols and in vitro protein digestibility of horse gram and moth bean. *Plant Foods Hum. Nutr.* **1981**, *31*, 71–76. [CrossRef]

91. Giczewska, G.A.; Borowska, J. Physical properties of selected legume seeds as indicators of technological suitability of small-seed broad bean. *Pol. J. Food Nutr. Sci.* **2003**, *12*, 9–13.

92. Yeo, J.; Shahidi, F. Effect of hydrothermal processing on changes of insoluble-bound phenolics of lentils. *J. Funct. Foods* **2017**, *38*, 716–722. [CrossRef]

93. Alam, M.S.; Kaur, J.; Khaira, H.; Gupta, K. Extrusion and Extruded Products: Changes in Quality Attributes as Affected by Extrusion Process Parameters: A Review. *Crit. Rev. Food Sci. Nutr.* **2016**, *56*, 445–473. [CrossRef]

94. Castells, M.; Marín, S.; Sanchis, V.; Ramos, A.J. Fate of mycotoxins in cereals during extrusion cooking: A review. *Food Addit. Contam.* **2005**, *22*, 150–157. [CrossRef] [PubMed]

95. Dalbhagat, C.G.; Mahato, D.K.; Mishra, H.N. Effect of extrusion processing on physicochemical, functional and nutritional characteristics of rice and rice-based products: A review. *Trends Food Sci. Tech.* **2019**, *85*, 226–240. [CrossRef]

96. Camire, M.E.; Camire, A.; Krumhar, K. Chemical and nutritional changes in foods during extrusion. *Crit. Rev. Food Sci. Nutr.* **1990**, *29*, 35–57. [CrossRef] [PubMed]

97. Ciudad-Mulero, M.; Barros, L.; Fernandes, Â.; Berrios, J.; Cámara, M.; Morales, P.; Fernández-Ruiz, V.; Ferreira, I.C.F.R. Bioactive compounds and antioxidant capacity of extruded snack-type products developed from novel formulations of lentil and nutritional yeast flours. *Food Funct.* **2018**, *9*, 819–829. [CrossRef]

98. Allen, K.E.; Carpenter, C.E.; Walsh, M.K. Influence of protein level and starch type on an extrusion-expanded whey product. *Int. J. Food Sci. Technol.* **2007**, *42*, 953–960. [CrossRef]

99. Alonso, R.; Grant, G.; Dewey, P.; Marzo, F. Nutritional Assessment in vitro and in vivo of raw and extruded peas (*Pisum sativum* L.). *J. Agric. Food Chem.* **2000**, *48*, 2286–2290. [CrossRef] [PubMed]

100. Arribas, C.; Cabellos, B.; Cuadrado, C.; Guillamón, E.; Pedrosa, M.M. Bioactive Compounds, Antioxidant Activity, and Sensory Analysis of Rice-Based Extruded Snacks-Like Fortified with Bean and Carob Fruit Flours. *Foods* **2019**, *8*, 381. [CrossRef]

101. Arribas, C.; Pereira, E.; Barros, L.; Alves, M.J.; Calhelha, R.C.; Guillamón, E.; Pedrosa, M.M.; Ferreira, I.C.F.R. Healthy novel gluten-free formulations based on beans, carob fruit and rice: Extrusion effect on organic acids, tocopherols, phenolic compounds and bioactivity. *Food Chem.* **2019**, *292*, 304–313. [CrossRef] [PubMed]

102. Brennan, C.; Brennan, M.; Derbyshire, E.; Tiwari, B. Effects of extrusion on the polyphenols, vitamins and antioxidant activity of foods. *Trends Food Sci. Technol.* **2011**, *22*, 570–575. [CrossRef]

103. Ciudad-Mulero, M.; Fernández-Ruiz, V.; Cuadrado, C.; Arribas, C.; Pedrosa, M.M.; De, J. Berrios, J.; Pan, J.; Morales, P. Novel gluten-free formulations from lentil flours and nutritional yeast: Evaluation of extrusion effect on phytochemicals and non-nutritional factors. *Food Chem.* **2020**, *315*, 126175. [CrossRef]

104. El-Hady, E.A.; Habiba, R.A. Effect of soaking and extrusion conditions on antinutrients and protein digestibility of legume seeds. *LWT-Food Sci. Technol.* **2003**, *36*, 285–293. [CrossRef]

105. Pedrosa, M.M.; Cuadrado, C. The nutritional and nutraceutical values of some 'ready-to-eat' expanded legume products. In *Legumes for Global Food Security*; Clemente, A., Jimenez-Lopez, J.C., Eds.; Nova Science: Hauppauge, NY, USA, 2017; pp. 41–73.

106. Alonso, R.; Rubio, L.A.; Muzquiz, M.; Marzo, F. The effect of extrusion cooking on mineral bioavailability in pea and kidney bean seed meals. *Anim. Feed Sci. Technol.* **2001**, *94*, 1–13. [CrossRef]

107. Borejszo, Z.; Khan, k. Reduction of flatulence-causing sugars by high temperature extrusion of pinto bean high starch fractions. *J. Food Sci.* **1992**, *57*, 771–777. [CrossRef]

108. Kelkar, S.; Siddiq, M.; Harte, J.B.; Dolan, K.D.; Nyombaire, G.; Suniaga, H. Use of low-temperature extrusion for reducing phytohemagglutinin activity (PHA) and oligosaccharides in beans (*Phaseolus vulgaris* L.) cv. Navy and Pinto. *Food Chem.* **2012**, *133*, 1636–1639. [CrossRef]

109. Korus, J.; Gumul, D.; Czechowska, K. Effect of extrusion on the phenolic composition and antioxidant activity of dry beans of *Phaseolus vulgaris* L. *Food Technol. Biotechnol.* **2007**, *45*, 139–146.

110. Zamora, N. Effect of extrusion on the activity of anti-nutritional factors and in vitro digestibility of protein and starch in flours of *Canavalia ensiformis*. *Arch. Latinoam. Nutr.* **2003**, *53*, 293–298.

111. Berrios, J.; Morales, P.; Cámara, M.; Sánchez-Mata, M.C. Carbohydrate composition of raw and extruded pulse flours. *Food Res. Int.* **2010**, *40*, 531–536. [CrossRef]

112. Poltronieri, F.; Arêas, J.A.G.; Colli, C. Extrusion and iron bioavailability in chickpea (*Cicer arietinum* L.). *Food Chem.* **2000**, *70*, 175–180. [CrossRef]

113. Hegazy, H.S.; El-Bedawey, A.E.-F.A.; Rahma, E.-S.H.; Gaafar, A.M. Effect of Extrusion Processs on Nutritional, Functional Properties and Antioxidant Activity of Germinated Chickpea Incorporated Corn Extrudates. *Am. J. Food Sci. Nutr. Res.* **2017**, *4*, 59–66.

114. Rathod, R.P.; Annapure, U.S. Effect of extrusion process on antinutritional factors and protein and starch digestibility of lentil splits. *LWT-Food Sci. Technol.* **2016**, *66*, 114–123. [CrossRef]

115. Varela, A.; Guillamón, E.; Cuadrado, C.; Marzo, F.; Burbano, C.; Muzquiz, M.; Pedrosa, M.M. Changes in nutritional active factors and protein in different legumes by extrusion/cooking. In Proceedings of the 6th European Conference of Grain Legumes, Lisboa, Portugal, 12–17 November 2007.

116. Berrios, J.; Cámara, M.; Torija, M.E.; Alonso, M. Effect of extrusion cooking and sodium bicarbonate addition on the carbohydrate composition of black bean flours. *J. Food Process. Preserv.* **2002**, *26*, 113–128. [CrossRef]

117. Delgado-Licon, E.; Ayala, A.L.M.; Rocha-Guzman, N.E.; Gallegos-Infante, J.-A.; Atienzo-Lazos, M.; Drzewiecki, J.; Martínez-Sánchez, C.E.; Gorinstein, S. Influence of extrusion on the bioactive compounds and the antioxidant capacity of the bean/corn mixtures. *Int. J. Food Sci. Nutr.* **2009**, *60*, 522–532. [CrossRef] [PubMed]

118. Yağcı, S.; Altan, A.; Doğan, F. Effects of extrusion processing and gum content on physicochemical, microstructural and nutritional properties of fermented chickpea-based extrudates. *LWT-Food Sci. Technol.* **2020**, *124*, 109150. [CrossRef]

119. Morales, P.; Cebadera-Miranda, L.; Cámara, R.M.; Reis, F.S.; Barros, L.; Berrios, J.D.J.; Ferreira, I.C.F.R.; Cámara, M. Lentil flour formulations to develop new snack-type products by extrusion processing: Phytochemicals and antioxidant capacity. *J. Funct. Foods* **2015**, *19*, 537–544. [CrossRef]

120. Morales, P.; Berrios, J.; Varela, A.; Burbano, C.; Cuadrado, C.; Muzquiz, M.; Pedrosa, M.M. Novel fiber-rich lentil flours as snack-type functional foods: An extrusion cooking effect on bioactive compounds. *Food Funct.* **2015**, *6*, 3135–3143. [CrossRef]

121. Berrios, J.; Pan, J. Evaluation of extruded black bean (*Phaseolus vulgaris* L.) processed under different screw speeds and particle sizes. In Proceedings of the Annual meeting of the Institute of Food technologist, New Orleans, LA, USA, 23–27 June 2001; p. 30.

122. Nikmaram, N.; Leong, S.Y.; Koubaa, M.; Zhu, Z.; Barba, F.J.; Greiner, R.; Oey, I.; Roohinejad, S. Effect of extrusion on the anti-nutritional factors of food products: An overview. *Food Control* **2017**, *79*, 62–73. [CrossRef]

123. Lombardi-Boccia, G.; Lullo, G.D.; Carnovale, E. In-vitro iron dialysability from legumes: Influence of phytate and extrusion cooking. *J. Sci. Food Agric.* **1991**, *55*, 599–605. [CrossRef]

124. Butron, J. Effect of extrusion of legumes and enzyme supplementation on the animal growth and their metabolism/Efecto_de_la_suplementacion_enzimatica_y_extrusion_de_leguminosas_en_el_crecimiento_y_metabolismo_animal. Ph.D. Thesis, Universidad Pública de Navarra, Navarra, Spain, 2002. Available online: https://www.researchgate.net/publication/45219512_Efecto_de_la_suplementacion_enzimatica_y_extrusion_de_leguminosas_en_el_crecimiento_y_metabolismo_animal (accessed on 9 February 2021).

125. Adeleye, O.O.; Awodiran, S.T.; Ajayi, A.O.; Ogunmoyela, T.F. Effect of high-temperature, short-time cooking conditions on in vitro protein digestibility, enzyme inhibitor activity and amino acid profile of selected legume grains. *Heliyon* **2020**, *6*, e05419. [CrossRef]

126. Korus, J.; Gumul, D.; Folta, M.; Barton, H. Antioxidant and antiradical activity of raw and extruded commons beans. *Electron. J. Pol. Agric. Univ.* **2007**, *10*, 6. Available online: http://www.ejpau.media.pl/volume10/issue4/art-06.html (accessed on 9 February 2021).

127. Díaz-Batalla, L.; Hernández-Uribe, J.P.; Gutiérrez-Dorado, R.; Téllez-Jurado, A.; Castro-Rosas, J.; Pérez-Cadena, R.; Gómez-Aldapa, C.A. Nutritional characterization of *Prosopis Laevigata* legume tree (mesquite) seed flour and the effect of extrusion cooking on its bioactive components. *Foods* **2018**, *7*, 124. [CrossRef] [PubMed]

128. Marti, A.; Seetharaman, K.; Pagani, M.A. Rice-based pasta: A comparison between conventional pasta-making and extrusion-cooking. *J. Cereal Sci.* **2010**, *52*, 404–409. [CrossRef]

129. Onwulata, C.I.; Tunick, M.H.; Qi, P.X. Extrusion texturized dairy proteins: Processing and application. In *Advances in Food and Nutrition Research*; Elsevier: Amsterdam, The Netherlands, 2011; Volume 62, pp. 173–200.

130. Boyaci, B.B.; Han, J.Y.; Masatcioglu, M.T.; Yalcin, E.; Celik, S.; Ryu, G.H.; Koksel, H. Effects of cold extrusion process on thiamine and riboflavin contents of fortified corn extrudates. *Food Chem.* **2012**, *132*, 2165–2170. [CrossRef]

131. Di Stefano, V.; Pagliaro, A.; Nobile, M.; Conte, A.; Melilli, M. Lentil Fortified Spaghetti: Technological Properties and Nutritional Characterization. *Foods* **2020**, *10*, 4. [CrossRef]

132. Lorenzo, G.; Sosa, M.; Califano, A. Alternative Proteins and Pseudocereals in the Development of Gluten-Free Pasta. In *Alternative and Replacement Foods*; Elsevier: San Diego, CA, USA, 2018; pp. 433–458. [CrossRef]

133. Sissons, M. Pasta. In *Reference Module in Food Science*; Elsevier: Amsterdam, The Netherlands, 2016. [CrossRef]

134. De Pilli, T.; Alessandrino, O. Effects of different cooking technologies on biopolymers modifications of cereal-based foods: Impact on nutritional and quality characteristics review. *Crit. Rev. Food Sci. Nutr.* **2020**, *60*, 556–565. [CrossRef]

135. Murray, J.; Kiszonas, A.; Morris, C.F. Pasta Production: Complexity in Defining Processing Conditions for Reference Trials and Quality Assessment Methods. *Cereal Chem. J.* **2017**, *94*, 791–797. [CrossRef]
136. Petitot, M.; Boyer, L.; Minier, C.; Micard, V. Fortification of pasta with split pea and faba bean flours: Pasta processing and quality evaluation. *Food Res. Int.* **2010**, *43*, 634–641. [CrossRef]
137. Bahnassey, Y.; Khan, K. Fortification of spaghetti with edible legumes. II. Rheological, processing, and quality evaluation studies. *Cereal Chem.* **1986**, *63*, 216–219.
138. Bilgicli, N. Some chemical and sensory properties of gluten-free noodle prepared with different legume, pseudocereal and cereal flour blends. *J. Food Nutr. Res.* **2013**, *52*, 251–255.
139. Bouasla, A.; Wójtowicz, A.; Zidoune, M.N. Gluten-free precooked rice pasta enriched with legumes flours: Physical properties, texture, sensory attributes and microstructure. *LWT-Food Sci. Technol.* **2017**, *75*, 569–577. [CrossRef]
140. Bouasla, A.; Wojtowicz, A.; Zidoune, M.N.; Olech, M.; Nowak, R.; Mitrus, M.; Oniszczuk, A. Gluten-free precooked rice-yellow pea pasta: Effect of extrusion-cooking conditions on phenolic acids composition, selected properties and microstructure. *J. Food Sci.* **2016**, *81*, C1070–C1079. [CrossRef] [PubMed]
141. Chillo, S.; Monro, J.; Mishra, S.; Henry, C. Effect of incorporating legume flour into semolina spaghetti on its cooking quality and glycaemic impact measured in vitro. *Int. J. Food Sci. Nutr.* **2010**, *61*, 149–160. [CrossRef]
142. Flores-Silva, P.C.; Berrios, J.D.J.; Pan, J.; Agama-Acevedo, E.; Monsalve-Gonzalez, A.; Bello-Perez, L.A. Gluten-free spaghetti with unripe plantain, chickpea and maize: Physicochemical, texture and sensory properties. *CyTA-J. Food Sci.* **2015**, *13*, 159–166. [CrossRef]
143. Flores-Silva, P.C.; Berrios, J.D.J.; Pan, J.; Osorio-Diaz, P.; Bello-Pérez, L.A. Gluten-free spaghetti made with chickpea, unripe plantain and maize flours: Functional and chemical properties and starch digestibility. *Int. J. Food Sci. Technol.* **2014**, *49*, 1985–1991. [CrossRef]
144. Giuberti, G.; Gallo, A.; Cerioli, C.; Fortunati, P.; Masoero, F. Cooking quality and starch digestibility of gluten free pasta using new bean flour. *Food Chem.* **2015**, *175*, 43–49. [CrossRef] [PubMed]
145. Schmiele, M.; Jaekel, L.Z.; Ishida, P.M.G.; Chang, Y.K.; Steel, C.J. Massa alimenticia sem gluten com elevado teor proteico obtida por processo convencional. *Ciencia Rural* **2013**, *43*, 908–914. [CrossRef]
146. Segura-Campos, M.R.; Garcia-Rodriguez, K.; Ruiz-Ruiz, J.C.; Chel-Guerrero, L.; Betancur-Ancona, D. In vitro bioactivity, nutritional and sensory properties of semolina pasta added with hard-to-cook bean (*Phaseolus vulgaris* L.) protein hydrolysate. *J. Funct. Foods* **2014**, *8*, 1–8. [CrossRef]
147. Susanna, S.; Prabhasankar, P. A study on development of Gluten free pasta and its biochemical and immunological validation. *LWT-Food Sci. Technol.* **2013**, *50*, 613–621. [CrossRef]
148. Gao, Y.; Janes, M.E.; Chaiya, B.; Brennan, M.A.; Brennan, C.S.; Prinyawiwatkul, W. Gluten-free bakery and pasta products: Prevalence and quality improvement. *Int. J. Food Sci. Technol.* **2018**, *53*, 19–32. [CrossRef]
149. Rani, S.; Singh, R.; Kamble, D.B.; Upadhyay, A.; Kaur, B.P.J.F.B. Structural and quality evaluation of soy enriched functional noodles. *Food Biosci.* **2019**, *32*, 100465. [CrossRef]
150. Sofi, S.A.; Singh, J.; Chhikara, N.; Panghal, A. Effect of incorporation of germinated flour and protein isolate from chickpea on different quality characteristics of rice-based noodle. *Cereal Chem.* **2020**, *97*, 85–94. [CrossRef]
151. Frias, J.; Kovacs, E.; Sotomayor, C.; Hedley, C.; Vidal-Valverde, C. Processing peas for producing macaroni. *LWT-Food Sci. Technol.* **1997**, *204*, 66–71. [CrossRef]
152. Yoshimoto, J.; Kato, Y.; Ban, M.; Kishi, M.; Horie, H.; Yamada, C.; Nishizaki, Y. Palatable Noodles as a Functional Staple Food Made Exclusively from Yellow Peas Suppressed Rapid Postprandial Glucose Increase. *Nutrients* **2020**, *12*, 1839. [CrossRef]
153. Torres, A.; Frias, J.; Granito, M.; Vidal-Valverde, C.n. Fermented pigeon pea (*Cajanus cajan*) ingredients in pasta products. *J. Agric. Food Chem.* **2006**, *54*, 6685–6691. [CrossRef]
154. Herken, E.N.; Äbanolu, Å.; Aner, M.D.; Bilgili, N.; Gazel, S. Effect of storage on the phytic acid content, total antioxidant capacity and organoleptic properties of macaroni enriched with cowpea flour. *J. Food Eng.* **2007**, *78*, 366–372. [CrossRef]
155. Laleg, K.; Cassan, D.; Barron, C.; Prabhasankar, P.; Micard, V. Structural, culinary, nutritional and anti-nutritional properties of high protein, gluten free, 100% legume pasta. *PLoS ONE* **2016**, *11*, e0160721. [CrossRef]
156. Carcea, M.; Narducci, V.; Turfani, V.; Giannini, V. Polyphenols in raw and cooked cereals/pseudocereals/legume pasta and couscous. *Foods* **2017**, *6*, 80. [CrossRef]
157. Arribas, C.; Cabellos, B.; Guillamón, E.; Pedrosa, M.M. Cooking and sensorial quality, nutritional composition and functional properties of cold-extruded rice/white bean gluten-free fettuccine fortified with whole carob fruit flour. *Food Funct.* **2020**, *11*, 7913–7924. [CrossRef]
158. Drago, S.R.; Franco-Miranda, H.; Cian, R.l.E.; Betancur-Ancona, D.; Chel-Guerrero, L. Bioactive properties of *Phaseolus lunatus* (Lima Bean) and *Vigna unguiculata* (Cowpea) hydrolyzates incorporated into pasta. Residual activity after pasta cooking. *Plant Foods Hum. Nutr.* **2016**, *71*, 339–345. [CrossRef] [PubMed]
159. Turco, I.; Bacchetti, T.; Bender, C.; Zimmermann, B.; Oboh, G.; Ferretti, G. Polyphenol content and glycemic load of pasta enriched with faba bean flour. *Funct. Foods Health Dis.* **2016**, *6*, 291–305. [CrossRef]
160. Torres, A.; Frias, J.; Granito, M.; Vidal-Valverde, C. Germinated *Cajanus cajan* seeds as ingredients in pasta products: Chemical, biological and sensory evaluation. *Food Chem.* **2007**, *101*, 202–211. [CrossRef]
161. Turco, I.; Bacchetti, T.; Morresi, C.; Padalino, L.; Ferretti, G. Polyphenols and glycaemic index of legume pasta. *Food Funct.* **2019**, *10*. [CrossRef]

162. Gallegos-Infante, J.A.; Rocha-Guzman, N.E.; Gonzalez-Laredo, R.F.; Ochoa-Martínez, L.A.; Corzo, N.; Bello-Perez, L.A.; Medina-Torres, L.; Peralta-Alvarez, L.E. Quality of spaghetti pasta containing Mexican common bean flour (*Phaseolus vulgaris* L.). *Food Chem.* **2010**, *119*, 1544–1549. [CrossRef]
163. Biernacka, B.; Dziki, D.; Gawlik-Dziki, U.; Rayao, R.; Siasta, M. Physical, sensorial, and antioxidant properties of common wheat pasta enriched with carob fiber. *LWT-Food Sci. Technol.* **2017**, *77*, 186–192. [CrossRef]
164. Seczyk, Å.U.; Åšwieca, M.; Gawlik-Dziki, U. Effect of carob (*Ceratonia siliqua* L.) flour on the antioxidant potential, nutritional quality, and sensory characteristics of fortified durum wheat pasta. *Food Chem.* **2016**, *194*, 637–642. [CrossRef]

Article

Combinations of Legume Protein Hydrolysates Synergistically Inhibit Biological Markers Associated with Adipogenesis

Cecilia Moreno [1], Luis Mojica [1], Elvira González de Mejía [2], Rosa María Camacho Ruiz [3] and Diego A. Luna-Vital [4,*]

[1] Tecnología de Alimentos, Centro de Investigación y Asistencia en Tecnología y Diseño del Estado de Jalisco, A.C., Guadalajara 44270, Mexico; cemoreno_al@ciatej.edu.mx (C.M.); lmojica@ciatej.mx (L.M.)

[2] Department of Food Science and Human Nutrition, University of Illinois at Urbana-Champaign, Champaign, IL 61801, USA; edemejia@illinois.edu

[3] Biotecnología Industrial, Centro de Investigación y Asistencia en Tecnología y Diseño del Estado de Jalisco, A.C., Guadalajara 44270, Mexico; rcamacho@ciatej.net.mx

[4] Tecnologico de Monterrey, School of Engineering and Science, Ave. Eugenio Garza Sada 2501, Monterrey 64849, N.L., Mexico

* Correspondence: dieluna@tec.mx; Tel.: +52-272-105-6868

Received: 2 October 2020; Accepted: 12 November 2020; Published: 17 November 2020

Abstract: The objective was to investigate the anti-adipogenesis potential of selected legume protein hydrolysates (LPH) and combinations using biochemical assays and in silico predictions. Black bean, green pea, chickpea, lentil and fava bean protein isolates were hydrolyzed using alcalase (A) or pepsin/pancreatin (PP). The degree of hydrolysis ranged from 15.5% to 35.5% for A-LPH and PP-LPH, respectively. Antioxidant capacities ranged for ABTS$^{\bullet+}$ IC$_{50}$ from 0.3 to 0.9 Trolox equivalents (TE) mg/mL, DPPH$^{\bullet}$ IC$_{50}$ from 0.7 to 13.5 TE mg/mL and nitric oxide (NO) inhibition IC$_{50}$ from 0.3 to 1.3 mg/mL. LPH from PP–green pea, A–green pea and A–black bean inhibited pancreatic lipase (PL) (IC$_{50}$ = 0.9 mg/mL, 2.2 mg/mL and 1.2 mg/mL, respectively) ($p < 0.05$). For HMG-CoA reductase (HMGR) inhibition, the LPH from A–chickpea (0.15 mg/mL), PP–lentil (1.2 mg/mL), A–green pea (1.4 mg/mL) and PP–green pea (1.5 mg/mL) were potent inhibitors. Combinations of PP–green pea + A–black bean (IC$_{50}$ = 0.4 mg/mL), A–green pea + PP–green pea (IC$_{50}$ = 0.9 mg/mL) and A–black bean + A–green pea (IC$_{50}$ = 0.6 mg/mL) presented synergistic effects to inhibit PL. A–chickpea + PP–lentil (IC$_{50}$ = 0.8 mg/mL) and PP–lentil + A–green pea (IC$_{50}$ = 1.3 mg/mL) interacted additively to inhibit HMGR and synergistically in the combination of A–chickpea + PP–black bean (IC$_{50}$ = 1.3 mg/mL) to block HMGR. Peptides FEDGLV and PYGVPVGVR inhibited PL and HMGR in silico, showing predicted binding energy interactions of −7.6 and −8.8 kcal/mol, respectively. Combinations of LPH from different legume protein sources could increase synergistically their anti-adipogenic potential.

Keywords: legumes; protein hydrolysates; anti-adipogenic potential; antioxidant capacity; peptide synergism

---

1. Introduction

Obesity is defined as excessive fat accumulation, characterized by increased visceral white adipose tissue mass and abnormalities in lipid metabolism that present a risk to health [1,2]. However, its prevalence has doubled around the world and steadily increased over the past 50 years, reaching pandemic levels [3]. This pathological condition results from complex interactions between genes and environmental factors, such as calorie-dense food intake, sedentary lifestyle and stress [4,5].

Adipose cells store energy in the form of triglycerides as well as controlling lipid mobilization and its distribution in the body [6–8]. As a result of increased fat storage, excessive adipose tissue expansion alters its histology and function. Consequently, interactions among adipocytes and immune cells at different stages of this process trigger adipocyte lipolysis, increasing circulating free fatty acids, as well as the production of multiple proinflammatory factors [9].

Several pharmacological agents have been developed to influence eating behavior, food intake, energy expenditure and nutrient absorption [10]. Currently, lorcaserin, phentermine/topiramate, naltrexone/bupropion, liraglutide and orlistat are anti-obesity drugs that have been approved by the U.S. Food and Drug Administration (FDA) [11]. For instance, tetrahydrolipstatin (orlistat) is marketed for the long-term regulation of energy intake; this works by inhibiting pancreatic lipase, an enzyme that breaks down triglycerides in the intestinal lumen. Once this enzyme is inactivated, it is unable to hydrolyze fats into fatty acids and monoglycerides, leading to their elimination through the feces [12].

Statins are another class of drugs prescribed as a lipid-lowering medication. Their mechanism of action is through the inhibition of the enzyme 3-hydroxy-3-methylglutaryl coenzyme A (HMG-CoA) reductase (HMGR). This enzyme catalyzes the rate-limiting step of the mevalonate pathway that leads to cholesterol biosynthesis. Inhibition of its enzymatic activity lowers endogenous cholesterol and serum low-density lipoprotein levels [13].

Pharmacological management of obesity has been unsuccessful due to adverse side effects generating safety concerns. For example, the use of orlistat has been associated with several gastrointestinal adverse effects, such as oily stools, diarrhea, abdominal pain and fecal spotting, and even a few cases of serious hepatic adverse effects [14]. Statins have also shown concerning side effects including myositis, myalgia, rhabdomyolysis muscle pain, fatigue and weakness [15].

Legumes are an excellent source of proteins and, when they are enzymatically digested, they become peptides with multiple sizes that can exert a wide spectrum of biological potentials [16–18]. Hence, the incorporation of bioactive compounds from legumes into diets could exert beneficial effects by regulating lipid metabolism, resulting in a potential alternative in the prevention or as an adjuvant in the treatment of obesity [19,20]. Furthermore, studies with legume-derived hydrolysates have shown beneficial effects on immunity, inflammation, infection, hypertension, hypercholesterolemia, type 2 diabetes and some types of cancer [21–23].

Therefore, this study aimed to evaluate the anti-adipogenic potential and antioxidant properties of selected legume protein hydrolysates by comparing their potential to inhibit pancreatic lipase and HMG-CoA reductase using biochemical assays and in silico approaches. Synergistic, additive and antagonistic effects between the combinations of legume protein hydrolysates were also evaluated through isobolographic analyses.

2. Materials and Methods

2.1. Materials

Raw varieties of *Phaseolus vulgaris* L., *Pisum sativum* L., *Cicer arietinum* L., *Lens culinaris* L. and *Vicia faba* L. were obtained from local farmers of Guadalajara, Mexico. The dry grains were stored at 4 °C until use. Commercial proteases alcalase (EC 3.4.21.62), porcine pepsin (EC 3.4.23.1) and pancreatin (8xUSP, 232-468-9) were purchased from Sigma-Aldrich (St. Louis, MO, USA). DC protein assay was purchased from Bio-Rad Laboratories. Molecular weight protein standard (10 to 250 kDa) and SimplyBlue Safe Stain were purchased from Amersham Pharmacia Biotech (Carlsbad, CA, USA). All other chemicals and reagents were obtained from Sigma-Aldrich (St. Louis, MO, USA).

2.2. Legume Protein Isolate Extraction

The extraction of proteins from legumes was conducted using a previously established methodology [24]. Common black bean seeds were soaked in water at room temperature for 16 h. Black bean hulls were manually removed; then, bean cotyledons, green pea, chickpea, lentil and fava bean were ground separately in a commercial blender in a 1:10 grain/water ratio. The pH of the supernatant was adjusted to 8.0 with 1 M sodium hydroxide, and protein extraction was carried out at 35 °C with stirring for 1 h. The mixture was centrifuged at 5000× g for 15 min at 25 °C. Then, the pH was adjusted to the isoelectric point of the different legumes with 1 M hydrochloric acid to precipitate proteins, followed by centrifugation at 10,000× g for 20 min at 4 °C. The supernatant was discarded and the pellet was freeze-dried in a Lab Conco Freeze Dryer 4.5 (Kansas, MO, USA). Legume protein isolates (LPI) were stored at −20 °C until further analysis.

2.3. Protein Hydrolysis

2.3.1. Simulated Gastrointestinal Digestion

Legume protein hydrolysates (LPH) were obtained after simulated gastrointestinal digestion with pepsin/pancreatin (PP) following the method described by Mojica et al. (2015) [25]. Briefly, LPI was suspended in water (1:20 *w/v*) and autoclaved for 20 min at 121 °C to denature proteins and improve hydrolysis. Sequential enzyme digestion was carried out with pepsin/substrate 1:20 (*w/w*) (pH 2.0) followed by pancreatin/substrate 1:20 (*w/w*) at pH 7.5 at 37 °C for 2 h each. The hydrolysis was stopped by heating at 75 °C for 20 min, and the resulting LPI hydrolysates were centrifuged at 20,000× g for 15 min at 4 °C. LPH were dialyzed to eliminate salts using a 500 Da molecular weight cutoff membrane and then freeze-dried in a LabConco FreeZone Freeze Dry System. Hydrolysates were stored at −20 °C until further analysis.

2.3.2. Alcalase Enzymatic Digestion

Alcalase (A) was used for the hydrolysis of LPI. First, a portion of LPI was suspended in water (1:20 *w/v*) and autoclaved for 20 min at 121 °C. Then, enzymatic digestion was carried out using a protease/substrate ratio of 1:20 (*w/w*), time of hydrolysis 2 h, with pH and temperature optimal for alcalase activity (pH 7.0, T: 50 °C, respectively). Protein hydrolysis was stopped by heating at 75 °C for 20 min, and the resulting LPH were centrifuged at 20,000× g for 15 min at 4 °C. LPH were dialyzed to eliminate salts using a 500 Da molecular weight cutoff membrane and then freeze-dried in a LabConco FreeZone Freeze Dry System. Hydrolysates were stored at −20 °C until further analysis [24].

2.4. Degree of Hydrolysis (DH)

An aliquot of 64 µL of sample solution was placed in 1 mL of 0.2125 M sodium phosphate buffer pH 8.2. A 0.05% TNBS (trinitrobenzene sulfonic acid) solution was sequentially added to light-protected test tubes and then the reaction mixture was placed in a water bath at 50 °C for 30 min. The reaction was stopped by adding 0.1 M sodium sulfite. After cooling for 15 min at room temperature, the absorbance was measured at 420 nm. The total hydrolysis of the sample was carried out using 6 N HCl at 110 °C for 24 h. A calibration curve was produced using leucine (0 to 10 mM) as standard. DH was calculated according to the following equation:

$$\%DH = \frac{h}{h\ total} * 100$$

where h is the number of free amino groups in the hydrolyzed sample and $h\ total$ is the number of free amino groups after complete hydrolysis of the LPI.

2.5. Gel Electrophoresis Analysis SDS–PAGE (Sodium Dodecyl Sulfate–Polyacrylamide Gel Electrophoresis)

The freeze-dried LPI of the five legume types tested and their respective LPH were analyzed by SDS–PAGE, under reducing conditions (1:20 β-mercaptoethanol, β-ME). Precast (4% to 20%) gradient polyacrylamide, Tris–HCl gels were used with a Bio-Rad Criterion Cell under a constant voltage of 200 V for 35 min. Standards (10 to 250 kDa) were used to calculate molecular mass. After staining with Simply Blue Safe Stain overnight, and destaining with water, the gel was visualized using a Chemidoc^{TM} XRS$^+$ System (Bio-Rad Laboratories, Inc., Hercules, CA, USA).

2.6. Identification and Characterization of Potentially Bioactive Peptides

LPH were analyzed by liquid chromatography–electrospray ionization–mass spectrometry (LC–ESI–MS/MS) using a Q-tof Ultima mass spectrometer (Water, Milford, CT, USA), equipped with an Alliance 2795 HPLC system. Separation of the components was performed by using a mobile phase of Solvent A (95% H_2O, 5% ACN and 0.1% formic acid) and Solvent B (95% ACN, 5% H_2O and 0.1% formic acid) using a flow rate of 400 µL min^{-1}. The elution was in a linear gradient (0 min, 100% A; 1 min, 100% A; 2 min, 90% A, 10% B; 6 min, 60% A, 40% B; 10 min, 100% B; 12 min, 100% B; 14 min, 100% A; 15 min, 100% A). The temperature was kept at 20 °C during the whole procedure. A splitter with a split ratio of 1:10 was used, where one part was used by the mass spectrometer and ten parts were used for the waste. The Q-tof Ultima mass spectrometer was equipped with a Z-spray ion source. Using the positive ion electrospray mode (+ESI), the analysis on the Q-tof was carried out in V-mode with an instrument resolution between 9000 and 10,000 based on full width at half maximum, with a flow rate of 400 µL min^{-1}. The source temperature was set at 80 °C and desolvation temperatures were set at 250 °C. The Q-tof was operated at a capillary voltage of 3.5 kV and a cone voltage of 35 V. The final detector was a microchannel plate with high sensitivity. The MassLynx 4.1 V software (Waters, Milford, CT, USA) was used to control the instruments and to process the data to obtain the highest probability of the peptide's sequences. Confirmation of peptide sequences in each legume protein was performed using the BLAST$^®$ tool (http://www.blast.ncbi.nlm.nih.gov/Blast.cgi, accessed on 15 February 2019).

The potential biological activity of the peptides was predicted by using BIOPEP$^®$ database (http://www.uwm.edu.pl/biochemia, accessed on 16 February 2019). Peptide structures were predicted using the PepDraw tool (http://www.tulane.edu/biochem/WW/PepDraw/, accessed on 16 February 2019).

2.7. Antioxidant Capacity Assays

2.7.1. ABTS Radical Scavenging Activity

The ABTS$^{•+}$ radical cation was produced by reacting 7 mM 2,2′-azinobis (3-ethyl-benzothiazoline -6-sulfonic acid) diammonium salt (ABTS) with 2.45 mM potassium persulfate. The mixture was left to stand in the dark at room temperature for 20 h before use (overnight). The radical was stable in this form under these conditions for more than 48 h. The ABTS$^{•+}$ solution was diluted with 0.01 M phosphate buffer (pH 7.4) to an absorbance of 0.70 ± 0.02 at 734 nm. The scavenging activity was expressed as IC_{50} values Trolox equivalent per mg of the sample using the following equation y = 411.91x + 8.4207, R^2 = 0.99.

2.7.2. DPPH Radical Scavenging Activity

Antioxidant activity of legume hydrolysates was determined using DPPH (2,2-diphenyl-1 -picrylhydrazyl) free radical scavenging assay. Protected from light, 100 µL of each sample or buffer as control was added to 1.5 mL of methanolic DPPH solution (0.1 mM) (1,1-diphenyl-2-picrylhydrazyl) and stirred by vortex (3000 rpm) for 30 s. After 30 min of incubation, the absorbance of the solution was read at 517 nm. The analytical curve was constructed using 5–205 µg/mL of the Trolox solution to determine TE (y = 0.0016 x − 0.0294; R^2 = 0.94). The results were expressed as IC_{50} values of Trolox equivalent (TE) per mg of the sample.

2.7.3. Nitric Oxide (NO) Radical Scavenging

Nitric oxide production was estimated by the accumulation of nitrite (NO_2), a stable product of the nitric oxide (NO) reaction with oxygen in Griess reagent (1% sulfanilamide and 0.1% *N*-1-(naphtha-yl) ethylenediamine dihydrochloride in 2.5% H_3PO_4). Sodium nitroprusside (SNP) was used as the NO donor. NO reacts with oxygen to produce nitrate and nitrite. Briefly, SNP (10 mM) in phosphate-buffered saline was mixed with different concentrations of each LPH sample (0.1–5 mg/mL freeze-dried hydrolysate). The samples were incubated at 25 °C. After 120 min, 0.5 mL incubated solution was mixed with 0.5 mL of Griess reagent. Results were expressed as % inhibition related to PBS treatment [26].

2.8. Biochemical Analyses to Determine Anti-Adipogenic Potential

2.8.1. Lipase Activity Measurement Using pH Indicator-Based Lipase Assay

Inhibitory pancreatic lipase activity was determined according to the method implemented by Camacho-Ruiz et al. (2015) [27]. Briefly, to measure the formation of free fatty acids upon hydrolysis of TG (4:0) or TG (8:0), each substrate was prepared with one volume of the substrate (50 mM dissolved in tert-butanol), also containing the pH indicator, which was mixed vigorously on a vortex with nine volumes of buffer solution to reach a final substrate concentration of 5 mM. Buffer solution included 2.5 mM 3-Morpholinopropane-1-sulfonic acid (MOPS), 0.5 mM sodium taurodeoxycholate hydrate (NaTDC), 150 mM NaCl, 6 mM $CaCl_2$. Then, 20 µL of enzyme solution at appropriate dilution in buffer was added in each microplate well and 100 µL of substrate emulsion was quickly added using an eight-channel pipette. Subsequently, the plate was placed in a microtiter plate scanning spectrophotometer (x-Mark™, Bio-rad, Hercules, CA, USA) and shaken for 5 s before each reading. The decrease in absorbance at a wavelength corresponding to the λmax of the pH indicator was recorded every 30 s at 37 °C. Blanks without enzyme were performed and data were collected at least in triplicate for 15 min.

2.8.2. Hydroxy-3-methylglutaryl Coenzyme a Reductase (HMG-CoA Reductase) Activity Assay

The HMG-CoA reductase assay kit CS-1090 from Sigma-Aldrich (St. Louis, MO, USA) was carried out under conditions recommended by the manufacturer at 37 °C. Statin drug (pravastatin) was employed as a positive control. In summary, aliquots containing NADPH (4 µL), HMG-CoA substrate (12 µL) and a buffer pH 7.4 were placed into a UV compatible 96-well plate. The analyses were initiated by the addition of HMG-CoA reductase (2 µL) in each well and incubated in the presence or absence of pravastatin (250 nM) or 5–0.1 mg/mL hydrolysate concentration of each LPH. The rate of NADPH consumed was monitored by reading the decrease in absorbance at 340 nm by microtiter plate scanning spectrophotometer (x-Mark™, Bio-rad, Hercules, CA, USA). The HMG-CoA-dependent oxidation of NADPH and the inhibition properties of LPH were measured by the absorbance reduction, which is directly proportional to the enzyme activity.

2.9. Isobolographic Analysis of PL Inhibitory Activity by LPH

The interactions were validated by isobolographic analysis in which the combinations were comprised of equieffective doses of the individual components. Using the IC_{50} values of each LPH, the additive line was plotted and the equieffective dose was calculated. Subsequently, a dose–response curve of PL inhibition was obtained in a fixed-ratio for the mixture of LPH (1:1) that was based on the IC_{50} values of each protein hydrolysate. The experimental IC_{50} values for the LPH combinations were calculated. In an isobologram, when the protein hydrolysate combination IC_{50} lies on the theoretical IC_{50} add line, then the mixture is considered to be synergistic. An interaction index (γ) was calculated according to the following formula: IC_{50} combination/IC_{50} theoretical. Gamma values around 1 (γ = 1) indicated additive interaction; γ > 1 implied and antagonistic interaction and γ < 1 indicated a synergistic interaction.

2.10. Molecular Docking (In Silico Analysis)

The structural mechanism by which the peptides present in LPH interact with the enzymes, pancreatic lipase and HMG-CoA reductase was evaluated by in silico analysis through molecular docking, as described by Mojica et al. (2017) [28]. Molecular docking analysis was performed to predict individual peptide biological potential using DockingServer® [29]. Peptides were designed using the software Instant MarvinSketch (Chem Axon Ltd., Budapest, Hungary). The MMFF94 force field [30] was used for the energy minimization of ligand molecules, peptides, orlistat and pravastatin. Gasteiger partial charges were added to the peptide ligand atoms (peptides). Non-polar hydrogen atoms were merged, and rotatable bonds were defined. Docking calculations were carried out on PL (1LPB) and HMG-CoA reductase (1DQ9) protein crystal structures. Essential hydrogen atoms, Kollman united atom type charges and solvation parameters were added with the aid of AutoDock tools [31]. Affinity maps and spacing were generated using the Autogrid program. AutoDock parameter set- and distance-dependent dielectric functions were used in the calculation of the van der Waals and the electrostatic terms, respectively. Initial position, orientation and torsions of the ligand molecules were set randomly. Each docking experiment was derived from 100 different runs that were set to terminate after a maximum of 250,000 energy evaluations. The population size was set up to 150. During the search, a translational step of 0.2 Å, and quaternions and torsion steps of 5 were applied.

2.11. Statistical Analysis

The results were expressed as the mean ± SD of at least two independent experiments with three repetitions each and analyzed through ANOVA. Statistical significance ($p < 0.05$) was determined using Student's *t*-test for comparing mean pairs and Tukey's test for multiple mean comparisons using software JMP version 8.0 (SAS Institute, Cary, NC, USA).

3. Results

3.1. Protein Profile for LPI and LPH by Gel Electrophoresis Analysis

The SDS–PAGE analysis was conducted to observe the effect of enzymatic hydrolysis with alcalase and pepsin/pancreatin in the molecular weight distribution of LPI (Figure 1A). Three lanes per legume are depicted. The first lane represents LPI before hydrolysis; the second lane contains the LPI after hydrolysis with the commercial enzyme alcalase (A); and the third lane contains the LPI after simulated gastrointestinal digestion with pepsin and pancreatin (PP). Protein components of the LPI ranged from 10 to 100 kDa. Enzymatic hydrolysis exerted an effect on molecular weight distribution, mostly to the high molecular weight fractions. Black bean protein isolate (first lane) reveals several major bands which are 47 and 44 kDa phaseolin bands [25] (Figure 1A). Phytohemagglutinin is also visible, which corresponds to the 31 kDa band. Major identified proteins in common black bean were phaseolin, lectin, protease and α-amylase inhibitors, Kunitz trypsin inhibitor and Bowman–Birk inhibitor [32]. Green pea LPI revealed convicilin (72 kDa), legumin (25, 39 kDa) and vicilin (44, 32, 16 kDa) fractions [33]. In the case of lentil LPI, the most intense bands observed in the SDS–PAGE correspond to subunits of vicilin (48 kDa) and convicilin (63 kDa). Other lower molecular mass bands were observed which can represent gamma-vicilin and a mixture of albumin polypeptides [34]. The protein profile of chickpea and fava bean was also influenced by the hydrolysis process.

3.2. Degree of Hydrolysis

The effect of the digestion conditions on the degree of hydrolysis (DH) was evaluated. Figure 1B shows the comparison between the LPI hydrolyzed with alcalase and by a simulation of gastrointestinal digestion with PP. The average yield of hydrolysis ranged from 2.72% to 26.61% and 21.50% to 45.26% for A-LPH and PP-LPH, respectively. In general, across all the legume hydrolysates, DH showed great variability; however, significant differences between the enzymes alcalase and PP were observed in the black bean, chickpea and fava bean hydrolysates. In the case of green pea and lentil hydrolysates, no statistical differences were found using alcalase and PP enzyme. The lowest values of DH were observed in chickpea and fava bean LPI hydrolyzed with alcalase.

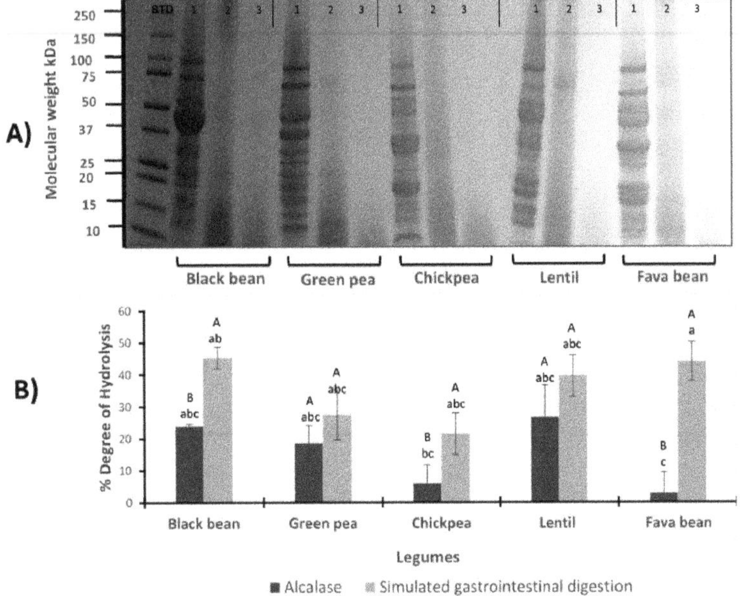

Figure 1. (**A**) Electrophoretic SDS–PAGE profile of legumes. Protein profile of black bean, green pea, chickpea, lentil and fava bean, before and after hydrolysis with alcalase and simulated gastrointestinal digestion. Each sample is presented in 3 wells: 1st well belongs to LPI profile (1), 2nd belongs to protein isolate after hydrolysis with alcalase (2), and the 3rd belongs to protein isolate after pepsin/pancreatin digestion (3). STD: standard. (**B**) Degree of hydrolysis (%) of legume protein isolates hydrolyzed with alcalase (A-LPH) and after simulated gastrointestinal digestion with pp (PP-LPH). Different uppercase letters indicate significant differences between alcalase and pepsin/pancreatin digestion ($p < 0.05$); different lowercase letters indicate significant differences ($p < 0.05$) among legume protein hydrolysates. Results represent the mean ± SD of at least two independent experiments.

3.3. Peptide Sequences and Predicted Bioactivity

The peptide sequences resulting from the PP and alcalase digestion were identified by high-performance–liquid chromatography–electrospray ionization–mass spectrometry (LC–ESI–MS/MS). Twenty-seven peptide sequences were obtained by alcalase enzymatic digestion and thirty-four peptides were sequenced from PP. Peptide identity was confirmed by Blast tool and most peptides were identified in legume parental proteins. Moreover, the physicochemical properties of the peptides are listed in Table 1. A-hydrolysates and PP-hydrolysates showed peptides with around six and eight amino acids, respectively. Most of their amino acids were aliphatic (glycine, proline, valine and alanine), which are non-polar and hydrophobic. Molecular weight for both A and PP peptides ranged from 440 (SPPE) to 1408 Da (VNPDPAGGPTSGRAL). The isoelectric point ranged from the acidic pI 2.78 (DLVLDVPS) to alkaline pI 11.52 (KPSSAAGAVR). The net charge was neutral in 46% of the peptides sequenced; 36% of the peptides presented negative charge and 18% were positively charged. Hydrophobicity ranged from 4.88 (TKAGGTAF) to 22.78 kcal/mol (VELVGPK).

The biological potential of peptide sequences generated by A and PP is shown in Figure 2. The percentage of biological potential is relative to the total bioactive peptides produced by each enzymatic system used. Most peptides in all the legumes evaluated potentially exert activity by blocking dipeptidyl peptidase-IV (DPP-IV) and angiotensin-converting enzyme (ACE). Potential biological activities such as stomach mucosa regulator, antiamnestic, antithrombotic, antioxidative and glucose uptake promoters were also observed.

Table 1. Physicochemical properties and their origin protein of the peptides obtained from the legume protein hydrolysates (LPH).

Sample	Molecular Mass (Da)	Peptide	Bioactive Sequence	pI	Net Charge	Hydrophobicity (kcal/mol)	Parental Protein [1]
A–Black bean	526	PVALK	PV, VA, AL, LK	9.8	−1	15.2	Photosystem I P700 chlorophyll a apoprotein A2
	565	DYRL	DY, YR, RL	6.6	−2	13.0	Phaseolin alpha-type
	596	VEGHGV	VE, GH, GV	5.0	1	9.6	Protein kinase PVPK−1
	650	FEELN	EE, EL, LN	2.9	0	11.3	Putative resistance protein TIR 17
	751	THGPVGAN	TH, HG, GP, GPV, PV, VG, GA	7.3	0	10.2	Nitrate reductase
	757	ANGSPGGAGA	NG, GS, SP, PG, GG, GA, AG	5.6	0	16.1	Inositol-3-phosphate synthase
	834	KPSASCSR	KP, KPS, PS, AS	9.8	0	18.0	Glycerol-3-phosphate acyltransferase
	872	NVGPGSLET	NV, VG, VGP, GP, PG, GS, SL, ET	3.1	−1	13.8	Serine/threonine-protein phosphatase PP1
	983	PKEDLRLL	PK, KE, LR, LL	6.9	0	15.4	Phaseolin, alpha-type
	983	PSVADLRLL	PS, SV, VA, AD, LR, LL	6.7	0	13.8	DNA-directed RNA polymerase subunit beta
	1408	VNPDPAGGPTSGRAL	VN, VNP, NP, DP, PA, AG, GG, GP, PT, TS, SG, GR, RA, AL	6.7	2	14.5	Phaseolin, beta-type
PP–Black bean	656	DEGEAH	EG, GE, EA, AH	3.7	0	16.4	Phaseolin, alpha-type
	740	VELVGPK	VE, EL, LV, VG, VGP, GP, PK	6.5	−3	22.7	Phaseolin, beta-type
	742	VELTGPK	VE, LT, LTGP, TGP, TG, GP, PK	6.5	0	14.1	Phaseolin, alpha-type
	850	SGNGGGGGASM	SG, NG, GG, GA, AS	5.4	1	15.3	Glycine-rich cell wall structural protein
	855	SKPGGGSPVA	SK, KP, PG, GG, GS, SP, PV, VA	10.2	0	13.4	9-cis-epoxycarotenoid dioxygenase NCED1
	943	KPTTGKGALA	KP, PT, KPT, TT, TG, GK, GK, KG, AL, LA	10.6	2	16.1	9-cis-epoxycarotenoid dioxygenase NCED1
A–Green pea	539	GPAAGPA	GP, GPA, AA, AG, PA	5.6	1	13.29	Preprotein translocase subunit SECY
	541	TKGGAV	TK, KG, GG, GA, AV	10.1	0	11.98	Aminomethyltransferase, mitochondrial
	543	NPEGQ	NP, EG, GQ	3.1	0	5.7	Not reported
	544	TLSPGA	TL, TLS, LSP, SP, PG, GA	5.3	−2	12.3	Photosystem II CP43 reaction center protein

Table 1. *Cont.*

Sample	Molecular Mass (Da)	Peptide	Bioactive Sequence	pI	Net Charge	Hydrophobicity (kcal/mol)	Parental Protein [1]
PP–Green pea	620	SPGDVF	SP, PG, GD, VF	3.0	0	10.8	Mitochondrial Type Ii Peroxiredoxin
	627	LTAVPAG	LT, TA, AV, AVP, VP, PA, AG	5.5	−1	14.4	ATP synthase subunit alpha
	678	HALLLL	HA, AL, LL, LLL	7.8	0	11.0	Photosystem II D2 protein
	683	SHLGAVT	SH, HL, LG, GA, AV, VT	7.5	0	9.1	Protein translocase subunit SecA, chloroplastic
	687	GRSAAGVA	GR, AA, AG, GV, VA	11.1	0	8.7	Asparagine synthetase, root
	780	HSLPGVAT	HS, SL, LP, LPG, PG, GV, VA, AT	7.5	−1	17.1	Dihydrolipoyl dehydrogenase, mitochondrial
	785	RDTAGLGP	TA, AG, GL, LG, LGP, GP	7.0	2	15.7	NAD(P)H-quinone oxidoreductase subunit 5
A–Chickpea	856	DLVLDVPS	LVL, LV, VL, VP, PS	2.7	0	15.2	Tubulin beta chain
	943	KPSSAAGAVR	KP, KPS, PS, AA, AG, GA, AV, VR	11.5	−1	11.1	Non-specific lipid-transfer protein
	1091	TAPHGGLPAGDV	TA, TAP, AP, PH, PHG, GG, GL, LP, PA, GD	4.9	1	13.5	Acidic endochitinase
PP–Chickpea	428	SPPE	SP, PP	3.1	−1	12.4	Not reported
	526	CSSSSG	SSS, SG	4.9	0	10.4	Alpha-amylase inhibitor
	620	SPGDV	SP, PG, GD	3.0	−1	11.1	Not reported
	812	TPSGLNPQ	TP, PS, SG, GL, LN, LNP, NP, PQ	5.2	1	8.5	Not reported
	812	TPEKNPQ	TP, EK, NP, PQ	6.5	0	10.8	Not reported
	815	EPNGGLVM	EP, PN, NG, GG, GL, LV, VM	3.0	0	10.4	Not reported
	900	HGAESAGGDT	HG, GA, AE, ES, AG, GG, GD	3.9	0	16.4	Non-specific lipid-transfer protein
	949	RTPVPPGLL	TP, PV, VP, VPP, PP, PPG, PG, PGL, GL, LL	11.1	−1	12.2	Acetyl-coenzyme A carboxylase carboxyl transferase subunit beta

Table 1. *Cont.*

Sample	Molecular Mass (Da)	Peptide	Bioactive Sequence	pI	Net Charge	Hydrophobicity (kcal/mol)	Parental Protein [1]
A–Lentil	467	VVPGP	VV, VP, PG, PGP, GP	5.6	−1	10.6	Not reported
	533	PGDVF	PG, GD, VF	2.9	−2	12.9	Not reported
	596	DGHLR	DG, GH, HL, LR	7.5	0	15.5	Not reported
	652	EVGTFT	EV, VG, GT, TF, FT	3.0	1	13.8	Not reported
	678	FEDGLV	DG, DGL, GL, LV	2.9	−1	11.0	Not reported
	715	TPVSAGGK	TP, PV, VS, AG, GG, GK	9.8	0	8.4	Not reported
	428	SPPE	SP, PP	3.1	−1	12.8	Not reported
PP–Lentil	473	SPGDV	SP, PG, GD	3.0	0	8.6	Not reported
	552	VPPGAL	VP, VPP, PPG, PP, PG, GA, AL	5.6	1	12.6	Not reported
	627	LSVPGGV	SV, VP, PG, GG, GGV, GV	5.5	0	13.1	Not reported
	630	KGGLGVT	KG, GG, GL, LG, LGV, GV, VT	9.8	−1	13.6	Not reported
	758	TSPSPGDV	TS, SP, PS, PG, GD	3.0	−1	12.2	Not reported
	942	KTDVLPTGL	KT, TD, VL, VLP, LP, PT, TG, GL	6.7	0	8.1	Linoleate 9S-lipoxygenase
A–Fava bean	677	TPVHPQ	TP, PV, VH, HP, PQ	7.5	−1	15.2	Legumin type B alpha chain
	682	NLLAPR	NL, LL, LA, LAP, LLAP, AP, PR	10.7	1	8.7	Probable sucrose-phosphate synthase
	706	SFGGGGLL	SF, FG, GG, FGG, GL, LL	5.4	−1	18.3	14-3-3-like protein B
	715	FGGLLPL	FG, FGG, GG, GL, LL, LLP, LP, LPL, PL	5.4	0	11.0	NAD(P)H-quinone oxidoreductase subunit 5
PP–Fava bean	751	TKAGGTAF	TK, KA, AG, GG, GT, TA, AF	9.9	0	4.8	14-3-3-like protein B
	807	GPPVDVPQ	GP, GPP, PV, VD, VP, PQ	3.1	−1	12.9	Photosystem II protein D1
	810	PPNGPSEN	PP, PN, NG, GP, PS, SE	3.0	1	10.2	Acid beta-fructofuranosidase
	869	PPRSDSDP	PP, PR, DP	3.9	1	12.7	Not reported
	942	PYGVPVGVR	PY, YG, GV, VP, PV, VG, GV, VR	9.5	0	8.7	Elongation factor 1-alpha

[1] As determined by BLAST®; A: enzymatic hydrolysis with alcalase; PP: enzymatic hydrolysis with pepsin/pancreatin; peptides obtained from the HPLC elution profile with intensity of at least 30% using LC–ESI–MS/MS, liquid chromatography–electrospray ionization tandem mass–spectrometry. Potential bioactivities were obtained from the BIOPEP database. Sequences were confirmed by BLAST tool, according to "UnitProtKB". pI: isoelectric point; ACE: angiotensin-converting enzyme; DPP-IV: dipeptidyl peptidase IV; amino acid nomenclature: C, cysteine; H, histidine; I, isoleucine; M, methionine; S, serine; V, valine; A, alanine; G, glycine; L, leucine; P, proline; T, threonine; F, phenylalanine; R, arginine; Y, tyrosine; W, tryptophan; D, aspartic acid; N, asparagine; E, glutamic acid; Q, glutamine; K, lysine.

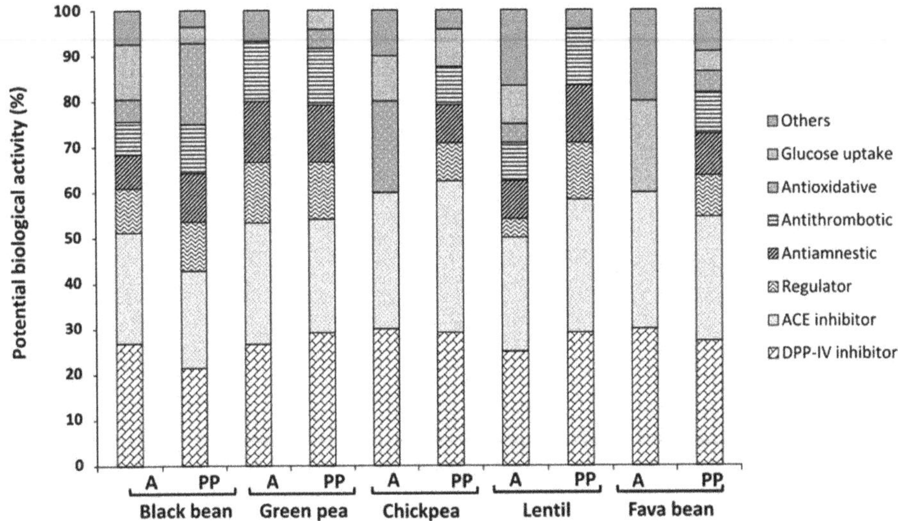

Figure 2. Potential biological activity (%) of different legume-generated peptides using Biopep database. PP, pepsin/pancreatin; A, alcalase; DPP-IV, dipeptidyl peptidase IV inhibitor; ACE, angiotensin-converting enzyme inhibitor; regulator, stomach mucosal membrane activity regulator. Results represent the mean ± SD of at least two independent experiments.

3.4. Antioxidant Capacity

3.4.1. ABTS Radical Scavenging

The antioxidant capacity of the whole hydrolysate (LPH) was measured spectrophotometrically by the disappearance of the blue/green stable ABTS$^{\bullet+}$ radical, caused by scavenging (Figure 3A). The concentration of inhibitor required to produce 50% inhibition was calculated (IC$_{50}$) and the results of ABTS$^{\bullet+}$ radical assay were presented as Trolox equivalent (TE) antioxidant capacity using Trolox as standard. PP–lentil hydrolysate was demonstrated to have the highest ability to scavenge the ABTS$^{\bullet+}$ radical (IC$_{50}$ = 0.30 ± 0.13 TE/mL), which was followed by PP–black bean (IC$_{50}$ = 0.60 ± 0.06 TE/mL) and both PP and A–green pea hydrolysates (IC$_{50}$ = 0.61 ± 0.03 and 0.60 ± 0.03 TE mg/mL, respectively) ($p < 0.05$).

3.4.2. DPPH Inhibition Capacity

The antioxidant activity of LPH measured by DPPH$^{\bullet}$ scavenging capacity is shown in Figure 3B. Hydrolysates obtained from alcalase digestion showed higher DPPH$^{\bullet}$ scavenging activity compared to PP-LPH ($p < 0.05$). As the lower the IC$_{50}$ value, the most potent scavenging capacity, A–chickpea (IC$_{50}$ = 0.68 ± 0.18 mg/mL) had the lowest IC$_{50}$ value, followed by A–lentil (IC$_{50}$ = 3.40 ± 0.30 mg/mL), A–black bean (IC$_{50}$ = 3.60 ± 1.50 mg/mL) and A–green pea (IC$_{50}$ = 4.45 ± 1.03 mg/mL).

3.4.3. Nitric Oxide (NO) Scavenging Capacity

LPH were assessed for NO radical inhibitory activity (Figure 3C). It was found that NO radical inhibitory activity showed no significant differences among IC$_{50}$ values of PP–black bean (0.23 ± 0.01 mg/mL), A and PP–green pea (0.20 ± 0.00 mg/mL and 0.25 ± 0.01 mg/mL, respectively), PP–lentil (0.24 ± 0.04 mg/mL) and A and PP–fava bean (0.16 ± 0.01 mg/mL and 0.28 mg/mL ± 0.01, respectively) hydrolysates.

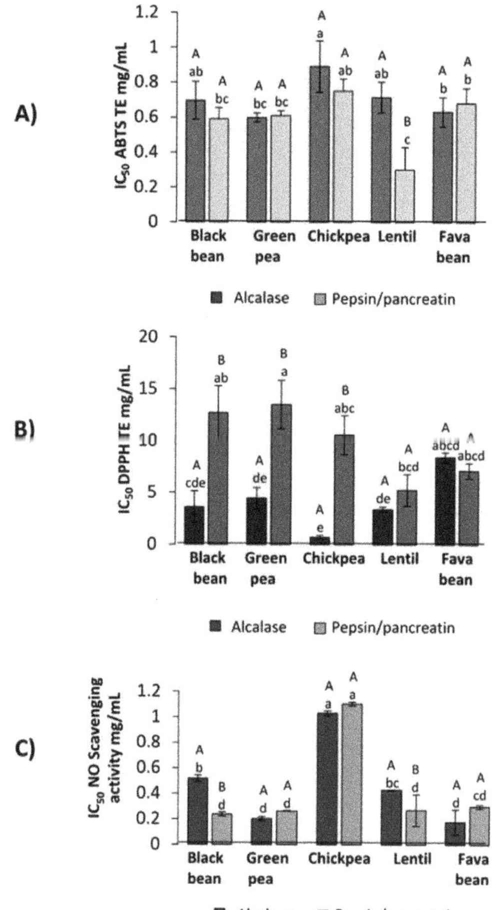

Figure 3. Antioxidant potential of LPH against different radicals. (**A**) ABTS scavenging activity (IC$_{50}$ TE mg/mL), (**B**) DPPH scavenging activity (IC$_{50}$ TE mg/mL), (**C**) nitric oxide scavenging activity (IC$_{50}$ mg/mL). Different capital letters indicate significant differences between alcalase and pepsin/pancreatin digestion ($p < 0.05$); different lowercase letters indicate significant differences ($p < 0.05$) among legume protein hydrolysates. Results represent the mean ± SD of at least two independent experiments.

3.5. Anti-Adipogenic Potential

3.5.1. Pancreatic Lipase Inhibitory Activity

LPH were tested for their ability to inhibit PL. All the hydrolysates demonstrated to inhibit the enzyme in a dose-dependent manner, as shown in Figure 4A. There were no significant differences between PP and A–black bean hydrolysates and PP and A–chickpea, ($p < 0.05$). PP–green pea hydrolysate presented the lowest IC$_{50}$ value (0.8 ± 0.06 mg/mL) among all the legume PP hydrolysates. Nevertheless, the IC$_{50}$ values of the PP–lentil hydrolysate (7.7 ± 0.64 mg/mL) and A–fava bean hydrolysate (11.8 ± 1.75 mg/mL) were significantly higher (less potent inhibition) compared to the rest of the hydrolysates.

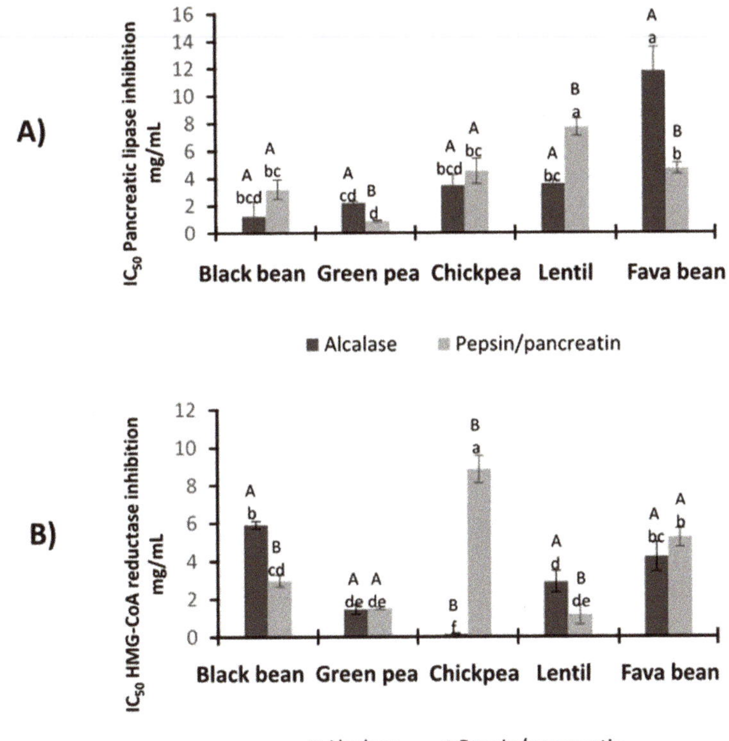

Figure 4. (**A**) Inhibition potential of LPH to pancreatic lipase (IC_{50} mg/mL). Orlistat synthetic inhibitor IC_{50} = 2.73 µg/mL [31]. Different capital letters indicate significant differences between alcalase and pepsin/pancreatin digestion ($p < 0.05$); different lowercase letters indicate significant differences ($p < 0.05$) among legume protein hydrolysates. (**B**) Inhibition potential of LPH to HMG-CoA (IC_{50} mg/mL). Simvastatin synthetic inhibitor IC_{50} = 0.08 µg/mL. Different capital letters indicate significant differences between alcalase and pepsin/pancreatin digestion ($p < 0.05$); different lowercase letters indicate significant differences ($p < 0.05$) among legume protein hydrolysates. Results represent the mean ± SD of at least two independent experiments.

3.5.2. HMG-CoA Reductase Inhibitory Activity

HMGR catalyzes the rate-limiting step of the mevalonate pathway that leads to cholesterol biosynthesis; thus, inhibition of its enzymatic activity plays a significant role in lowering endogenous cholesterol levels during hypercholesterolemia [35]. Figure 4B shows the inhibition of HMGR activity by PP and A hydrolysates. A–chickpea was the most effective hydrolysate with an IC_{50} = 0.15 ± 0.04 mg/mL, followed by PP–lentil (IC_{50} = 1.17 ± 0.52 mg/mL), A–green pea (IC_{50} = 1.45 ± 0.25 mg/mL) and PP–green pea (IC_{50} = 1.50 ± 0.07 mg/mL). By contrast, the highest IC_{50} value was observed in PP–chickpea (IC_{50} = 8.84 ± 0.74 mg/mL).

3.5.3. Isobolograms of the LPH Interactions

The results of the isobolographic PL assay are presented in Figure 5A,B. It was observed that three combinations of LPH showed synergistic potential, namely PP–green pea and A–black bean, PP–green pea and A–green pea, A–black bean and A–green pea, with IC_{50} values of 0.40 ± 0.00 mg/mL, 0.95 ± 0.02 mg/mL and 0.65 ± 0.04 mg/mL, respectively.

For the inhibition of HMGR, interactions among LPH presented synergistic and additive potential (Figure 5B). A synergistic effect was observed with the combination of A–chickpea and PP–black bean (IC_{50} = 1.33 ± 0.07), while additive interactions were observed between A–chickpea and PP–green pea (IC_{50} = 0.84 ± 0.04 mg/mL) and PP–lentil and A–green pea (IC_{50} = 1.34 ± 0.00 mg/mL).

A)

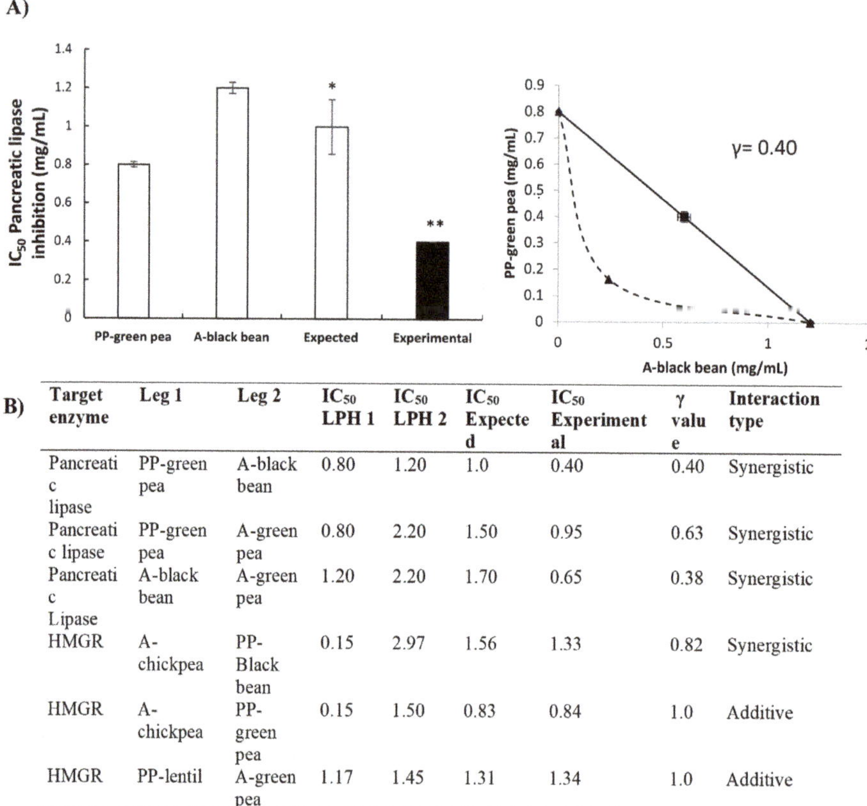

B)

Target enzyme	Leg 1	Leg 2	IC$_{50}$ LPH 1	IC$_{50}$ LPH 2	IC$_{50}$ Expected	IC$_{50}$ Experimental	γ value	Interaction type
Pancreatic lipase	PP-green pea	A-black bean	0.80	1.20	1.0	0.40	0.40	Synergistic
Pancreatic lipase	PP-green pea	A-green pea	0.80	2.20	1.50	0.95	0.63	Synergistic
Pancreatic Lipase	A-black bean	A-green pea	1.20	2.20	1.70	0.65	0.38	Synergistic
HMGR	A-chickpea	PP-Black bean	0.15	2.97	1.56	1.33	0.82	Synergistic
HMGR	A-chickpea	PP-green pea	0.15	1.50	0.83	0.84	1.0	Additive
HMGR	PP-lentil	A-green pea	1.17	1.45	1.31	1.34	1.0	Additive

Figure 5. (**A**) Synergistic interaction of PP–green pea and A–black bean hydrolysates to block pancreatic lipase (IC_{50} mg/mL). Bar plots represent the mean ± standard deviation of at least two independent experiments, with only the expected and combined values compared statistically. Different number of stars means significant difference ($p < 0.05$). The lower the values, the more potency. Results represent the mean ± SD of at least two independent experiments. (**B**) Interactions between legume hydrolysate combinations to block pancreatic lipase and 3-hydroxy-3-methylglutaryl coenzyme A reductase HMGR. Results are expressed in IC_{50} mg/mL; γ value < 1 synergism, γ value = 1 additive effect, γ value > 1 antagonism.

3.6. Molecular Docking Study of Peptides Inhibiting Pancreatic Lipase and HMG-CoA Reductase

Molecular docking analysis was performed to predict the potential of the LPH to interact with PL and HMG-CoA reductase enzymes. Peptide and macromolecular target docking shows theoretical affinity, type of interactions and distances [36,37]. Table 2 shows the minimum estimated free energy for peptides sequenced from LPH, with PL and HMGR. Estimated free energy indicates that compounds with the most negative value present higher potential to interact with the target enzyme. The peptides studied had free energy values ranging from −5.5 (CSSSSG) to −7.6 (FEDGLV) kcal/mol for PL. Peptide FEDGLV was obtained from A–lentil and CSSSSG from PP–chickpea. In the case of HMGR, binding affinities of peptides ranged from −6.1 (CSSSSG and GPPVDVPQ) to −8.8 (PYGVPVGVR) kcal/mol, obtained from PP–chickpea and PP–fava bean, respectively. Orlistat and pravastatin were also evaluated as a control, presenting free energy values of −5.6 and −6.2 kcal/mol for pancreatic lipase and HMGR, respectively. The most stabilized pose of the peptide bonds with PL and HMG-CoA reductase can be observed in Figure 6A,B, respectively. All the peptides identified in the different legume protein hydrolysates were able to interact with amino acid residues of PL and HMGR catalytic site. These peptides interacted with the enzymes mainly through hydrogen bonds and hydrophobic, polar and cation π interactions.

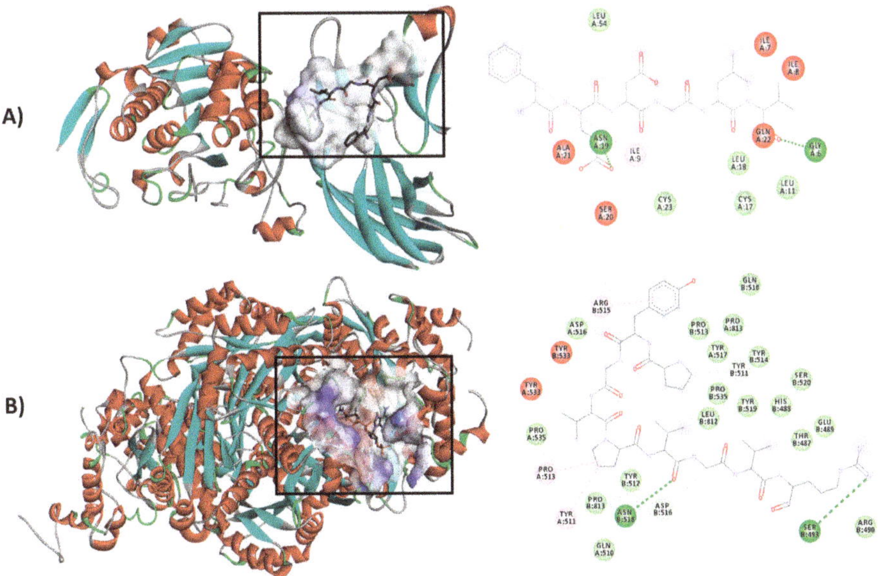

Interaction type	Color	Target enzyme	Peptide/source	Binding affinity (kcal/mol)
	Van der Waals	A) LP	FEDGLV	-7.6
	Conventional hydrogen bond		Orlistat	-5.6
	Carbon hydrogen bond	B) HMGR	PYGVPVGVR	-8.8
	Pi-Alkyl		Pravastatin	-6.2

Figure 6. Molecular docking diagrams, examples of the best pose of the most potent legume peptides in the catalytic site of pancreatic lipase (**A**) and HMGR (**B**). Results represent the mean ± SD of at least two independent experiments.

Table 2. Estimated free energy binding and chemical interactions among peptides, present in legumes hydrolyzed with alcalase and simulated gastrointestinal digestion, and the catalytic site of the pancreatic lipase and HMG-CoA reductase.

	Black Bean	Binding Affinity (kcal/mol)		Green Pea	Binding Affinity (kcal/mol)		Chickpea	Binding Affinity (kcal/mol)		Lentil	Binding Affinity (kcal/mol)		Fava bean	Binding Affinity (kcal/mol)	
		PL	HMG		PL	HMG		PL	HMG		PL	HMG		PL	HMG
A	FEELN	−6.1	−6.7	GPAAGPA	−7.0	−7.6	DLVLDVPS	−6.1	−6.8	VVPGL	−7.1	−7.4	TPVHPQ	−6.9	−7.6
	THGPVGAN	−6.3	−7.0	TKGGAV	−6.0	−6.8	KPSSAAGAVR	−6.1	−6.3	PGDVL	−7.0	−7.9	NLLAPR	−6.3	−7.6
	ANGSPGGAGA	−6.5	−8.1	NPEGQ	−6.3	−6.6	TAPHGGLPAGDV	−7.1	−7.7	DGHLM	−6.5	−6.9	SFGGGGLL	−5.8	−7.5
	KPSASCSR	−5.9	−6.7	TLSPGA	−6.3	−6.8				EVGTF	−7.3	−8.2			
	NVGPGSLET	−6.9	−7.3							FEDGLV	−7.6	−7.0			
	PKEDLRLL	−5.6	−7.4							TPVSAGGK	−6.0	−6.2			
	PSVADLRLL	−5.9	−7.7				SPPE	−6.6	−7.7						
	VNPDPAGGPTSGRAL	−7.0	−8.0	SPGDVF	−6.7	−7.3	CSSSSG	−5.5	−6.1	SPPE	−6.8	−7.5			
PP	DEGEAH	−6.3	−7.6	LTAVPAG	−6.3	−6.3	SPGDV	−7.0	−6.8	SPGDV	−6.4	−6.6	FGGLLPL	−6.6	−7.6
	VELVGPK	−6.2	−6.9	HALLLL	−5.9	−6.9	TPSGLNPQ	−6.9	−7.8	VPPGAL	−6.4	−7.7	TKAGGTAF	−5.5	−7.2
	VELTGPK	−6.4	−6.7	SHLGAVT	−6.5	−6.7	TPEKNPQ	−6.9	−7.1	LSVPGCV	−6.2	−7.2	GPPVDVPQ	−5.8	−6.1
	SGNGGGGASM	−6.0	−6.8	GRSAAGVA	−5.8	−6.7	EPNGGLVM	−5.7	−7.0	KGGLGFT	−6.4	−6.7	PPNGPSEN	−5.9	−8.1
	SKPGGGSPVA	−5.6	−7.9	HSLPGVAT	−6.8	−7.1	HGAESAGGDT	−5.8	−6.8	TSPSPGIV	−7.0	−7.4	PPRSDSDP	−7	−7.7
	KPTTGKGALA	−6.4	−7.0	RDTAGLGP	−6.5	−7.3	RTPVPPGLL	−6.8	−7.9	KTDVLPGGL	−6.3	−6.7	PYGVPVGVR	−6.9	−8.8

A: enzymatic hydrolysis with alcalase; PP: enzymatic hydrolysis with pepsin/pancreatin; PL: pancreatic lipase; HMG: HMG-CoA reductase. Amino acid nomenclature: C, cysteine; H, histidine; I, isoleucine; M, methionine; S, serine; V, valine; A, alanine; G, glycine; L, leucine; P, proline; T, threonine; F, phenylalanine; R, arginine; Y, tyrosine; W, tryptophan; D, aspartic acid; N, asparagine; E, glutamic acid; Q, glutamine; K, lysine.

4. Discussion

Vicilin and legumin-like proteins comprise more than 70% of the total proteins found in the legumes used for this study. However, differences in the amino acid sequence of homologous proteins contained in the legume species used can lead to changes in the protein profile obtained after the hydrolysis treatment.

A higher DH is indicative of more peptide bonds cleaved, resulting in lower molecular weight peptides [38]. The application of the enzymatic hydrolysis with A and PP notably changed the structure of the legume native proteins. These changes can be noted in the disappearance of protein bands higher than 10 kDa after treatment and the formation of polypeptides with lower molecular mass. Similar findings were obtained in a study of legume isolate proteins [39].

The enzymatic action of pepsin and pancreatin in a sequential way to perform the simulated gastrointestinal digestion increases the enzyme selectivity and specificity, enabling the production of a larger number of small protein fractions or free amino acids [40]. Increasing cleavage sites are available for the second enzyme after the first one has acted, thus forming a more diverse mixture of amino acid residues in comparison to alcalase. According to these results, the action of the gastrointestinal enzymes was more effective to cleave distinct peptide bonds in black bean, chickpea and fava bean LPI. Even though enzymatic hydrolysis with PP treatment exhibited higher DH in the mentioned legumes, the action of alcalase showed the same effect as PP in green pea and lentil. These results suggest that fractions of vicilin commonly found in plants such as peas or lentils were susceptible to digestion with the different enzymatic treatments [41,42]. On the other hand, due to the very broad substrate specificity, alcalase can hydrolyze most peptide bonds within a protein molecule. Therefore, its use on substrates like legume proteins causes a high degree of hydrolysis, producing many peptides of small sizes [43]. The low DH values obtained in fava bean and chickpea LPH with alcalase could be related to vicilin, as this protein is known for its emulsifying, foaming and gelling properties [44,45]. Moreover, legumes rich in glutamic acid could have led to lower degradation by alcalase. This enzyme has a negative charge in physiological pH, which may diminish the DH due to the net negative charge present in glutamic acid [46]. In addition, resistance toward enzymatic activity may be a consequence of structural differences and the compactness of proteins [33]. Similar results were found by Ghribi et al. (2015) [47] in chickpea isolates hydrolyzed with alcalase, where it was found that a small degree of chickpea hydrolysis (DH = 4%) could enhance the emulsifying properties of chickpea protein. In another study performed with fava bean by Liu et al. (2019) [48], this moderate alcalase hydrolysis suggests the production of suitable lower molecular mass that could result in a more flexible peptide structure, increased surface charge and hydrophobicity, which could positively affect the emulsifying activity [41,49].

All the hydrolysates demonstrated the potential to inhibit pancreatic lipase and HMG-CoA reductase in a dose-dependent manner. According to the literature, Lee et al. (2015) [50] performed an inhibition assay for pancreatic lipase with phenolic compounds from methanolic extracts of seven selected legumes. The results showed pancreatic lipase inhibition in a dose-dependent manner, with no significant differences among red bean, chickpea, black soybean, yellow soybean and black-eyed pea extracts. The lowest reported IC_{50} values were for red bean (5.90 ± 0.59 mg/mL), chickpea (6.97 ± 2.19 mg/mL) and black soybean (6.65 ± 0.62 mg/mL). These values are higher (less potent) compared to most of the hydrolysates obtained in this work. The variation may be attributed to the differences in the extraction system, the varieties used and the bioactive compounds present in legume protein hydrolysates. Another report associated the peptide fragments released after protein enzymatic hydrolysis with the potential to act as HMGR inhibitors. For instance, the authors identified peptide sequences below 3 kDa from amaranth protein with hypocholesterolemic potential (GGV, IVG and VGVL) [51]. Recent studies performed on soybean glycinin and β-Conglycinin protein-derived peptides have also reported that these peptides could inhibit HMGR activity in vitro and in silico studies [52,53].

From all evaluated legume protein hydrolysate combinations used to block PL and HMG-CoA reductase activities, four combinations acted synergistically. PP–green pea with A–black bean, PP–green pea with A–green pea and A–black bean with A–green pea were demonstrated to inhibit more efficiently the activity of PL. Furthermore, the A–chickpea and PP–black bean combination acted synergistically against HMG-CoA reductase activity. Previous studies have demonstrated the efficacy of legume protein hydrolysates to inhibit markers related to obesity (black bean, green pea and chickpea hydrolysates) [19,54–56]. Several reports have demonstrated synergistic interactions among diverse bioactive compounds. For instance, synergistic and additive interactions of peptides produced from black bean for ACE inhibition were reported by Luna-Vital et al. (2015) [57]. The combination of peptides GLTSK and MTEEY showed a synergistic interaction, reducing the concentration needed to inhibit half of the enzymatic activity by approximately 30%. Further studies focusing on each peptide sequence are essential to confirm the possible effects of their interaction. These interactions can be significant and lead to lower effective dosages associated with the effects of pure compounds [58–60]. There is evidence that legume protein hydrolysates could interact with enzymes related to obesity. Exploring the possibilities to improve their performance, along with increasing the desired effect, is important.

The peptide profile of the protein hydrolysates obtained was different due to the legume source and enzyme used. As a result of the enzymatic hydrolysis and the specificity of the enzymes used, the biological activity of the legume protein hydrolysates was conferred by the specific combination of both protease used and source of proteins. The amino acid profile in peptide sequences with higher biological activity was associated with aliphatic amino acids, sulfur-containing and aromatic amino acids. Through molecular docking, it was possible to predict the binding affinity to the target enzymes (LP and HMG-CoA reductase). Peptide sequences presented lower predicted free energy values compared to the available drugs orlistat and pravastatin. These results suggest that they have a higher affinity for the catalytic site, thus exerting significant potential anti-adipogenesis activity. Non-polar interactions among peptide sequences and evaluated enzymes were more common due to the frequency of the aliphatic amino acids.

In addition to the effect of the enzymes related to lipid metabolism (LP and HMG-CoA reductase), obesity is associated with the production of reactive oxygen species and increased oxidative stress [8]. Peptides are able to contribute to the antioxidant defense in the body, being able to rapidly scavenge reactive oxygen species before cellular damage, therefore inactivating them [61]. The antioxidant capacity of the legume hydrolysates was measured using free radicals generated by ABTS and DPPH. Even though the meaning of the results obtained with these assays is limited as they use no physiological radicals, it is still possible to achieve representative data evaluating antioxidant activity [62]. Lentil protein hydrolysate generated with PP showed higher potential to scavenge the radical ABTS$^\bullet$ compared to the lentil protein hydrolysate generated using alcalase. Different legumes have demonstrated antioxidant capacities through this method. For instance, Ngoh et al. (2016) [63] determined the antioxidant activity presented in pinto bean peptides; in this study, peptide fractions < 3 kDa exhibited 42.2% inhibition of ABTS$^{\bullet+}$ scavenging activity. The results for DPPH$^\bullet$ suggest that lower DH in A-LPH may enhance the potential to scavenge the radical DPPH$^\bullet$. This is consistent with previous reports by Evangelho et al. (2017) [64], where it was observed that alcalase protein hydrolysates obtained from black bean showed higher DPPH$^\bullet$ scavenging capacities at lower DH. Likewise, similar findings were observed in a study performed by Kou et al. (2013) [65] showing that chickpea peptides obtained by alcalase exhibited 41.3% DPPH$^\bullet$ radical scavenging activity. Another free radical scavenging evaluation was performed with NO. This radical plays an important role in inflammatory processes; high levels of NO and its oxidized derivatives are known to be toxic, resulting in vascular damage and other ailments [66]. Legume protein hydrolysates showed the potential to scavenge NO, with important results. In the case of the NO scavenging capacity, the most potent legume hydrolysates were the ones hydrolyzed with PP. Pepsin/pancreatin protein hydrolysates presented a more extensive degree of hydrolysis, leading to more diversity of peptides in the protein hydrolysate. In a study performed by Oseguera-Toledo et al. (2015) [26], potent hydrolysate fractions of 5–10 kDa were obtained with the enzyme alcalase. These hydrolysates from black bean demonstrated the capacity to scavenge the radical NO with a range of inhibition from 57.46% to 68.26%.

Protein hydrolysates from selected legumes could participate in the inhibition of the enzymatic activity of LP and HMG-CoA reductase. Besides this, legume protein hydrolysates show the potential to exert antioxidant activity.

5. Conclusions

Legumes are an important source of ingredients for the formulation of healthy foods. Legume protein hydrolysates were able to block pancreatic lipase and HMG-CoA reductase using in silico and biochemical assays. Furthermore, legume protein hydrolysates showed important radical scavenging activities against ABTS, DPPH and NO radicals. These results shed light on the potential antioxidant activity of the peptides. More importantly, the combination of different legume protein hydrolysates inhibited synergistically the adipogenesis-related enzymes evaluated. Further studies are needed to determine the anti-obesity potential and the antioxidant capacity of pure synthesized peptides and their combinations. Mixtures of legume protein hydrolysates could be used as functional ingredients in the formulation of foods with the potential to prevent or treat non-communicable diseases such as obesity.

Author Contributions: C.M.: conceptualization; formal analysis; investigation; methodology; writing—original draft, review and editing. E.G.d.M.: conceptualization; funding acquisition; investigation; methodology; writing—review and editing. D.A.L.-V.: conceptualization; funding acquisition; methodology; writing—review and editing. R.M.C.R.: conceptualization; methodology; writing—review and editing. L.M.: conceptualization; funding acquisition; project administration; methodology; writing—review and editing. All authors have read and agreed to the published version of the manuscript.

Funding: This work was supported by the Consejo Nacional de Ciencia y Tecnologia CONACyT Mexico (grant number 847861) and the research was supported by CONACyT (Problemas Nacionales 2016–2018 grant).

Acknowledgments: Author Cecilia Alejandra Moreno Valdespino was supported by a scholarship from the Consejo Nacional de Ciencia y Tecnologia CONACyT Mexico, number 847861.

Conflicts of Interest: The authors declare no conflict of interest.

Abbreviations

A: alcalase; A-LPH: legume protein hydrolyzed with alcalase; ABTS: 2,2′-azinobis (3-ethyl-benzothiazoline -6-sulfonic acid) diammonium salt; ACE: angiotensin-converting enzyme; DH: degree of hydrolysis; DPP-IV: dipeptidyl peptidase-IV; DPPH: 2,2-diphenyl-1-picrylhydrazyl; FDA: U.S. Food and Drug Administration; HMGR: 3-hydroxy-3-methylglutaryl coenzyme A reductase; LC–ESI–MS/MS: liquid chromatography–electrospray ionization–mass spectrometry; LPH: legume protein hydrolysates; LPI: legume protein isolates; MOPS: 3-Morpholinopropane-1-sulfonic acid; NaTDC: sodium taurodeoxycholate hydrate; NO: nitric oxide; PL: pancreatic lipase; PP: pepsin and pancreatin; PP-LPH: legume protein hydrolyzed with pepsin/pancreatin; SDS–PAGE: sodium dodecyl sulfate–polyacrylamide gel electrophoresis: SNP: sodium nitroprusside; TE: Trolox equivalent; TNBS: trinitrobenzene sulfonic acid.

References

1. Feingold, K.R.; Grunfeld, C. *Obesity and Dyslipidemia*; MDText.com, Inc.: South Darmouth, MA, USA, 2000.
2. World Health Organization. Obesity and Overweight. Available online: https://www.who.int/en/news-room/fact-sheets/detail/obesity-and-overweight (accessed on 3 April 2019).
3. Blüher, M. Obesity: Global epidemiology and pathogenesis. *Nat. Rev. Endocrinol.* **2019**, *15*, 288–298. [CrossRef] [PubMed]
4. Waalen, J. The genetics of human obesity. *Transl. Res.* **2014**, *164*, 293–301. [CrossRef] [PubMed]
5. Ades, P.A.; Savage, P.D. Obesity in coronary heart disease: An unaddressed behavioral risk factor. *Prev. Med.* **2017**, *104*, 117–119. [CrossRef] [PubMed]
6. Fernández-Sánchez, A.; Madrigal-Santillán, E.; Bautista, M.; Esquivel-Soto, J.; Morales-González, Á.; Esquivel-Chirino, C.; Durante-Montiel, I.; Sánchez-Rivera, G.; Valadez-Vega, C.; Morales-González, J.A. Inflammation, oxidative stress, and obesity. *Int. J. Mol. Sci.* **2011**, *12*, 3117–3132. [CrossRef] [PubMed]

7.   Heymsfield, S.B.; Wadden, T.A. Mechanisms, pathophysiology, and management of obesity. *N. Engl. J. Med.* **2017**, *376*, 254–266. [CrossRef] [PubMed]

8.   Marseglia, L.; Manti, S.; D'Angelo, G.; Nicotera, A.; Parisi, E.; Di Rosa, G.; Gitto, E.; Arrigo, T. Oxidative stress in obesity: A critical component in human diseases. *Int. J. Mol. Sci.* **2014**, *16*, 378–400. [CrossRef] [PubMed]

9.   Dannenberg, A.J.; Berger, N.A. *Obesity, Inflammation and Cancer*; Dannenberg, A.J., Berger, N.A., Eds.; Springer: New York, NY, USA, 2013; ISBN 978-1-4614-6818-9.

10.   Eckel, R.H.; Kahn, S.E.; Ferrannini, E.; Goldfine, A.B.; Nathan, D.M.; Schwartz, M.W.; Smith, R.J.; Smith, S.R. Obesity and Type 2 diabetes: What can be unified and what needs to be individualized? *Diabetes Care* **2011**, *34*, 1424–1430. [CrossRef]

11.   Daneschvar, H.L.; Aronson, M.D.; Smetana, G.W. FDA-approved anti-obesity drugs in the United States. *Am. J. Med.* **2016**, *129*, 879.e1–879.e6. [CrossRef]

12.   Leung, W.Y.S.; Neil Thomas, G.; Chan, J.C.N.; Tomlinson, B. Weight management and current options in pharmacotherapy: Orlistat and sibutramine. *Clin. Ther.* **2003**, *25*, 58–80. [CrossRef]

13.   Gupta, R. Traumatic brain injury. In *Rapid Review Anesthesiology Oral Boards*; Gupta, R., Tran, M.C.J., Eds.; Cambridge University Press: Cambridge, UK, 2018; pp. 73–78. ISBN 9780128040645.

14   Filippatos, T.D.; Derdemezis, C.S.; Gazi, I.F.; Nakou, E.S.; Mikhailidis, D.P.; Elisaf, M.S. Orlistat-associated adverse effects and drug interactions. *Drug Saf.* **2008**, *31*, 53–65. [CrossRef]

15.   Golomb, B.A.; Evans, M.A. Statin adverse effects. *Am. J. Cardiovasc. Drugs* **2008**, *8*, 373–418. [CrossRef] [PubMed]

16.   Clemente, A.; Olias, R. Beneficial effects of legumes in gut health. *Curr. Opin. Food Sci.* **2017**, *14*, 32–36. [CrossRef]

17.   Bennetau-Pelissero, C. *Plant Proteins from Legumes*; Springer: Cham, Switzerland, 2019; pp. 223–265. ISBN 978-3-319-78029-0.

18.   López-Barrios, L.; Gutiérrez-Uribe, J.A.; Serna-Saldívar, S.O. Bioactive peptides and hydrolysates from pulses and their potential use as functional ingredients. *J. Food Sci.* **2014**, *79*, R273–R283. [CrossRef] [PubMed]

19.   Moreno-Valdespino, C.A.; Luna-Vital, D.; Camacho-Ruiz, R.M.; Mojica, L. Bioactive proteins and phytochemicals from legumes: Mechanisms of action preventing obesity and type-2 diabetes. *Food Res. Int.* **2020**, *130*, 108905. [CrossRef] [PubMed]

20.   Callcott, E.T.; Santhakumar, A.B.; Luo, J.; Blanchard, C.L. Therapeutic potential of rice-derived polyphenols on obesity-related oxidative stress and inflammation. *J. Appl. Biomed.* **2018**, *16*, 255–262. [CrossRef]

21.   Milán-Noris, A.K.; Gutiérrez-Uribe, J.A.; Santacruz, A.; Serna-Saldívar, S.O.; Martínez-Villaluenga, C. Peptides and isoflavones in gastrointestinal digests contribute to the anti-inflammatory potential of cooked or germinated desi and kabuli chickpea (*Cicer arietinum* L.). *Food Chem.* **2018**, *268*, 66–76. [CrossRef]

22.   Luna-Vital, D.; Mojica, L.; de Mejía, E.G.; Mendoza, S.; Loarca-Piña, G. Biological potential of protein hydrolysates and peptides from common bean (*Phaseolus vulgaris* L.): A review. *Food Res. Int.* **2015**, *76*, 39–50. [CrossRef]

23.   Luna-Vital, D.; de Mejía, E.G. Peptides from legumes with antigastrointestinal cancer potential: Current evidence for their molecular mechanisms. *Curr. Opin. Food Sci.* **2018**, *20*, 13–18. [CrossRef]

24.   Mojica, L.; de Mejía, E.G. Optimization of enzymatic production of anti-diabetic peptides from black bean (*Phaseolus vulgaris* L.) proteins, their characterization and biological potential. *Food Funct.* **2016**, *7*, 713–727. [CrossRef]

25.   Mojica, L.; Chen, K.; de Mejía, E.G. Impact of commercial precooking of common bean (*Phaseolus vulgaris*) on the generation of peptides, after pepsin-pancreatin hydrolysis, capable to inhibit Dipeptidyl Peptidase-IV. *J. Food Sci.* **2015**, *80*, H188–H198. [CrossRef]

26.   Oseguera-Toledo, M.E.; de Mejia, E.G.; Amaya-Llano, S.L. Hard-to-cook bean (*Phaseolus vulgaris* L.) proteins hydrolyzed by alcalase and bromelain produced bioactive peptide fractions that inhibit targets of type-2 diabetes and oxidative stress. *Food Res. Int.* **2015**, *76*, 839–851. [CrossRef]

27.   de los Angeles Camacho-Ruiz, M.; Mateos-Díaz, J.C.; Carrière, F.; Rodriguez, J.A. A broad pH range indicator-based spectrophotometric assay for true lipases using tributyrin and tricaprylin. *J. Lipid Res.* **2015**. [CrossRef]

28.   Mojica, L.; de Mejia, E.G.; Granados-Silvestre, M.Á.; Menjivar, M. Evaluation of the hypoglycemic potential of a black bean hydrolyzed protein isolate and its pure peptides using in silico, in vitro and in vivo approaches. *J. Funct. Foods* **2017**, *31*, 274–286. [CrossRef]

29. Bikadi, Z.; Hazai, E. Application of the PM6 semi-empirical method to modeling proteins enhances docking accuracy of AutoDock. *J. Cheminform* **2009**, *1*, 15. [CrossRef]

30. Halgren, T. A merck molecular force field. II. MMFF94 van der Waals and electrostatic parameters for intermolecular interactions. *J. Comput. Chem.* **1996**, *17*, 520–552. [CrossRef]

31. Huey, R.; Morris, G.M.; Olson, A.J.; Goodsell, D.S. A semiempirical free energy force field with charge-based desolvation. *J. Comput. Chem.* **2007**, *28*, 1145–1152. [CrossRef]

32. Mojica, L.; Luna-Vital, D.A.; de Mejía, E.G. Characterization of peptides from common bean protein isolates and their potential to inhibit markers of type-2 diabetes, hypertension and oxidative stress. *J. Sci. Food Agric.* **2017**, *97*, 2401–2410. [CrossRef]

33. Barac, M.; Cabrilo, S.; Stanojevic, S.; Pesic, M.; Pavlicevic, M.; Zlatkovic, B.; Jankovic, M. Functional properties of protein hydrolysates from pea (*Pisum sativum*, L) seeds. *Int. J. Food Sci. Technol.* **2012**. [CrossRef]

34. Barbana, C.; Boye, J.I. Angiotensin I-converting enzyme inhibitory properties of lentil protein hydrolysates: Determination of the kinetics of inhibition. *Food Chem.* **2011**. [CrossRef]

35. Rasmussen, T.P. Genomic medicine and lipid metabolism. In *Translational Cardiometabolic Genomic Medicine*; Elsevier: Amsterdam, The Netherlands, 2016; pp. 99–118. ISBN 9780127999616.

36. Hakes, L.; Lovell, S.C.; Oliver, S.G.; Robertson, D.L. Specificity in protein interactions and its relationship with sequence diversity and coevolution. *Proc. Natl. Acad. Sci. USA* **2007**, *104*, 7999–8004. [CrossRef]

37. Gane, P.J.; Dean, P.M. Recent advances in structure-based rational drug design. *Curr. Opin. Struct. Biol.* **2000**, *10*, 401–404. [CrossRef]

38. Man, D. Enzymic hydrolysis of food proteins. *J. Food Eng.* **1989**, *9*, 165–166. [CrossRef]

39. Ladjal-Ettoumi, Y.; Boudries, H.; Chibane, M.; Romero, A. Pea, chickpea and lentil protein isolates: Physicochemical characterization and emulsifying properties. *Food Biophys.* **2016**. [CrossRef]

40. Arana-Peña, S.; Carballares, D.; Berenguer-Murcia, Á.; Alcántara, A.R.; Rodrigues, R.C.; Fernandez-Lafuente, R. One pot use of combilipases for full modification of oils and fats: Multifunctional and heterogeneous substrates. *Catalysts* **2020**, *10*, 605. [CrossRef]

41. Jarpa-Parra, M. Lentil protein: A review of functional properties and food application. An overview of lentil protein functionality. *Int. J. Food Sci. Technol.* **2018**, *53*, 892–903. [CrossRef]

42. Lam, A.C.Y.; Can Karaca, A.; Tyler, R.T.; Nickerson, M.T. Pea protein isolates: Structure, extraction, and functionality. *Food Rev. Int.* **2018**, *34*, 126–147. [CrossRef]

43. Ahmadifard, N.; Murueta, J.H.C.; Abedian-Kenari, A.; Motamedzadegan, A.; Jamali, H. Comparison the effect of three commercial enzymes for enzymatic hydrolysis of two substrates (rice bran protein concentrate and soy-been protein) with SDS-PAGE. *J. Food Sci. Technol.* **2016**, *53*, 1279–1284. [CrossRef]

44. Kimura, A.; Fukuda, T.; Zhang, M.; Motoyama, S.; Maruyama, N.; Utsumi, S. Comparison of physicochemical properties of 7S and 11S globulins from pea, fava bean, cowpea, and french bean with those of soybean french bean 7S globulin exhibits excellent properties. *J. Agric. Food Chem.* **2008**, *56*, 10273–10279. [CrossRef]

45. Rangel, A.; Domont, G.B.; Pedrosa, C.; Ferreira, S.T. Functional properties of purified vicilins from cowpea (Vigna unguiculata) and pea (Pisum sativum) and cowpea protein isolate. *J. Agric. Food Chem.* **2003**, *51*, 5792–5797. [CrossRef]

46. Tacias-Pascacio, V.G.; Morellon-Sterling, R.; Siar, E.-H.; Tavano, O.; Berenguer-Murcia, Á.; Fernandez-Lafuente, R. Use of Alcalase in the production of bioactive peptides: A review. *Int. J. Biol. Macromol.* **2020**. [CrossRef]

47. Ghribi, A.M.; Sila, A.; Przybylski, R.; Nedjar-Arroume, N.; Makhlouf, I.; Blecker, C.; Attia, H.; Dhulster, P.; Bougatef, A.; Besbes, S. Purification and identification of novel antioxidant peptides from enzymatic hydrolysate of chickpea (*Cicer arietinum* L.) protein concentrate. *J. Funct. Foods* **2015**, *12*, 516–525. [CrossRef]

48. Liu, C.; Bhattarai, M.; Mikkonen, K.S.; Heinonen, M. Effects of enzymatic hydrolysis of fava bean protein isolate by alcalase on the physical and oxidative stability of oil-in-water emulsions. *J. Agric. Food Chem.* **2019**. [CrossRef] [PubMed]

49. Avramenko, N.A.; Low, N.H.; Nickerson, M.T. The effects of limited enzymatic hydrolysis on the physicochemical and emulsifying properties of a lentil protein isolate. *Food Res. Int.* **2013**, *51*, 162–169. [CrossRef]

50. Lee, S.S.; Mohd Esa, N.; Loh, S.P. In vitro inhibitory activity of selected legumes against pancreatic lipase. *J. Food Biochem.* **2015**, *39*, 485–490. [CrossRef]

51. Soares, R.; Mendonça, S.; de Castro, L.Í.; Menezes, A.; Arêas, J. Major peptides from Amaranth (Amaranthus cruentus) protein inhibit HMG-CoA reductase activity. *Int. J. Mol. Sci.* **2015**, *16*, 4150–4160. [CrossRef]

52. Lammi, C.; Zanoni, C.; Arnoldi, A.; Vistoli, G. Two peptides from Soy β-conglycinin induce a hypocholesterolemic effect in HepG2 cells by a statin-Like mechanism: Comparative in vitro and in silico modeling studies. *J. Agric. Food Chem.* **2015**, *63*, 7945–7951. [CrossRef]

53. Lammi, C.; Zanoni, C.; Arnoldi, A. IAVPGEVA, IAVPTGVA, and LPYP, three peptides from soy glycinin, modulate cholesterol metabolism in HepG2 cells through the activation of the LDLR-SREBP2 pathway. *J. Funct. Foods* **2015**, *14*, 469–478. [CrossRef]

54. Jakubczyk, A.; Karaś, M.; Złotek, U.; Szymanowska, U. Identification of potential inhibitory peptides of enzymes involved in the metabolic syndrome obtained by simulated gastrointestinal digestion of fermented bean (*Phaseolus vulgaris* L.) seeds. *Food Res. Int.* **2017**, *100*, 489–496. [CrossRef]

55. Oseguera Toledo, M.E.; de Mejia, E.G.; Sivaguru, M.; Amaya-Llano, S.L. Common bean (*Phaseolus vulgaris* L.) protein-derived peptides increased insulin secretion, inhibited lipid accumulation, increased glucose uptake and reduced the phosphatase and tensin homologue activation in vitro. *J. Funct. Foods* **2016**, *27*, 160–177. [CrossRef]

56. Gomes, M.J.C.; Lima, S.L.S.; Alveo, N.E.C.; Assis, A.; Moreira, M.E.C.; Toledo, R.C.L.; Rosa, C.O.B.; Teixeira, O.R.; Bassinello, P.Z.; De Mejía, E.G.; et al. Common bean protein hydrolysate modulates lipid metabolism and prevents endothelial dysfunction in BALB/c mice fed an atherogenic diet. *Nutr. Metab. Cardiovasc. Dis.* **2020**, *30*, 141–150. [CrossRef]

57. Luna-Vital, D.A.; De Mejía, E.G.; Mendoza, S.; Loarca-Piña, G. Peptides present in the non-digestible fraction of common beans (*Phaseolus vulgaris* L.) inhibit the angiotensin-I converting enzyme by interacting with its catalytic cavity independent of their antioxidant capacity. *Food Funct.* **2015**, *6*, 1470–1479. [CrossRef]

58. Rayalam, S.; Della-Fera, M.A.; Baile, C.A. Synergism between resveratrol and other phytochemicals: Implications for obesity and osteoporosis. *Mol. Nutr. Food Res.* **2011**, *55*, 1177–1185. [CrossRef]

59. Leena, M.M.; Silvia, G.; Kannadasan, V.; Jeyan, M.; Anandharamakrishnan, C. Synergistic potential of nutraceuticals: Mechanisms and prospects for futuristic medicine. *Food Funct.* **2020**. [CrossRef]

60. Zhang, L.; Virgous, C.; Si, H. Synergistic anti-inflammatory effects and mechanisms of combined phytochemicals. *J. Nutr. Biochem.* **2019**, *69*, 19–30. [CrossRef]

61. Elias, R.J.; Kellerby, S.S.; Decker, E.A. Antioxidant activity of proteins and peptides. *Crit. Rev. Food Sci. Nutr.* **2008**, *48*, 430–441. [CrossRef]

62. Floegel, A.; Kim, D.-O.; Chung, S.-J.; Koo, S.I.; Chun, O.K. Comparison of ABTS/DPPH assays to measure antioxidant capacity in popular antioxidant-rich US foods. *J. Food Compos. Anal.* **2011**, *24*, 1043–1048. [CrossRef]

63. Ngoh, Y.-Y.; Gan, C.-Y. Enzyme-assisted extraction and identification of antioxidative and α-amylase inhibitory peptides from Pinto beans (Phaseolus vulgaris cv. Pinto). *Food Chem.* **2016**, *190*, 331–337. [CrossRef]

64. do Evangelho, J.A.; Vanier, N.L.; Pinto, V.Z.; De Berrios, J.J.; Dias, A.R.G.; da Zavareze, E.R. Black bean (*Phaseolus vulgaris* L.) protein hydrolysates: Physicochemical and functional properties. *Food Chem.* **2017**, *214*, 460–467. [CrossRef]

65. Kou, X.; Gao, J.; Zhang, Z.; Wang, H.; Wang, X. Purification and identification of antioxidant peptides from chickpea (*Cicer arietinum* L.) albumin hydrolysates. *LWT Food Sci. Technol.* **2013**, *50*, 591–598. [CrossRef]

66. Habu, J.B.; Ibeh, B.O. In vitro antioxidant capacity and free radical scavenging evaluation of active metabolite constituents of Newbouldia laevis ethanolic leaf extract. *Biol. Res.* **2015**, *48*, 16. [CrossRef]

Publisher's Note: MDPI stays neutral with regard to jurisdictional claims in published maps and institutional affiliations.

 *foods*

Article

The Effect of Steaming and Fermentation on Nutritive Values, Antioxidant Activities, and Inhibitory Properties of Tea Leaves

Chaowanee Chupeerach [1,2], Amornrat Aursalung [1], Thareerat Watcharachaisoponsiri [1], Kanyawee Whanmek [1], Parunya Thiyajai [1], Kachakot Yosphan [1], Varittha Sritalahareuthai [1], Yuraporn Sahasakul [1,2], Chalat Santivarangkna [1,2] and Uthaiwan Suttisansanee [1,2,*]

[1] Institute of Nutrition, Mahidol University, Salaya, Phuttamonthon, Nakhon Pathom 73170, Thailand; chaowanee.chu@mahidol.ac.th (C.C.); amornrat.aur@mahidol.ac.th (A.A.); thareerat.wat@hotmail.com (T.W.); kanyaweebiosci@gmail.com (K.W.); parunya.thy@mahidol.ac.th (P.T.); kachakot.yp@gmail.com (K.Y.); varittha.sri@hotmail.com (V.S.); yuraporn.sah@mahidol.ac.th (Y.S.); chalat.san@mahidol.ac.th (C.S.)
[2] Food and Nutrition Academic and Research Cluster, Institute of Nutrition, Mahidol University, Salaya, Phuttamonthon, Nakhon Pathom 73170, Thailand
* Correspondence: uthaiwan.sut@mahidol.ac.th; Tel.: +66-(0)-2800-2380 (ext. 422)

Citation: Chupeerach, C.; Aursalung, A.; Watcharachaisoponsiri, T.; Whanmek, K.; Thiyajai, P.; Yosphan, K.; Sritalahareuthai, V.; Sahasakul, Y.; Santivarangkna, C.; Suttisansanee, U.et al. The Effect of Steaming and Fermentation on Nutritive Values, Antioxidant Activities, and Inhibitory Properties of Tea Leaves. *Foods* **2021**, *10*, 117. https://doi.org/10.3390/foods10010117

Received: 24 November 2020
Accepted: 2 January 2021
Published: 8 January 2021

Publisher's Note: MDPI stays neutral with regard to jurisdictional clai-ms in published maps and institutio-nal affiliations.

Abstract: Fermented tea (Cha-miang in Thai) is a local product made by traditional food preservation processes in Northern Thailand that involve steaming fresh tea leaves followed by fermenting in the dark. Information on changes in nutritive values, bioactive compounds, antioxidant activities, and health properties that occur during the steaming and fermenting processes of tea leaves is, however, limited. Changes in nutritive values, phenolics, antioxidant activities, and in vitro health properties through inhibition of key enzymes that control obesity (lipase), diabetes (α-amylase and α-glucosidase), hypertension (angiotensin-converting enzyme (ACE)), and Alzheimer's disease (cholinesterases (ChEs) and β-secretase (BACE-1)) of fermented tea were compared to the corresponding fresh and steamed tea leaves. Results showed that energy, carbohydrate, and vitamin B1 increased after steaming, while most nutrients including protein, dietary fiber, vitamins (B2, B3, and C), and minerals (Na, K, Ca, Mg, Fe, and Zn) decreased after the steaming process. After fermentation, energy, fat, sodium, potassium, and iron contents increased, while calcium and vitamins (B1, B2, B3, and C) decreased compared to steamed tea leaves. However, the contents of vitamin B1 and iron were insignificantly different between fresh and fermented tea leaves. Five flavonoids (quercetin, kaempferol, cyanidin, myricetin, and apigenin) and three phenolic acids (gallic acid, caffeic acid, and *p*-coumaric acid) were identified in the tea samples. Total phenolic content (TPC) and antioxidant activities increased significantly after steaming and fermentation, suggesting structural changes in bioactive compounds during these processes. Steamed tea exhibited high inhibition against lipase, α-amylase, and α-glucosidase, while fermented tea possessed high anti-ChE and anti-ACE activities. Fresh tea exhibited high BACE-1 inhibitory activity. Results suggest that tea preparations (steaming and fermentation) play a significant role in the amounts of nutrients and bioactive compounds, which, in turn, affect the in vitro health properties. Knowledge gained from this research will support future investigations on in vivo health properties of fermented tea, as well as promote future food development of fermented tea as a healthy food.

Keywords: Cha-miang; fermented tea; bioactive compounds; nutritive values; enzyme inhibitory activities

1. Introduction

Tea (*Camellia sinensis*) is a globally popular drink with a higher consumption rate than coffee, wine, beer, and soft drinks [1]. Tea is also used in both pharmaceutical and industrial applications. The key chemical compounds in tea that control the flavor and taste consist of volatile terpenes, organic acids, caffeine, and polyphenols (present predominantly as catechins; others are gallic acid, caffeic acid, chlorogenic acid, quercetin, kaempferol, and myricetin) [1,2]. Tea has health benefits, with positive effects in terms of antihypertensive,

81

antiarteriosclerotic, hypocholesterolemic, and hypolipidemic activities [3]. Fermented tea is also popular among tea drinkers as a new flavor with diverse bioactive compounds. Thai fermented tea (or Cha-miang) is a local product in Northern Thailand, which serves as an alternative application to consume tea [4]. Conventional methods of fermentation vary among regions; however, the basic procedures include steaming for 1–2 h before fermenting in the dark for several days or up to a year [2]. Instead of drinking tea, the final product of Cha-miang is chewing or eating tea with a sour–bitter taste, consumed as a snack alone or served with salt and other ingredients such as roasted coconut, shredded ginger, and garlic [2]. Previously, chewing Cha-miang was only popular in the elderly. However, chewing has become a normal habit for all ages. Therefore, it is of great interest to investigate the nutrients and health properties of this historical food product for the future development of other functional foods and drinks from Cha-miang.

Cha-miang possesses many beneficial attributes in food, pharmaceutical, and nutraceutical applications. It is rich in phenolic compounds, with several health benefits including antimicrobial and antioxidant activities [2,5] due to the nutritional biotransformation which occurs during the fermenting process [6]. Cha-miang is considered to be a unique cultural food with limited consumption among locals; nevertheless, beneficial health properties resulting from variations in bioactive compounds and nutritional changes have not been previously investigated. Currently, regulations on the activity of key enzymes that can ameliorate the incidence of noncommunicable diseases (NCDs) such as obesity, diabetes, Alzheimer's disease (AD), and hypertension have received attention due to enzyme characteristics in enzyme–substrate specificity [7]. Obesity is a major global public health problem that is continuously increasing due to lifestyle changes. Obesity can be controlled through the triglyceride-hydrolyzing enzyme, lipase. Inhibition of lipase decreases the rate of fat digestion and, thus, absorption [8]. Likewise, a new approach for diabetic management involves inhibition of the carbohydrate-hydrolyzing enzymes, α-amylase and α-glucosidase [9,10], while prevention of hypertension is associated with control of the angiotensin-converting enzyme (ACE) [11]. Recent pathways to control AD include termination of the physiological role of cholinergic synapses and β-amyloid formation via inhibition of the key enzymes, acetylcholinesterase (AChE), butyrylcholinesterase (BChE), and β-secretase (β-site amyloid precursor protein cleaving enzyme 1 or BACE-1) [12–15]. However, despite this information, limited data exist regarding the effects of steaming and fermenting processes on nutritive values, bioactive compounds, antioxidant activities, and prevention of obesity, diabetes, hypertension, and AD via inhibition of the key enzymes controlling these diseases. Therefore, here, changes in nutritive values, bioactive compounds, and antioxidant activities during Cha-miang preparation (fresh tea leaves compared to steamed and fermented tea leaves) were investigated. Changes in in vitro health properties through inhibition of the key enzymes that control obesity (lipase), diabetes (α-glucosidase and α-amylase), hypertension (ACE), and AD (AChE, BChE, and BACE-1) of fresh tea leaves compared to the corresponding steamed and fermented tea leaves were also examined. Knowledge gained from these experiments can be used to support the consumption of Cha-miang as a healthy food and promote the future development of other functional foods from Cha-miang.

2. Materials and Methods

2.1. Sample Preparation, Microbiological Analysis, and Extraction

Fresh tea leaves (*Camellia sinensis* var. assamica), as well as steamed and fermented teas, were collected from Chiang Rai Province, Thailand in October 2017. Briefly, fermented tea was prepared by collecting the tea leaves into small bundles using bamboo strips, followed by steaming at 100 °C for 1 h before cooling. The steamed tea leaves underwent nonfilamentous fungal-based fermentation (or a single-step fermentation process) [2]. The steamed tea leaves were packed in a cement container and tightly wrapped in the dark at ambient temperature to minimize air content and to initiate the fermentation in an anaerobic environment. Fermentation occurred for 1 month. The fresh, steamed, and fermented tea leaves (1 kg each, $n = 3$) were collected throughout the preparation to ensure homogeneity of the leaves.

To perform the microbiological analysis, 10 g of fermented tea leaves were aseptically weighed and homogenized with 90 mL of 0.85% (w/v) NaCl solution. The mixture was subjected to 10-fold serial dilution and analyzed using the spread plate technique. Media used for microbial enumeration included plate count agar (PCA) (Merck, Darmstadt, Germany) incubated at 37 °C for 48 h for total viable bacteria, de Man Rogosa and Sharpe (MRS) agar (Difco, Franklin Lakes, NJ, USA) incubated at 37 °C for 48 h for total lactic acid bacteria, and yeast mold (YM) agar (Difco, Franklin Lakes, NJ, USA) incubated at 25 °C for 3–5 days for yeast and mold. Microbial numbers were calculated and expressed as log colony-forming units (CFU) per gram of sample.

The samples were prepared for antioxidant analysis and in vitro health properties by freeze-drying for 3 days in a freeze-dryer (Heto PowerDry PL9000 series from Heto Lab Equipment, Allerod, Denmark). Dry samples were then ground into fine powder using a grinder (Philips 600W series from Philips Electronics Co., Ltd., Jakarta, Indonesia) before packing in vacuum aluminum foil bags and keeping at -20 °C in a freezer until required for further analysis. Moisture content was analyzed utilizing a Halogen moisture analyzer (HE53 series, Mettler-Toledo AG, Greifensee, Switzerland).

Extractions to determine total phenolic contents, antioxidant activities, and enzyme inhibitory activities were performed as follows: the dry samples (400 mg) were dissolved in 10 mL of distilled water and incubated at 70 °C for 1 h using a water-bath shaker (WNE45 series from Memmert GmBh, Eagle, WI, USA). The supernatant was collected via centrifugation at $3800\times g$ (a Hettich® Rotina 38R refrigerated centrifuge from Andreas Hettich GmbH, Tuttlingen, Germany) for 15 min and filtered through a 0.45 µM polyethersulfone (PES) membrane syringe filter. All extracted samples were stored at -20 °C until required for further analysis.

2.2. Determination of Nutritive Values

The nutritive values of fresh samples were determined utilizing the standard protocols of the Association of Official Analytical Chemists (AOAC) [16] at the Institute of Nutrition, Mahidol University (ISO/IEC 17025:2005). Nutritional values including moisture content, energy, protein, fat, carbohydrate, total dietary fiber, soluble dietary fiber, insoluble dietary fiber, ash, minerals (Na, K, Ca, Mg, Fe, and Zn), and vitamins (vitamin C, B1, B2, and B3) were reported as per 100 g fresh weight, as shown in Table S1 (Supplementary Materials). To accurately determine the effect of the steaming and fermenting processes, nutritional contents were calculated and reported per 100 g dry weight.

Moisture content was determined by drying the fresh sample in a 100 °C hot-air oven (Memmert Model UNE 500, Eagle, WI, USA) until the sample weight was constant (AOAC 930.04, 934.01).

Protein content was determined via the Kjeldahl method using digestion and distillation units (Buchi Model K-435 and B-324, Flawil, Switzerland, respectively) and then calculated using a conversion factor of 6.25 (AOAC 992.23).

Crude fat content was analyzed via acidic digestion and extracted with petroleum ether using a Soxtec System (Tecator Model 1043, Hoganas, Sweden) (AOAC 948.15, 945.16).

Total dietary fiber, soluble dietary fiber, and insoluble dietary fiber were determined via the enzymatic gravimetric method (AOAC 991.42 for soluble dietary fiber and AOAC 991.43 for insoluble dietary fiber). The sample was sequentially digested with α-amylase, amyloglucosidase, and protease. Soluble and insoluble dietary fibers were calculated separately, while total dietary fiber was indirectly calculated as the sum of both dietary fibers.

Ash content was determined via incineration in a muffle furnace (Carbolite Model CWF 1100, Hope, UK) at 550 °C (AOAC 930.30, 945.46).

Carbohydrate and energy were derived via calculation using the following equations:

$$\text{Total carbohydrate (g)} = 100 - \text{moisture} - \text{protein} - \text{total fat} - \text{ash}.$$

$$\text{Energy (kcal)} = (\text{total carbohydrate} \times 4) + (\text{protein} \times 4) + (\text{total fat} \times 9).$$

Ash residue was used to determine calcium, sodium, and potassium contents using a flame atomic absorption spectrophotometer (AAS, Thermo S series, Waltham, MA, USA) (AOAC 985.35), while magnesium, iron, and zinc contents were analyzed using inductively coupled plasma optical emission spectroscopy (ICP-OES) (AOAC 984.27).

Vitamin C was determined by extracting the samples in 10% (v/v) metaphosphoric acid (MPA) before filtering through a 0.22 µM polytetrafluorothylene (PTFE) membrane syringe filter and loading to a high-performance liquid chromatography (HPLC) system utilizing a Waters 515 pump (Waters Corporation, Milford, MA, USA), a Zorbax original ODS column (5 µm, $250 \times 4.6 \times 10^{-3}$ m, Agilent Technologies, Santa Clara, CA, USA), and an isocratic solvent system (0.5% (v/v) KH_2PO_4, adjusted to pH 2.5 with H_3PO_4) with a flow rate of 0.8 mL/min [17]. Vitamin C was visualized at 254 nm using an ultraviolet/visible light (UV/Vis)-975 detector (JASCO International Co., Ltd., Tokyo, Japan).

Vitamin B1 (thiamine) and B2 (riboflavin) were extracted following the protocol of AOAC method no. 942.23 and 970.65, respectively [16], and detected utilizing an HPLC system with a pump (LC-20AT pump, Shimadzu Scientific Instrument, Columbia, MD, USA), and fluorescence detector (FP-920, JASCO International Co., Ltd., Tokyo, Japan). Vitamin B1 and B2 were separated by a Luna® C18(2) 100A column (5 µm, $250 \times 4.6 \times 10^{-3}$ m, Phenomenex, Torrance, CA, USA) under an isocratic solvent system (50% (v/v) methanol) with a flow rate of 1.0 mL/min [18,19].

Vitamin B3 (niacin) was extracted following the protocol of AOAC method no. 961.14 [16] and detected utilizing an HPLC system with a pump (1200 series G1310A isocratic pump, Agilent Technologies, Santa Clara, CA, USA), and variable wavelength detector (VWD) UV detector (1100 series G1314B, Agilent Technologies, Santa Clara, CA, USA). Vitamin B3 was separated by a Luna® C8(2) 100A column (5 µm, $250 \times 4.6 \times 10^{-3}$ m, Phenomenex, Torrance, CA, USA) under an isocratic solvent system (15% (v/v) methanol) with a flow rate of 1.0 mL/min [20,21].

2.3. Determination of Phenolic Profiles

Phenolic profiles were determined utilizing the Agilent 1100 HPLC system with a photodiode array detector (Agilent Technologies, Santa Clara, CA, USA), a 5 µm Zorbax Eclipse XDB-C_{18} column (150×4.6 mm from Agilent Technologies), and a solvent system, as described previously [22]. Phenolics were identified at 280 nm and 325 nm using a ChemStation software (Agilent Technologies, Santa Clara, CA, USA) by comparing the retention time (t_R) and spectral fingerprint with standards. The standards for phenolic acids were caffeic acid (>98.0% HPLC, T), chlorogenic acid (>98.0% HPLC, T), *p*-coumaric acid (>98.0% GC, T), ferulic acid (>98.0% GC, T), 4-hydroxybenzoic acid (>99.0% GC, T), sinapic acid (>99.0% GC, T), and syringic acid (>97.0% T) from Tokyo Chemical Industry (Tokyo, Japan), and gallic acid (97.5–102.5% T) from Sigma-Aldrich (St. Louis, MO, USA). Standards for flavonoids were apigenin (>98.0% HPLC), kaempferol (>97.0% HPLC), hesperidin (>90.0% HPLC, T), myricetin (>97.0% HPLC), luteolin (>98.0% HPLC), quercetin (>98.0% HPLC, E), and naringenin (>93.0% HPLC, T) from Tokyo Chemical Industry (Tokyo, Japan), and isorhamnetin (≥99.0% HPLC) from Extrasynthese (Genay, France). Anthocyanins were visualized at 524 nm, in which the t_R and spectral fingerprint were compared with standards including cyanidin (≥96.0% HPLC), malvidin (≥97.0% HPLC), delphinidin (≥97.0% HPLC), petunidin (≥95.0% HPLC), and peonidin (≥97.0% HPLC) from Extrasynthese (Genay, France). The validation parameters for HPLC analysis are shown in Table S2 (Supplementary Materials) and HPLC chromatograms are shown in Figure S1 (Supplementary Materials).

Total phenolic content (TPC) was analyzed utilizing Folin–Ciocalteu's phenol as a reagent and gallic acid (0–200 µg/mL) as a standard, as previously reported [23,24]. Results were monitored at 765 nm using a Synergy™ HT 96-well UV/visible microplate reader (BioTek Instruments, Inc., Winooski, VT, USA) and Gen 5 data analysis software.

2.4. Determination of Antioxidant Activities

Antioxidant activities were analyzed via 2,2-diphenyl-1-picrylhydrazyl (DPPH) radical scavenging, ferric ion reducing antioxidant power (FRAP), and oxygen radical absorbance capacity (ORAC) assays using well-established protocols [25–28]. The DPPH radical scavenging assay utilizing DPPH in 95% (*v/v*) aqueous ethanol and the Trolox standard (0.01–0.64 mM) was performed as previously described [25,28]. The FRAP assay utilizing FRAP reagent and the Trolox standard (7.81–250.00 µM) was determined according to the previously reported procedures [26,28]. The ORAC assay utilizing fluorescein reagent and the Trolox standard (3.12–100.00 µM) was analyzed as previously described [27,28].

2.5. Determination of Enzyme Inhibitory Activities

Lipase inhibitory activity was determined using a well-established protocol as previously described [29]. Briefly, the reaction consisted of 100 µL of 0.01 mg/mL *Candida rugosa* lipase (type VII, ≥700 unit/mg) in 50 mM Tris (pH 8.0) containing 0.1% (*w/v*) bovine serum albumin (BSA), 50 µL of 0.2 mM 5-5′-dithiobis(2-nitrobenzoic-*N*-phenacyl-4,5-dimethyyhiazolium bromide) (DMPTB) in 50 mM Tris (pH 7.2) containing 10 mMKCl and 1 mM ethylenediaminetetraacetic acid (EDTA), 10 µL of 16 mM 5,5′-dithiobis(2-nitrobenzoic acid) (DTNB) in 50 mM Tris (pH 7.2) containing 10% (*v/v*) Triton X-100, 10 mM KC1,and 1 mM EDTA, and 40 µL of extract. Lipase inhibitory activity was monitored at 412 nm using the microplate reader. Results were calculated as percentage inhibition using the following equation:

$$\% \text{ inhibition } = \left(1 - \frac{B - b}{A - a}\right) \times 100,$$

where A is the initial velocity of the control reaction with enzyme (control), a is the initial velocity of the control reaction without enzyme (control blank), B is the initial velocity of the enzyme reaction with extract (sample), and b is the initial velocity of the reaction with extract but without enzyme (sample blank).

The α-glucosidase inhibitory activity was determined using a well-established protocol as previously described [22]. Briefly, the reaction consisted of 50 µL of 2 mM *p*-nitrophenyl-α-D-glucopyranoside in a 50 mM phosphate buffer (pH 7.0), 100 µL of 0.1 U/mL *Saccharomyces cerevisiae* α-glucosidase (type 1, ≥10 U/mg protein), and 50 µL of extract. The α-glucosidase inhibitory activity was monitored at 405 nm using the microplate reader, and inhibition percentage was calculated as described above.

The α-amylase inhibitory activity was determined using a well-established protocol as previously described [22]. Briefly, the reaction consisted of 50 µL of 30 mM *p*-nitrophenyl-α-D-maltohexaoside in a 50 mM phosphate buffer (pH 7.0) containing 200 mM KCl, 100 µL of 30 mg/mL of porcine pancreatic α-amylase (typeVII, ≥10 unit/mg), and 50 µL of extract. The α-amylase inhibitory activity was monitored at 405 nm using the microplate reader, and inhibition percentage was calculated as described above.

Acetylcholinesterase (AChE) inhibitory activities were determined using a well-established protocol as previously described [22]. Briefly, the reaction consisted of 100 µL of 20 ng of *Electrophorus electricus* AChE (1000 units/mg) in 50 mM potassium phosphate buffer (KPB) (pH 7.0), 10 µL of 16 mM 5,5-dithio-bis-(2-nitrobenzoic acid) (DTNB), 40 µL of 0.8 mM acetylthiocholine in 50 mM KPB (pH 7.0), and 40 µL of extract. Likewise, the butyrylcholinesterase (BChE) assay consisted of 100 µL of 0.5 µg/mL equine serum BChE (≥10 units/mg protein), 50 µL of 0.4 mM butylthiocholine (BTCh) in 50 mM KPB (pH 7.0) containing MgCl$_2$ (1 mM), 10 µL of 16 mM DTNB in 50 mM KPB (pH 7.0), and 40 µL of extract. Inhibitory activity was monitored at 412 nm using the microplate reader. The inhibition percentage was then calculated as above.

Beta-secretase (BACE-1) inhibitory activity was measured using a BACE1 FRET Assay Kit (Sigma-Aldrich, St. Louis, MO, USA) with slight modifications [22]. The assay consisted of 2 µL of 0.3 U/µL BACE-1, 20 µL of 50 mM BACE-1 substrate (Rh-EVNLDAEFK-Quencher in 50 mM ammonium bicarbonate) in BACE-1 assay buffer (50 nM sodium acetate), and 20 µL of extract. The enzyme reaction was monitored at an excitation wavelength of 545 nm and an emission wavelength of 585 nm using the microplate reader. The result was calculated as percentage inhibitory activity using the equation below.

$$\% \text{ inhibition } = \left(1 - \frac{B - b}{A - a}\right) \times 100,$$

where A is the absorbance of reaction with enzyme (control), a is the absorbance of reaction without enzyme (control blank), B is the absorbance of the enzyme reaction with extract (sample), and b is the absorbance of the reaction with extract but without enzyme (sample blank).

High-throughput angiotensin-converting enzyme (ACEX) inhibitory activity was determined according to a previously described assay with slight modifications [30]. The reaction consisted of 3 µL of 0.5 U/mL rabbit lung ACE ($\geq$2 unit/mg), 30 µL of 3 mM hippuryl-histidyl-leucine (HHL) in 50 mMKPB (pH 7.0) containing 0.025 NaOH and 3 M NaCl, and 50 µL of extract. The mixture was then incubated at 37 °C for 30 min in the dark. To stop the enzyme reaction, 177 µL of 0.28 M NaOH was added to the mixture. Then, 15 µL of 20 mg/mL *o*-phthaldialdehyde in aqueous methanol was added to visualize the enzyme reaction, and 25 µL of 3 M HCl was added to neutralize the reaction. The reaction was monitored at an excitation wavelength of 360 nm and an emission wavelength of 485 nm using the microplate reader. Results were calculated as percentage inhibitory activity, similar to the BACE-1 assay.

All enzymes, chemicals, and reagents used in the enzyme inhibitory assays were purchased from Sigma-Aldrich (St. Louis, MO, USA).

2.6. Statistical Analysis

The experiments were performed in triplicate ($n = 3$). The results were expressed as mean $\pm$ standard deviation (SD). Statistical analysis was performed using one-way analysis of variance (ANOVA) and a Duncan's multiple comparison test with significant differences between values at $p < 0.05$.

3. Results

3.1. Microbial Numbers of Fermented Tea Leaves

Microbiological analysis of fermented tea leaves gave microbial numbers of total viable bacteria, lactic acid bacteria, and yeast and mold as 6.57, 6.66, and 6.49 log colony-forming units (CFU)/g sample, respectively (Table 1). Similar results were reported by Ketwal et al. (2014), suggesting that total viable bacteria and yeast and mold detected in fermented tea leaves (Cha-miang) were in the range of 6–10 log CFU/g sample [31], while lactic acid bacteria detected in Cha-miang were in the range of 6–8 CFU/g sample [6].

Table 1. Microbial numbers of viable bacteria, lactic acid bacteria, and yeast and mold in fermented tea (log colony-forming units (CFU)/g sample).

Sample	Microbial Numbers (log CFU/g Sample)		
	Total Viable Bacteria	**Lactic Acid Bacteria**	**Yeast and Mold**
Fermented tea	6.57 ± 0.07	6.66 ± 0.07	6.49 ± 0.01

All data are shown as the mean $\pm$ standard deviation (SD) of triplicate experiments ($n = 3$).

3.2. Nutritive Values

Nutritive values per 100 g dry weight (DW) suggested that fermented tea leaves possessed significantly higher energy levels (389.89 kcal) than steamed and fresh tea leaves (386.49 and 372.38 kcal, respectively) (Table 2). The high energy detected in fermented tea leaves was contributed from the high fat and carbohydrate contents. Low fat content was detected in fresh and steamed tea leaves (1.76 and 1.75 g, respectively); however, a significantly higher amount was detected in fermented tea leaves (3.20 g). High carbohydrate content was found in steamed and fermented tea leaves (67.99 and 67.69 g, respectively), while a significantly lower amount was found in fresh tea leaves (38.74 g). Fresh tea leaves contained the highest content of total dietary fiber, as well as soluble and insoluble dietary fibers (88.06, 15.23, and 72.83 g, respectively). Insoluble dietary fiber in fresh samples was approximately 4.8 times higher than soluble dietary fiber. Steamed and fermented tea leaves possessed 2.6–2.9 times higher insoluble dietary fiber than soluble dietary fiber. Fresh tea leaves also expressed a significantly higher amount of protein (50.40 g) than steamed and fermented tea leaves (24.70 and 22.58 g, respectively). Ash content was found in small amounts in all samples (5.57–9.10 g).

Table 2. Nutritional compositions of fresh, steamed, and fermented tea leaves.

Nutrients	Nutritive Values (per 100 g Dry Weight)		
	Fresh Tea Leaves	Steamed Tea Leaves	Fermented Tea Leaves
Energy (kcal)	372.38 ± 1.32 [c]	386.49 ± 0.67 [b]	389.89 ± 0.73 [a]
Protein (g)	50.40 ± 1.08 [a]	24.70 ± 1.32 [b]	22.58 ± 1.44 [b]
Fat (g)	1.76 ± 0.04 [b]	1.75 ± 0.13 [b]	3.20 ± 0.23 [a]
Carbohydrate (g)	38.74 ± 1.33 [b]	67.99 ± 1.78 [a]	67.69 ± 2.14 [a]
Total dietary fiber (g)	88.06 ± 4.24 [a]	51.99 ± 3.22 [b]	50.11 ± 3.56 [b]
Soluble dietary fiber (g)	15.23 ± 0.88 [a]	13.48 ± 0.91 [a]	13.82 ± 1.67 [a]
Insoluble dietary fiber (g)	72.83 ± 3.35 [a]	38.52 ± 2.31 [b]	36.28 ± 1.89 [b]
Ash (g)	9.10 ± 0.29 [a]	5.57 ± 0.33 [c]	6.53 ± 0.47 [b]
Minerals			
Calcium (mg)	587.69 ± 14.86 [a]	511.31 ± 32.49 [b]	453.65 ± 31.60 [c]
Sodium (mg)	227.15 ± 15.31 [b]	106.98 ± 2.67 [c]	706.04 ± 54.82 [a]
Potassium (mg)	2726.64 ± 82.13 [a]	1432.62 ± 87.39 [c]	1745.96 ± 172.99 [b]
Magnesium (mg)	244.32 ± 13.33 [a]	146.54 ± 7.65 [b]	127.82 ± 5.53 [b]
Iron (mg)	7.11 ± 0.07 [a]	5.96 ± 0.35 [b]	6.72 ± 0.42 [a]
Zinc (mg)	8.65 ± 0.43 [a]	1.75 ± 0.03 [b]	2.01 ± 0.11 [b]
Vitamins			
Vitamin B1 (mg)	2.52 ± 0.13 [b]	4.17 ± 0.42 [a]	2.56 ± 0.18 [b]
Vitamin B2 (mg)	1.45 ± 0.11 [a]	0.67 ± 0.06 [b]	0.32 ± 0.02 [c]
Niacin (mg)	10.78 ± 0.17 [a]	3.70 ± 0.32 [b]	2.69 ± 0.19 [c]
Vitamin C (mg)	58.69 ± 3.56 [a]	46.08 ± 4.00 [b]	ND

All data are shown as the mean ± standard deviation (SD) of triplicate experiments ($n = 3$). ND: not detected. The letters indicate significant differences ($p < 0.05$) of the same nutrients in different tea samples using one-way analysis of variance (ANOVA) and Duncan's multiple comparison test.

Major minerals found in all tea samples were calcium, sodium, potassium, and magnesium. Fresh tea leaves contained the highest amounts of calcium (587.69 mg), potassium (2726.64 mg), and magnesium (244.32 mg), while the steaming process significantly reduced quantities to 511.31 mg of calcium, 1432.62 mg of potassium, and 146.54 mg of magnesium. Fermentation further decreased calcium and magnesium contents to 453.65 mg and 127.82 mg, respectively. However, compared to the steaming process, potassium increased after fermentation (1745.96 mg). Interestingly, the highest sodium content (706.04 mg) was observed in fermented tea leaves, followed by fresh (227.15 mg) and steamed (106.98 mg) tea leaves. Iron and zinc were found in trace amounts in all samples. Fresh tea leaves contained 7.11 mg of iron and 8.65 mg of zinc. Steaming and fermenting processes reduced the contents of both minerals. Steamed tea leaves contained 0.96 mg of iron and 1.75 mg of zinc, while fermented tea leaves contained 6.72 mg of iron and 2.01 mg of zinc.

For vitamins, vitamin B1 in steamed tea leaves (4.17 mg) increased significantly compared to fresh tea leaves (2.52 mg), which possessed similar vitamin B1 content to fermented tea leaves (2.56 mg). Other vitamins, namely, vitamins B2, B3, and C, were found to be the highest in fresh tea leaves (1.45, 10.78, and 58.69 mg, respectively). Steaming and fermenting processes further reduced the contents of these vitamins, while no vitamin C was detected in fermented tea leaves.

3.3. Phenolic Profiles and Antioxidant Activities

Five detected flavonoids including cyanidin, myricetin, quercetin, kaempferol, and apigenin, and three detected phenolic acids including gallic acid, caffeic acid, and *p*-coumaric acid were identified in tea samples using high-performance liquid chromatography (HPLC) and the set of standards indicated in Section 2.3 (Table 3). Among flavonoids, quercetin was detected at the highest content (184.30–280.47 mg/100 g dry weight (DW)), followed by kaempferol (116.71–164.47 mg/100 g DW), cyanidin (34.22–54.41 mg/100 g DW), apigenin (31.90 mg/100 g DW), and myricetin (9.01–28.86 mg/100 g DW). All flavonoids except apigenin were detected in all tea samples, while apigenin was only detected in fermented tea leaves. Fresh and steamed tea leaves contained the two most abundant flavonoids including quercetin and kaempferol at higher contents than detected in fermented tea leaves. On the other hand, fermented tea leaves contained higher contents of cyanidin than fresh and steamed tea leaves, while myricetin content was higher in steamed and fermented tea leaves than detected in fresh tea leaves.

Table 3. Identification of flavonoids and phenolic acids in fresh, steamed, and fermented tea leaves.

Phenolics (mg/100 g DW)	Tea Leaves		
	Fresh	**Steamed**	**Fermented**
Flavonoids			
Cyanidin	44.54 ± 3.15 [b]	34.22 ± 0.45 [c]	54.41 ± 3.65 [a]
Myricetin	9.01 ± 0.06 [b]	25.48 ± 2.24 [a]	28.86 ± 2.51 [a]
Quercetin	280.47 ± 5.22 [a]	247.91 ± 18.09 [b]	184.30 ± 12.25 [c]
Kaempferol	164.47 ± 4.79 [a]	156.56 ± 9.12 [a]	116.71 ± 9.77 [b]
Apigenin	ND	ND	31.90 ± 2.68
Phenolic acids			
Gallic acid	237.18 ± 17.24 [b]	395.08 ± 21.08 [a]	83.13 ± 5.73 [c]
Caffeic acid	9.20 ± 0.56 [a]	7.86 ± 0.80 [b]	3.75 ± 0.32 [c]
p-Coumaric acid	14.79 ± 0.60 [a]	8.70 ± 0.12 [b]	ND
TPCs (mg GAE/g DW)	25.74 ± 1.21 [c]	97.11 ± 2.83 [b]	102.60 ± 1.40 [a]

All data are shown as the mean ± standard deviation (SD) of triplicate experiments ($n = 3$). DW: dry weight; GAE: gallic acid equivalent; ND: not detected. The letters indicate significant differences ($p < 0.05$) of the same phenolics in different tea samples using one-way analysis of variance (ANOVA) and Duncan's multiple comparison test.

Among phenolic acids detected in fresh, steamed, and fermented tea leaves, gallic acid was found at the highest content (83.13–395.08 mg/100 g DW), followed by caffeic acid (3.75–9.20 mg/100 g DW) and *p*-coumaric acid (8.70–14.79 mg/100 g DW). Interestingly, gallic acid and caffeic acid were detected in all tea samples, while *p*-coumaric acid was only detected in fresh and steamed tea leaves. Fresh and steamed tea leaves contained higher amounts of all detected phenolic acids than fermented tea leaves.

The HPLC analysis indicated that the identified phenolics were found at higher quantities in fresh and steamed tea leaves; however, spectrophotometric analysis of total phenolic content (TPC) suggested otherwise (Table 3). Fermented tea leaves exhibited a TPC of 102.60 mg gallic acid equivalent (GAE)/g DW, significantly higher than steamed (97.11 mg GAE/g DW) and fresh tea leaves (25.74 mg GAE/g DW).

The TPCs concurred with the antioxidant activities determined by 2,2-diphenyl-1-picrylhydrazyl (DPPH) radical scavenging, ferric ion reducing antioxidant power (FRAP), and oxygen radical absorbance capacity (ORAC) assays (Table 4). The DPPH radical scavenging activity of fermented tea leaves (690.26 µmol Trolox equivalent (TE)/100 g DW) was significantly higher than steamed (649.91 µmol TE/100 g DW) and fresh (410.91 µmol TE/100 g DW) tea leaves. Likewise, the same trend was observed with FRAP and ORAC activities, in which fermented tea leaves exhibited significantly higher antioxidant activities (311.45 and 1539.22 µmol TE/g DW, respectively) than steamed (235.42 and 1508.49 µmol TE/g DW, respectively) and fresh (30.45 and 499.11 µmol TE/g DW, respectively) tea leaves.

Table 4. Total phenolic content (TPC) and antioxidant activities detected by 2,2-diphenyl-1-picrylhydrazyl (DPPH) radical scavenging, ferric ion reducing antioxidant power (FRAP), and oxygen radical absorbance capacity (ORAC) assays of fresh, steamed, and fermented teas.

	Antioxidant Activities		
Tea Leaves	DPPH Radical Scavenging Assay (µmol TE/100 g DW)	FRAP Assay (µmol TE/g DW)	ORAC Assay (µmol TE/g DW)
Fresh	410.91 ± 9.62 [c]	30.45 ± 2.62 [c]	499.11 ± 9.91 [c]
Steamed	649.91 ± 7.07 [b]	235.42 ± 5.60 [b]	1508.49 ± 85.52 [b]
Fermented	690.26 ± 10.66 [a]	311.45 ± 9.38 [a]	1539.22 ± 83.55 [a]

All data are shown as the mean ± standard deviation (SD) of triplicate experiments (n = 3). DW: Dry weight; TE: Trolox equivalent. The letters indicate significant differences (p < 0.05) of different tea samples in the same antioxidant assay using one-way analysis of variance (ANOVA) and Duncan's multiple comparison test.

3.4. In Vitro Enzyme Inhibitory Activities

Enzyme inhibitory assays were performed using lipase as the key enzyme to control obesity, α-amylase and α-glucosidase to test for diabetes, angiotensin-converting enzyme (ACE) to control hypertension, and acetylcholinesterase (AChE), butyrylcholinesterase (BChE), and β-secretase (BACE-1) to control Alzheimer's disease (AD). Results for in vitro enzyme inhibitory activities were compared among fresh, steamed, and fermented tea leaves, as shown in Table 5.

Table 5. In vitro enzyme inhibitory activities against lipase, α-amylase, α-glucosidase, angiotensin-converting enzyme (ACE), acetylcholinesterase (AChE), butyrylcholinesterase (BChE), and β-secretase (BACE-1) in fresh, steamed, and fermented tea leaves.

Tea Leaves	% Inhibition						
	[1] Lipase	[2] α-Amylase	[2] α-Glucosidase	[3] ACE	[1] AChE	[1] BChE	[4] BACE-1
Fresh	12.27 ± 0.92 [c]	20.65 ± 1.95 [a]	72.72 ± 1.17 [b]	77.43 ± 2.61 [c]	59.13 ± 1.03 [c]	17.80 ± 1.22 [c]	90.70 ± 0.49 [a]
Steamed	50.37 ± 3.12 [a]	21.02 ± 1.16 [a]	84.41 ± 3.31 [a]	90.91 ± 0.68 [b]	61.75 ± 2.34 [b]	37.12 ± 2.12 [b]	80.47 ± 0.99 [b]
Fermented	39.06 ± 0.21 [b]	15.60 ± 1.30 [b]	62.04 ± 2.73 [c]	92.54 ± 2.70 [a]	90.19 ± 2.84 [a]	66.58 ± 6.52 [a]	67.89 ± 4.26 [c]

All data are shown as the mean ± standard deviation (SD) of triplicate experiments (n = 3). The letters indicate significant differences (p < 0.05) of different samples in the same enzyme assay using one-way analysis of variance (ANOVA) and Duncan's multiple comparison test. [1] Final concentration = 1.00 mg/mL; [2] final concentration = 1.25 mg/mL; [3] final concentration = 1.11 mg/mL; [4] final concentration = 8.00 mg/mL.

Results suggested that steamed tea leaves exhibited significantly higher lipase inhibition (50.37%) than fermented (39.06%) and fresh (12.27%) tea leaves, using the extract concentration of 1 mg/mL. Fresh and steamed tea leaves exhibited higher anti-α-amylase activities (20.65–21.02% inhibition) than fermented tea leaves (15.60% inhibition) using the same extract concentration (1.25 mg/mL). However, steamed tea leaves exhibited the highest inhibition against α-glucosidase activity (84.41%), followed by fresh (72.72%) and fermented (62.04%) tea leaves, using the same extract concentration (1.25 mg/mL). On the other hand, fermented tea leaves exhibited higher ACE inhibitory activity (92.54%) than steamed (90.91%) and fresh (77.43%) tea leaves, using the extract concentration of 1.11 mg/mL. Similar results were observed with AChE and BChE inhibitory activities. Fermented tea leaves exhibited higher inhibitory activities (90.19% AChE inhibition and 66.58% BChE inhibition) than steamed (61.75% AChE inhibition and 37.12% BChE inhibition) and fresh (59.13% AChE inhibition and 17.80% BChE inhibition) tea leaves, using the extract concentration of 1 mg/mL. However, opposite results were observed with BACE-1, another key enzyme in controlling AD. Results of BACE-1 inhibition suggested that fresh tea leaves exhibited higher BACE-1 inhibitory activity (90.70%) than steamed (80.47%) and fermented (67.89%) teas, using the same extract concentration of 8 mg/mL.

4. Discussion

Cha-miang (*Camellia sinensis* var. assamica) is a traditional fermented tea that is locally consumed in Northern Thailand. Two major procedures are required for preparation of fermented tea, i.e., steaming and fermenting processes. Fermented tea leaves were previously reported to possess different amounts of phenolic compounds compared to fresh and steamed tea leaves [32], with nutritional biotransformation detected during the steaming and fermenting processes [6], leading to variations in health properties. However, limited information exists on the comparison of nutritive values, phenolic profiles, antioxidant activities, and health properties among fresh, steamed, and fermented tea leaves. Our results showed that (i) fermentation caused nutritional changes in tea leaves that were compatible with other nutrient-rich plants, (ii) fermentation possibly caused structural changes and varied the quantities of bioactive compounds, leading to increased antioxidant activities, and (iii) these changes also caused variations in in vitro health properties through inhibitions of key enzymes that control obesity (lipase), diabetes (α-amylase and α-glucosidase), hypertension (angiotensin-converting enzyme (ACE)), and Alzheimer's disease (AD) (acetylcholinesterase (AChE), butyrylcholinesterase (BChE), and β-secretase (BACE-1)) compared to fresh and steamed tea leaves. This information will be useful to promote this cultural food as an alternative source of nutrients, phenols, and antioxidants with health benefits.

Nutritional biotransformation was observed during the steaming and fermenting processes. Steaming caused an increase in energy, carbohydrate, and vitamin B1, while fermentation elevated fat, sodium, and potassium contents, compared to fresh tea leaves. Similar results were previously observed, and comparisons of nutritional components in fresh and fermented tea leaves suggested an increase in fat and sugar after the fermentation process [6]. An increase in fat might be due to the lipase activity of microorganisms to break down fat compounds (such as fatty acids and glycerols) [33], thus increasing the availability of fat content in the fermented product. Furthermore, these microorganisms could digest fibers to sugars, causing the decrease in fiber content and alternatively affecting the total carbohydrate content [34,35]. The decrease in protein content might also be the result of amino-acid usage in these microorganisms [34,35]. Compared to fresh tea leaves, the decrease in most mineral contents detected in steamed and fermented tea leaves might be the result of heat treatment at 100 °C and possibly the usage of these macrominerals in enzyme activities and energy production of microorganisms [36]. However, the increase in Na and K after fermentation might be due to the digestion of covalent bonds in mineral–food matrix complexes, leading to increased bioavailability [35]. Similarly, the decrease in most vitamin contents might be due to the application of 100 °C during the steaming process and the usage of microorganisms [36]. Nutritional values per 100 g fresh weight (FW) of fermented tea leaves with 84.32% moisture content consisted of 10.63 g of carbohydrate, 3.53 g of protein, 0.50 g of fat, 1.02 g of ash, and 7.84 g of total dietary fiber (2.16 and 5.68 g of soluble and insoluble fibers, respectively). The protein content of fermented tea leaves was comparable to leaves of star gooseberry (*Phyllanthus acidus*), *Lasia spinosa*, *Cuminum cyminum*, *Tamarindus indica*, *Polygonum odoratum*, *Diplazium esculentum*, *Anethum graveolens*, and *Coccinia grandis* at 3.3–3.7 g/100 g FW [33]. The carbohydrate content of fermented tea leaves was comparable to leaves of *Ipomoea batatas*, *Senna siamea*, *Erythrina subumbrans*, and *Tamarindus indica* at 9.1–10.4 g/100 g FW [37], while the fat content of fermented tea leaves was comparable to leaves of *Lasia spinosa*, *Polygonum odoratum*, *Anethum graveolens*, *Acacia pennata* subsp. *Insuavis*, and *Spinacia oleracea* at 0.4–0.6 g/100 g FW [37]. The total dietary fiber content of fermented tea leaves was comparable to leaves of *Cuminum cyminum*, *Sesbania grandiflora*, *Piper sarmentosum*, *Marsilea crenata*, and *Leucaena leucocephala* at 5.9–7.9 g/100 g FW [37]. Moreover, these nutrients could affect the bioactivities of tea samples. For example, vitamin C could act as antioxidant; thus, antioxidant activities observed in tea samples might be a result of this vitamin (other than the biological function of phenolics). Fiber, especially soluble fiber, could also retard the absorption of sugar; thus, it helps supporting the antidiabetic property.

Interestingly, compared to fresh tea leaves, most nutrients declined after fermentation such as protein, total dietary fiber, calcium, potassium, magnesium, zinc, vitamin B2, niacin, and vitamin C. Compared to steamed tea leaves, most vitamins were also decreased after fermentation. These results were in contrast with some bioactive compounds, in which previous reports indicated that some bioactive compounds increased (i.e., (−)-epicatechin content was increased after 20 days of fermentation), while others decreased (i.e., (+)-catechin and (−)-epicatechin-3-gallate) significantly [32]. Furthermore, even though catechins are the main phenolic compounds in tea, previous studies reported that catechins were drastically reduced during tea fermentation [32,38]. Therefore, we focused on other phenolics and used many standards to identify changes during steaming and fermenting processes in an attempt to explain the bioactivities of tea leaves. Fermentation was previously reported to cause a structural change in bioactive compounds [39,40]. A previous study suggested that fermentation reduced the phenolic content as these compounds broke down or polymerized into new strands of polymers [39]. Dihydrochalcone, as one of the major flavonoids detected in tea, also noticeably decreased during fermentation due to partial oxidization [40]. Likewise, the absence of *p*-coumaric acid in fermented tea leaves in our experiment indicated biodegradation during fermentation, while detection of apigenin that was only found in fermented tea leaves also possibly resulted from structural changes of bioactive compounds during fermentation. Interestingly, an increase in phenolic acid content was observed after the steaming process. Our results concurred with a previous study, suggesting that citrus fruit (*Citrus sinensis* (L.) Osbeck) peels exhibited higher total phenolic contents (TPCs) after heat treatment at 100 °C [41]. A previous study also suggested that the high temperature weakened plant cell walls, thus releasing more phenolics into the solvent [42]. Phenolics are normally present in bound form with the food matrix [43]; thus, the high temperature was able to break down interactions between the phenolics and food matrix, leading to higher phenolic content detected in steamed tea leaves than in the fresh form. Phenolic content is often related to antioxidant activity [44]. This finding concurred with our results showing that both TPCs and antioxidant activities increased significantly after fermentation. A previous study also supported our findings that fermented tea exhibited higher antioxidant activity than fresh-brewed green tea [45].

To control obesity through limitation of fat absorption, the lipase inhibitory activities of tea samples were investigated. Our results suggested that steamed and fermented tea leaves exhibited higher lipase inhibitory activities than fresh tea leaves, correlating with a previous report indicating that fermented green tea showed in vitro inhibitory activity against lipase enzyme, with a half maximal inhibitory concentration (IC_{50}) of 0.48 mg/mL and in vivo ameliorated property against postprandial plasma lipid in rats fed with fermented green tea (500 mg/kg body weight) for 2 weeks [46]. Black tea and its fermented form, kombucha (5 mL/kg body weight), were reported to reduce lipase enzyme activity in alloxan-induced diabetic rats within 30 days [47]. Interestingly, the lipase inhibitory action of these tea leaves might result from the biological functions of their bioactive compounds. A previous study revealed that quercetin, as the most abundant flavonoid detected in our tea leaves, exhibited strong inhibitory potential against lipase enzymes with an IC_{50} value of 6.1 μM [48], while the most abundant phenolic acid as gallic acid exhibited stronger lipase inhibitory activity with an IC_{50} value of 0.47 nM [49]. In addition to phenolics, short-chained peptides and free amino acids also act as lipase inhibitors [48]. Heptapeptides and amide tripeptides also effectively inhibited the pancreatic lipase reaction [50,51], while free amino acids including L-proline, L-alanine, and aspartic acid inhibited pancreatic lipase in a dose-dependent manner with IC_{50} values of 0.01, 0.14, and 0.02 μM, respectively [49]. Although fermented tea leaves possessed higher TPCs than steamed and fresh tea leaves, and the lipase inhibitory activity of steamed tea leaves was higher than that of fermented and fresh tea leaves, it is possible that the lipase inhibition observed in these tea leaves might result from the biological function of both phenolics and peptides/free amino acids degraded from protein during steaming and fermenting processes.

The control of diabetes was investigated through inhibition of the carbohydrate hydrolyzing enzymes, α-amylase and α-glucosidase. Bioactive compounds detected in tea leaves effectively inhibited these enzymes. Myricetin exhibited an IC_{50} value of 5 μM [41], while kaempferol, cyanidin, and quercetin exhibited IC_{50} values ≥500 μM against α-amylase [40]. Interestingly, these flavonoids effectively inhibited α-glucosidase, with the IC_{50} values ranging from 4 to 12 μM [42]. Gallic acid, caffeic acid, and *p*-coumaric acid exhibited the IC_{50} values of 737, 828, and >3000 μM, respectively, against α-glucosidase activities [52]. The effectiveness of these bioactive compounds led to higher α-amylase and α-glucosidase inhibitions being observed in fresh and steamed tea leaves with higher flavonoid and phenolic acid contents than fermented tea leaves.

The renin–angiotensin–aldosterone system (RAAS) hormonal cascade is one of the most significant pathways to preserve hemodynamic stability, with ACE as the key enzyme in this mechanism to control hypertension. Quercetin, kaempferol, and apigenin acted as ACE inhibitors with IC_{50} values of 43, 178, and 196 mM, respectively [53], while myricetin inhibited ACE at a similar rate to quercetin [54]. However, the IC_{50} value of caffeic acid was 430 μM [55], while gallic acid and *p*-coumaric acid exhibited 34.3% and 16.8% inhibitions, respectively, using a concentration of 4 mg/mL [56]. On the basis of this information, these phenolics were considered as weak to moderate ACE inhibitors. Interestingly, the ACE inhibitory activities of tea samples possibly resulted from other types of ACE inhibitors. ACE inhibitors have been identified in natural sources such as peptides from plants (corn, sesame, green beans, rice, and tea), animals (milk, eggs, and fish), and microorganisms [57,58], while ACE inhibitory peptides with particular amino-acid sequences were enzymatically released from protein precursors during food processing [59]. Fermented tea leaves contained lower phenolic contents than fresh and steamed tea leaves but exhibited higher ACE inhibitory activities. Thus, the high ACE inhibition observed in fermented tea leaves might result from the biological function of peptides degraded from protein during tea fermentation rather than phenolics.

Two main hypotheses of AD are proposed as the cholinergic and β-amyloid hypotheses. The former occurs from excessive cholinesterase enzymes (AChE and BChE), leading to neurotransmitter loss, while the latter leads to the formation of β-amyloid plaque by BACE-1, leading to β-amyloid aggregation [60]. A previous study suggested that quercetin was a strong AChE inhibitor with an IC_{50} value of 25.9 μM [61] but a weaker BChE inhibitor with an IC_{50} value of 177.8 μM [61]. Kaempferol and myricetin inhibited AChE with IC_{50} values of 92.8 and 37.8 μM, respectively [62], and inhibited BChE with IC_{50} values of 71.0 and 43.1 μM, respectively [62]. Caffeic acid inhibited AChE with an IC_{50} value of 23 μM and BChE with an IC_{50} value of 31 μM [63]. According to this information, fresh and steamed tea leaves should exhibit higher AChE and BChE inhibition than fermented tea leaves since they possessed higher flavonoid and phenolic acid contents. However, the results showed an opposing trend, whereby fermented tea leaves exhibited higher AChE and BChE inhibitions than fresh and steamed tea leaves. Currently, different types of peptides have been designed to act as potential AChE and BChE inhibitors, as a new approach to overcome AD occurrence [64–66]. Interestingly, some of these peptides are derived from the AD synthetic drug, galantamine [65], as well as from protein hydrolysates from natural sources such as hemp seed [66]. Other than peptides, a complex of polysaccharide–peptide conjugates was also reported to act as an AChE and BChE inhibitor [67]. Thus, high AChE and BChE inhibition in fermented tea leaves might be the result of peptides degraded from protein during tea fermentation. Results of BACE-1 inhibition also suggested that fresh tea leaves exhibited higher BACE-1 inhibitory activity than steamed and fermented tea leaves. Previous research suggested that food rich in flavonoids such as green tea, blueberry, and cocoa inhibited BACE-1 or disrupted amyloid β-aggregation [68]. A previous study suggested that quercetin, kaempferol, myricetin, and apigenin inhibited BACE-1 activities with

IC_{50} values of 5.4, 14.7, 2.8, and 38.5 µM, respectively [69]. Thus, quercetin and myricetin were considered as strong BACE-1 inhibitors. Moreover, *p*-coumaric acid inhibited BACE-1 in a noncompetitive manner with an IC_{50} value of 0.9 µM [70]. No report on the BACE-1 inhibitory activity of caffeic acid is currently available; however, caffeic acid was proven capable of improving spatial cognition and memory unit in $A\beta_{25-35}$-injected AD mice [71]. Gallic acid also reduced $A\beta_{1-42}$ aggregation and neurotoxicity in an APP/PS1 transgenic mouse model [72]. According to this information, higher BACE-1 inhibitory activities observed in fresh and steamed tea leaves than in fermented tea leaves resulted from the greater flavonoid and phenolic acid contents in nonfermented tea leaves.

The information received from this study would be useful for the promotion of Cha-miang consumption as a historical and unique product of northern Thailand with many health benefits. This information also suggests that Cha-miang could be used as a source of nutrients and bioactive compounds to develop other functional foods and drinks. Moreover, the knowledge gained from this research would support local tea fermentation as an alternative method of cultural food preservation, resulting in a fermented product with unique favor and taste, and leading to the sustainable conservation of local lifestyle.

Supplementary Materials: The following are available online at https://www.mdpi.com/2304-8158/10/1/117/s1: Table S1. Nutrient compositions of fresh, steamed, and fermented tea leaves per 100 g fresh weight; Table S2. The validation parameters used for HPLC analysis; Figure S1. High-performance liquid chromatograms of fresh, steamed, and fermented tea leaves detected at 280, 325, and 524 nm.

Author Contributions: Conceptualization, U.S.; methodology, U.S.; investigation, C.C., A.A., T.W., K.W., and P.T.; Writing—original draft, K.Y., V.S., Y.S., and C.S.; writing—review and editing, U.S. All authors have read and agreed to the published version of the manuscript.

Funding: This research received no external funding.

Institutional Review Board Statement: Not applicable.

Informed Consent Statement: Not applicable.

Data Availability Statement: Data is contained within this article and supplementary material.

Conflicts of Interest: The author, Varittha Sritalahareuthai, is currently an employee of MDPI; however, she did not work for the journal Foods at the time of submission and publication.

References

1. Pripdeevech, P.; Machan, T. Fingerprint of volatile flavour constituents and antioxidant activities of teas from Thailand. *Food Chem.* **2011**, *125*, 797–802. [CrossRef]
2. Khanongnuch, C.; Unban, K.; Kanpiengjai, A.; Seanjum, C. Recent research advances and ethno-botanical history of miang, a traditional fermented tea (*Camellia sinensis* var. *assamica*) of Northern Thailand. *J. Ethn. Foods.* **2017**, *4*, 135–144. [CrossRef]
3. Kim, Y.; Goodner, L.K.; Park, D.J.; Choi, J.; Talcott, T.S. Changes in antioxidant phytochemicals and volatile composition of *Camellia sinensis* by oxidation during tea fermentation. *Food Chem.* **2011**, *129*, 1331–1342. [CrossRef]
4. Tanasupawat, S.; Pakdeeto, A.; Thawai, C.; Yukphan, P.; Okada, S. Identification of lactic acid bacteria from fermented tea leaves (*miang*) in Thailand and proposals of *Lactobacillus thailandensis* sp. nov., *Lactobacillus camelliae* sp. nov., and *Pediococcus siamensis* sp. nov. *J. Gen. Appl. Microbiol.* **2007**, *53*, 7–15. [CrossRef] [PubMed]
5. Phromrukachat, S.; Tiengburanatum, N.; Meechui, J. Assessment of active ingredients in pickled tea. *Asian J. Food Agro-Ind.* **2010**, *3*, 312–318.
6. Unban, K.; Khatthongngam, N.; Shetty, K.; Khanongnuch, C. Nutritional biotransformation in traditional fermented tea (Miang) from north Thailand and its impact on antioxidant and antimicrobial activities. *J. Food Sci. Technol.* **2019**, *56*, 2687–2699. [CrossRef]
7. Aekplakorn, W.; Mo-suwan, L. Prevalence of obesity in Thailand: A review. *Obes. Rev.* **2009**, *10*, 589–592. [CrossRef]
8. Aekplakorn, W.; Klafter, A.J.; Premgamone, A.; Dhanmun, B.; Chaikittiporn, C.; Chongsuvatwong, V.; Suwanprapisa, T.; Chaipornsupaisan, W.; Tiptaradol, S.; Lim, S.S. Prevalence and management of diabetes and associated risk factors by regions of Thailand: Third National Health Examination Survey 2004. *Diabetes Care* **2007**, *30*, 2007–2012. [CrossRef]
9. Van de Laar, F.A. Alpha-glucosidase inhibitors in the early treatment of type 2 diabetes. *Vasc. Health Risk Manag.* **2008**, *4*, 1189–1195. [CrossRef]

10. De la garza, A.L.; Etexeberria, U.; Lostao, P.M.; Roman, S.B.; Barrenetxe, J.; Martinez, A.J.; Milagro, F.I. Helichrysum and grapefruit extracts inhibit carbohydrate digestion and absorption, improving postprandial glucose levels and hyperinsulinemia in rats. *J. Agric. Food Chem.* **2013**, *61*, 12012–12019. [CrossRef]

11. Balasuriya, B.N.; Rupasinghe, V.H. Plant flavonoids as angiotensin converting enzyme inhibitors in regulation of hypertension. *Funct. Food Health Dis.* **2011**, *1*, 172–188. [CrossRef]

12. Parihar, M.; Hemnani, T. Alzheimer's disease pathogenesis and therapeutic interventions. *J. Clin. Neurosci.* **2004**, *11*, 456–467. [CrossRef] [PubMed]

13. Rao, A.A.; Sridhar, G.R.; Das, N.U. Elevated butyrylcholinesterase and acetylcholinesterase may predict the development of type 2 Diabetes Mellitus and Alzheimer's disease. *Med. Hypotheses* **2007**, *69*, 1272–1276. [CrossRef]

14. Pákáski, M.; Hugyecz, M.; Sántha, P.; Jancsó, G.; Bjelik, A.; Domokos, Á. Capsaicin promotes the amyloidogenic route of brain amyloid precursor protein processing. *Neurochem. Int.* **2009**, *54*, 426–430. [CrossRef] [PubMed]

15. Huovila, A.P.; Turner, J.A.; Pelto-Huikko, M.; Kärkkäinen, I.; Ortiz, M.R. Shedding light on ADAM metalloproteinases. *Trends Biochem. Sci.* **2005**, *30*, 413–422. [CrossRef] [PubMed]

16. Latimer, G.W. *Official Method of Analysis of AOAC International*, 21st ed.; AOAC International: Rockville, ML, USA, 2019.

17. Puwastien, P.; Siong, T.E.; Kantasubrata, J.; Craven, G.; Feliciano, R.R.; Judprasong, K. *ASEAN Manual of Food Analysis*; Institute of Nutrition, Mahidol University: Salaya, Thailand, 2011.

18. Wehling, R.L.; Wetzel, D.L. Simultaneous determination of pyridoxine, riboflavin, and thiamin in fortified cereal products by high-performance liquid chromatography. *J. Agric. Food Chem.* **1984**, *32*, 1326–1331. [CrossRef]

19. Wimalasiri, P.; Wills, R.B.H. Simultaneous analysis of thiamin and riboflavin in foods by high-performance liquid chromatography. *J. Chromatogr.* **1985**, *318*, 412–416. [CrossRef]

20. Windahl, K.L.; Trenerry, V.C.; Ward, C.M. The determination of niacin in selected foods by capillary electrophoresis and high performance liquid chromatography: Acid extraction. *Food Chem.* **1998**, *65*, 263–270. [CrossRef]

21. Ward, C.M.; Trenerry, V.C. The determination of niacin in cereals, meat and selected foods by capillary electrophoresis and high performance liquid chromatography. *Food Chem.* **1997**, *60*, 667–674. [CrossRef]

22. Sritalahareuthai, V.; Temviriyanukul, P.; On-nom, N.; Charoenkiatkul, S.; Suttisansanee, U. Phenolic profiles, antioxidant, and inhibitory activities of *Kadsura heteroclita* (Roxb.) Craib and *Kadsura coccinea* (Lem.) A.C. Sm. *Foods* **2020**, *9*, 1222. [CrossRef] [PubMed]

23. Ainsworth, E.A.; Gillespie, K.M. Estimation of total phenolic content and other oxidation substrates in plant tissues using Folin–Ciocalteu reagent. *Nat. Protoc.* **2007**, *2*, 875–877. [CrossRef]

24. Sripum, C.; Kukreja, R.K.; Charoenkiatkul, S.; Kriengsinyos, W.; Suttisansanee, U. The effect of extraction conditions on antioxidant activities and total phenolic contents of different processed Thai jasmine rice. *Int. Food Res. J.* **2017**, *24*, 1644–1650.

25. Fukumoto, L.; Mazza, G. Assessing antioxidant and prooxidant activities of phenolic compounds. *J. Agric. Food Chem.* **2000**, *48*, 3597–3604. [CrossRef] [PubMed]

26. Benzie, I.F.F.; Strain, J.J. The ferric reduction ability of plasma (FRAP) as a measure of antioxidant power: The FRAP Assay. *Anal. Biochem.* **1996**, *239*, 70–76. [CrossRef] [PubMed]

27. Ou, B.; Huang, D.; Hampsch-Woodili, M.H.; Flanagan, J.A.; Deemer, E.K. Analysis of antioxidant activities of common vegetables employing oxygen radical absorbance capacity (ORAC) and ferric reducing power (FRAP) assays: A comparative study. *J. Agric. Food Chem.* **2002**, *50*, 3122–3128. [CrossRef]

28. Hinkaew, J.; Sahasakul, Y.; Tangsuphoom, N.; Suttisansanee, U. The effect of cultivar variation on total phenolic contents and antioxidant activities of date palm fruit (*Phoenix dactylifera* L.). *Curr. Res. Nutr. Food Sci.* **2020**, *8*, 155–163. [CrossRef]

29. Pongkunakorn, T.; Watcharachaisoponsiri, T.; Chupeerach, C.; On-Nom, N.; Suttisansanee, U. Inhibitions of key enzymes relevant to obesity and diabetes of Thai local mushroom extracts. *Curr. Appl. Sci. Technol. J.* **2017**, *17*, 181–190.

30. Schwager, L.S.; Carmona, K.A.; Sturrock, D.S. A high-throughput fluorimetric assay for angiotensin I-converting enzyme. *Nat. Protoc.* **2006**, *1*, 1961–1964. [CrossRef]

31. Ketwal, S.; Chueamchaitrakun, P.; Theppakorn, T.; Wongsakul, S. Contents of total polyphenol, microorganisms and antioxidant capacities of pickled tea (miang) commercially available in Chiang Rai, Thailand. In Proceedings of the 16th Food Innovation Asia Conference, Bitec Bangna, Thailand, 12–13 June 2014.

32. Bouphun, T.; Wei, X.; Dan, W.; Zhao, R.; Qu, Z. Dynamic changes in chemical constituents during processing of Miang (Thai fermented tea leaf) in various degree of tea leaf maturity. *Int. J. Food Eng.* **2018**, *4*, 178–185. [CrossRef]

33. Odunfa, S.A. Biochemical changes during production of ogiri, a fermented melon (*Citrullus vulgaris* Schrad) product. *Plant Foods Hum. Nutr.* **1983**, *32*, 11–18. [CrossRef]

34. Osman, M.A. Effect of traditional fermentation process on the nutrient and antinutrient contents of pearl millet during preparation of Lohoh. *J. Saudi Soc. Agric. Sci.* **2011**, *10*, 1–6. [CrossRef]

35. Pranoto, Y.; Anggrahini, S.; Efendi, Z. Effect of natural and *Lactobacillus plantarum* fermentation on in vitro protein and starch digestibilities of sorghum flours. *Food Biosci.* **2013**, *2*, 46–52. [CrossRef]

36. Karimian-Khosroshahi, N.; Hosseini, H.; Rezaei, M.; Khaksar, R.; Mahmoudzadeh, M. Effect of different cooking methods on minerals, vitamins, and nutritional quality indices of rainbow trout (*Oncorhynchus mykiss*). *Int. J. Food Prop.* **2016**, *19*, 2471–2480. [CrossRef]

37. Judprasong, K.; Puwastien, P.; Rojroongwasinkul, N.; Nitithamyong, A.; Sridonpai, P.; Somjai, A. *Thai Food Composition Database*, 2nd ed.; Institute of Nutrition, Mahidol University: Salaya, Thailand, 2015.
38. Tu, Y.Y.; Xia, H.L.; Watanabe, N. The changes of catechins during the fermentation of green tea. *Prikl. Biokhim. Mikrobiol.* **2005**, *41*, 652–655. [CrossRef]
39. Almajano, P.M.; Carbó, R.; Jiménez, L.A.; Gordon, H.M. Antioxidant and antimicrobial activities of tea Infusions. *Food Chem.* **2008**, *108*, 55–63. [CrossRef]
40. Joubert, E. HPLC quantification of the dihydrochalcones, aspalathin and nothofagin in rooibos tea (*Aspalathus linearis*) as affected by processing. *Food Chem.* **1996**, *55*, 403–411. [CrossRef]
41. Chen, L.M.; Yang, J.D.; Liu, C.S. Effects of drying temperature on the flavonoid, phenolic acid and antioxidative capacities of the methanol extract of citrus fruit (*Citrus sinensis* (L.) Osbeck) Peels. *Int. J. Food Sci. Technol.* **2011**, *46*, 1179–1185. [CrossRef]
42. Shi, J.; Yu, J.; Pohorly, J.; Young, J.C.; Bryan, M.; Wu, Y. Optimization of the extraction of polyphenols from grape seed meal by aqueous ethanol solution. *J. Food Agric. Environ.* **2003**, *1*, 42–47.
43. Jeong, S.M.; Kim, S.Y.; Kim, D.R.; Jo, S.C.; Nam, K.C.; Ahn, D.U.; Lee, S.C. Effect of heat treatment on the antioxidant activity of extracts from citrus peels. *J. Agric. Food Chem.* **2004**, *52*, 3389–3393. [CrossRef]
44. Lee, L.S.; Kim, S.H.; Kim, Y.B.; Kim, Y.C. Quantitative analysis of major constituents in green tea with different plucking periods and their antioxidant activity. *Molecules* **2014**, *19*, 9173–9186. [CrossRef]
45. Somsong, P.; Tiyayon, P.; Srichamnong, W. Antioxidant of green tea and pickle tea product, miang, from Northern Thailand. *Acta. Hortic.* **2018**, *1210*, 241–248. [CrossRef]
46. Seo, D.B.; Jeong, H.W.; Kim, Y.J.; Kim, S.K.; Kim, J.K.; Lee, J.H.; Joo, K.M.; Choi, J.K.; Shin, S.S.; Lee, S.J. Fermented green tea extract exhibits hypolipidaemic effects through the inhibition of pancreatic lipase and promotion of energy expenditure. *Br. J. Nutr.* **2017**, *117*, 177–186. [CrossRef] [PubMed]
47. Aloulou, A.; Hamden, K.; Elloumi, D.; Ali, M.B.; Hargafi, K.; Jaouadi, B.; Ayadi, F.; Elfeki, A.; Ammar, E. Hypoglycemic and antilipidemic properties of kombucha tea in alloxan-induced diabetic rats. *BMC Complementary Altern. Med.* **2012**, *12*, 1–9. [CrossRef] [PubMed]
48. Martinez-Gonzalez, A.I.; Alvarez-Parrilla, E.; Díaz-Sánchez, A.G.; De la Rosa, L.A.; Núñez-Gastélum, J.A.; Vazquez-Flores, A.A.; Gonzalez-Aguilar, G.A. In vitro inhibition of pancreatic lipase by polyphenols: A kinetic, fluorescence spectroscopy and molecular docking study. *Food Technol. Biotechnol.* **2017**, *55*, 519–530. [CrossRef] [PubMed]
49. De Pradhan, I.; Dutta, M.; Choudhury, K.; Bratati, D. Metabolic diversity and in vitro pancreatic lipase inhibition activity of some varieties of *Mangifera indica* L. fruits. *Int. J. Food Prop.* **2017**, *20*, S3212–S3223. [CrossRef]
50. Lunder, M.; Bratkovič, T.; Kreft, S.; Štrukelj, B. Peptide inhibitor of pancreatic lipase selected by phage display using different elution strategies. *J. Lipid Res.* **2005**, *46*, 1512–1516. [CrossRef] [PubMed]
51. Stefanucci, A.; Dimmito, P.M.; Zengin, G.; Luisi, G.; Mirzaie, S.; Novellino, E.; Mollica, A. Discovery of novel amide tripeptides as pancreatic lipase inhibitors by virtual screening. *New J. Chem.* **2019**, *43*, 3208–3217. [CrossRef]
52. Shobana, S.; Sreerama, Y.; Malleshi, N. Composition and enzyme inhibitory properties of finger millet (*Eleusine coracana* L.) seed coat phenolics: Mode of inhibition of α-glucosidase and pancreatic amylase. *Food Chem.* **2009**, *115*, 1268–1273. [CrossRef]
53. Yang, B.; Liu, P. Composition and health effects of phenolic compounds in hawthorn (*Crataegus* spp.) of different origins. *J. Sci. Food Agric.* **2012**, *92*, 1578–1590. [CrossRef]
54. Chen, S.Y.; Chu, C.C.; Jiang, C.L.; Duh, P.D. The vasodilating effect and angiotensin converting enzyme inhibition activity of three dietary flavonols: Comparsion between myricetin, quercetin and morin, in vitro. *J. Food Nutr. Res.* **2019**, *7*, 347–354. [CrossRef]
55. Bhullar, K.S.; Lassalle-Claux, G.; Touaibia, M.; Rupasinghe, H.V. Antihypertensive effect of caffeic acid and its analogs through dual renin–angiotensin–aldosterone system inhibition. *Eur. J. Pharmacol.* **2014**, *730*, 125–132. [CrossRef] [PubMed]
56. Nwaji, N.N.; Ojo, O.O.; Ayinla, Z.A.; Ajayi-Smith, A.F.; Mgbenka, U.R.; Nkwor, A.N.; Onuoha, M.O. Antioxidant and angiotensin converting enzyme inhibition activity of *Landophia owariensis*. *J. Pharmacogn. Phytochem.* **2016**, *8*, 871–880.
57. Iwaniak, A.; Minkiewicz, P.; Darewicz, M. Food-originating ACE inhibitors, including antihypertensive peptides, as preventive food components in blood pressure reduction. *Compr. Rev. Food Sci. Food Saf.* **2014**, *13*, 114–134. [CrossRef]
58. Persson, I.A.L.; Josefsson, M.; Persson, K.; Andersson, R.G. Tea flavanols inhibit angiotensin-converting enzyme activity and increase nitric oxide production in human endothelial cells. *J. Pharm. Pharmacol.* **2006**, *58*, 1139–1144. [CrossRef]
59. De Leo, F.; Panarese, S.; Gallerani, R.; Ceci, L.R. Angiotensin converting enzyme (ACE) inhibitory peptides: Production and implementation of functional food. *Curr. Pharm. Design.* **2009**, *15*, 3622–3643. [CrossRef]
60. Hardy, J.; Selkoe, D.J. The amyloid hypothesis of Alzheimer's disease: Progress and problems on the road to therapeutics. *Science* **2002**, *297*, 353–356. [CrossRef]
61. Min, S.W.; Ryu, S.N.; Kim, D.H. Anti-inflammatory effects of black rice, cyanidin-3-O-β-D-glycoside, and its metabolites, cyanidin and protocatechuic acid. *Int. Immunopharmacol.* **2010**, *10*, 959–966. [CrossRef]
62. Katalinič, M.; Rusak, G.; Barovič, J.D.; Šinko, G.; Jelić, D.; Antolović, R.; Kovarik, Z. Structural aspects of flavonoids as inhibitors of human butyrylcholinesterase. *Eur. J. Med. Chem.* **2010**, *45*, 186–192.
63. Oboh, G.; Agunloye, O.M.; Akinyemi, A.J.; Ademiluyi, A.O.; Adefegha, S.A. Comparative study on the inhibitory effect of caffeic and chlorogenic acids on key enzymes linked to Alzheimer's disease and some pro-oxidant induced oxidative stress in rats' brain-in vitro. *Neurochem. Res.* **2013**, *38*, 413–419. [CrossRef]

64. Mondal, P.; Gupta, V.; Das, G.; Pradhan, K.; Khan, J.; Gharai, P.K.; Ghosh, S. Peptide-based acetylcholinesterase inhibitor crosses the blood-brain barrier and promotes neuroprotection. *ACS Chem. Neurosci.* **2018**, *9*, 2838–2848. [CrossRef]
65. Yaneva, S.A.; Stoykova, I.I.; Ilieva, L.I.; Vezenkov, L.T.; Marinkova, D.A.; Yotova, L.K.; Raykova, R.N.; Danalev, D.L. Acetyl-cholinesterase inhibition activity of peptide analogues of galanthamine with potential application for treatment of Alzheimers disease. *Bulg. Chem. Commun.* **2017**, *49*, 90–94.
66. Malomo, S.A.; Aluko, R.E. Kinetics of acetylcholinesterase inhibition by hemp seed protein-derived peptides. *J. Food Biochem.* **2019**, *43*, 1–10. [CrossRef] [PubMed]
67. Thabthimsuk, S.; Panya, N.; Owatworakit, A.; Choksawangkarn, W. Acetylcholinesterase inhibition and antioxidant activities of polysaccharide-peptide complexes from edible mushrooms. In Proceedings of the 6th International Conference on BMB: Biochemitry and Molecular Biology, Rayong, Thailand, 20–22 January 2018.
68. Anand, P.; Singh, B. Flavonoids as lead compounds modulating the enzyme targets in Alzheimer's disease. *Med. Chem. Res.* **2013**, *22*, 3061–3075. [CrossRef]
69. Shimmyo, Y.; Kihara, T.; Akaike, A.; Niidome, T.; Sugimoto, H. Flavonols and flavones as BACE-1 inhibitors. Structure–activity relationship in cell-free, cell-based and in silico studies reveal novel pharmacophore features. *Biochim. Biophys. Acta. Gen. Subj.* **2008**, *1780*, 819–825. [CrossRef]
70. Youn, K.; Jun, M. Inhibitory effects of key compounds isolated from corni fructus on BACE1 activity. *Phytother. Res.* **2012**, *26*, 1714–1718. [CrossRef]
71. Kim, S.; Sato, Y.; Mohan, P.S.; Peterhoff, C.; Pensalfini, A.; Rigoglioso, A.; Jiang, Y.; Nixon, R.A. Evidence that the rab5 effector APPL1 mediates APP-βCTF-induced dysfunction of endosomes in down syndrome and Alzheimer's disease. *Mol. Psychiatr.* **2016**, *21*, 707–716. [CrossRef]
72. Yu, Y.; Jans, D.C.; Winblad, B.; Tjernberg, L.O.; Schedin-Weiss, S. Neuronal Aβ42 is enriched in small vesicles at the presynaptic side of synapses. *Life Sci. Alliance* **2018**, *1*, 1–11. [CrossRef]

Article

Thermal Degradation Kinetics of Anthocyanins Extracted from Purple Maize Flour Extract and the Effect of Heating on Selected Biological Functionality

Mioara Slavu (Ursu), Iuliana Aprodu, Ștefania Adelina Milea, Elena Enachi, Gabriela Râpeanu, Gabriela Elena Bahrim and Nicoleta Stănciuc *

Faculty of Food Science and Engineering, Dunarea de Jos University of Galati, 111 Domnească Street, 800201 Galați, Romania; ursumioara@ymail.com (M.S.); Iuliana.Aprodu@ugal.ro (I.A.); Adelina.Milea@ugal.ro (Ș.A.M.); Elena.Ionita@ugal.ro (E.E.); Gabriela.Rapeanu@ugal.ro (G.R.); Gabriela.Bahrim@ugal.ro (G.E.B.)
* Correspondence: Nicoleta.Stanciuc@ugal.ro

Received: 7 October 2020; Accepted: 31 October 2020; Published: 3 November 2020

Abstract: The thermal degradation of the anthocyanins and antioxidant activity in purple maize extracts was determined between 80 and 180 °C. The anthocyanins were found to be thermostable in the temperature range of 80 to 120 °C, whereas at higher temperatures the thermal degradation of both anthocyanins and antioxidant activity followed a first-order kinetic model. The z-values started from 61.72 ± 2.28 °C for anthocyanins and 75.75 ± 2.87 °C for antioxidant activity. The conformational space of pairs of model anthocyanin molecules at 25 and 180 °C was explored through a molecular dynamics test, and results indicated the occurrence of intermolecular self-association reactions and intramolecular co-pigmentation events, which might help explaining the findings of the degradation kinetics. The relationship between thermal degradation of anthocyanins and antioxidant activity and the in vitro release was further studied. The unheated extracts showed a high stability under gastric environment, whereas after heating at 180 °C, the digestion ended quickly after 60 min. After simulated intestinal digestion, the anthocyanins were slowly decreased to a maximum of 12% for the unheated extracts, whereas an 83% decrease was found after preliminary heating at 180 °C. The thermal degradation of anthocyanins was positively correlated with the in vitro decrease of antioxidant activity.

Keywords: purple maize; anthocyanins; antioxidant activity; thermal treatment; in vitro release; molecular modeling

---

1. Introduction

It has been suggested that food color is the one of the most important visual indicators of taste and flavor, as both are considered vital sensory properties of food [1]. In recent years, considerable effort has been made to identify alternatives for synthetic dyes in food and beverages with natural colorants due to the public concern regarding their potential adverse effects including potential adverse behavioral and neurological effects and hyperactivity in children [2]. Although these studies have not established a concrete link between the intake of synthetic dyes and detrimental health effects, still a general move toward "more natural ingredients" is forcing the food processing industry to explore natural replacements for synthetic dyes [3].

Anthocyanins (ANCs) are well known and currently intensely studied as sourced from dark-colored fruits and vegetables such as black grapes, blueberries, and purple carrots [4], in addition to their health promoting functions, such as: antioxidant and anti-cancer [5], and anti-obesity and anti-inflammation effects [6]. However, ANCs can be involved in chemical and enzymatic reactions that may degrade them to colorless products or transform them into new structures [7]. Besides this, their health promoting effects are also hindered by their low stability under the physicochemical conditions they are exposed to after oral consumption by humans, affecting their bioaccessibility and further bioavailability [8]. Thus, when considering the use of ANCs as food pigments and health promoting ingredients, their thermal treatment, enzymes, co-pigmentation, and the pH instability should be considered. These types of processing and factors might induce changes that are likely to significantly change ANCs' concentration and bioactivity, which further affects consumer acceptance of a product.

Thermal treatment is one of the most commonly used methods to preserve and extend the shelf life of foods in order to ensure food safety [9], involving heating in the temperatures range between 50 and 100 °C, depending on the characteristics of the product and the desired shelf life. Thus, it is expected that processing at different temperature–time combinations specific to unit operations affects the color, anthocyanin content, and the antioxidant capacity of the ANCs present in a food [10]. Additionally, associated with each thermal process, degradation of heat-sensitive vitamins, proteins, anthocyanins, conversion to colorless derivatives and subsequently to insoluble brown pigments, and changes in structure and other quality factors that are undesirable, should be considered. In general, the thermal processing must be drastic/strong enough to destroy the microorganisms and degrading enzymes, but at the same time, it should be mild enough to avoid chemical changes that might impair the food's flavor and nutritional value [11]. Therefore, accurate knowledge of the kinetic parameters is essential to predict the quality changes that occur during thermal processing.

Colored corn is one of the richest sources of ANCs, with concentrations ranging from 51 mg cyanidin-3-*O*-glucoside equivalent (C3G)/kg in red corn to 1300 mg C3G/kg fresh weight (fw) in purple corn [12]. Li et al. [13] suggested that large-scale economical cultivation of corn and its long shelf life makes it a very attractive source for efficient ANC extraction. Colored corn contains a number of different anthocyanin species including cyanidin-, pelargonidin-, and peonidin-based monoglucosides, malonyl and dimalonyl glucosides, and flavonol anthocyanin condensed forms.

Therefore, the aim of this study was to advance the knowledge on anthocyanins from the purple maize extract obtained by combining solvent- and ultrasound-assisted extraction, based on the total monomeric and individual anthocyanin profiles. Further, since accurate knowledge of the kinetic parameters is essential to predict the quality changes that occur during thermal processing of different foods, our study reports the degradation kinetic parameters of ANCs and related antioxidant activity during heating at various temperatures ranging from 80 and 180 °C for different heating times (0–40 min). Furthermore, studies involved establishing a link between ANCs' thermal degradation at different selected time–temperature combinations with the effect of in vitro digestion on ANCs and antioxidant activity.

2. Materials and Methods

2.1. Chemicals

Gallic acid, catechin, Trolox (6-hydroxy-2,5,7,8-tetramethylchromane-2-carboxylic acid), Folin–Ciocalteu's reagent, pepsin from porcine gastric mucosa (≥400 units/mg protein), and pancreatin from porcine pancreas (1400 FIP-U/g protease, 24,000 FIP-U/g lipase, 30,000 FIP-U/g amylase) were purchased from Sigma Aldrich Steinheim (Darmstadt, Germany). The standards used for the HPLC analysis, namely cyanidin-3-glucoside, pelargonidin-3-glucoside, and peonidin-3-glucoside, were acquired from Sigma Aldrich Steinheim (Darmstadt, Germany). All other chemicals and reagents used in the experiments were of analytical grade.

2.2. Plant Material

The purple maize flour (*Zea mays* L.) was generously supplied by a local producer (Brăila County, Romania), in September 2019. The flour with a water content <12% was sealed in a brown bottle and kept at 4 °C until analysis. The flour was sifted through a 100 mesh sieve and used for further anthocyanin extraction.

2.3. ANC Extraction

An amount of 10 g of sieved flour was extracted with 90 mL of 70% ethanol and 10 mL of HCL 1N. The extraction was performed using a sonication water (MRC Scientific Instruments, Holon, Israel) bath for at 40 °C for 30 min, followed by centrifugation at 5000× g for 10 min at 4 °C. The supernatant was collected, and the extraction was repeated three times. After extraction, the collected supernatants were pooled together and concentrated under reduced pressure at 40 °C, using a vacuum rotary evaporator (AVC 2-18, Christ, UK). The obtained extract was dissolved in ultrapure water, characterized, and used for kinetic experiments.

2.4. Total Monomeric ANC Content

The total monomeric anthocyanin content (TAC) was determined according the pH-differential method, as explained by Giusti and Worsltad [14]. An aliquot of the extract was dissolved in 1 mL of distillated water. Aliquots of 0.2 mL were added to 0.8 mL of buffers with of pH 1.0 and 4.5, respectively. Absorbance was measured by a spectrophotometer (Biochrom Libra S22 UV/Vis, Cambridge, UK) at 520 and 700 nm, respectively. Absorbance was calculated as Abs = $(A_{520} - A_{700})_{pH\,1.0} - (A_{520} - A_{700})_{pH\,4.5}$ with a molar extinction coefficient for cyanidin-3-glucoside of 26,900 L·mol^{-1}·cm^{-1}. TAC was calculated using the following equation and expressed as milligrams of cyanidin-3-O-glucoside equivalents per g dry weight (mg C3G/g DW) (Equation (1)).

$$TAC\left(mg\frac{C3G}{gDW}\right) = \frac{Abs}{eL}M_w \times D \times \frac{V}{M} \tag{1}$$

where *Abs* is absorbance, e is cyanidin-3-O-glucoside molar absorbance (26,900 L·mol^{-1}·cm^{-1}), L is the cell path length (1 cm), M_W is the molecular weight of anthocyanin (449.2 Da), D is the dilution factor, V is the final volume (mL), and M is the dry weight (mg) [15].

2.5. Chromatographic Analysis of ANCs

To obtain the chromatographic profile of the anthocyanins from purple maize flour extract, an HPLC analysis was conducted using a Thermo Finnigan HPLC system (Thermo Scientific, Waltham, MA, USA). In order to separate and identify the compounds, a Synergi 4u Fusion-RP 80A (150 × 4.6 mm, 4 μm) column was used, monitored at 520 nm wavelength, at an oven temperature of 27 °C. Firstly, the samples were filtered through a C18 Sep-Pak cartridge (Waters), in order to remove the unwanted compounds, and secondly through a 0.22 μm syringe filter (Bio Basic Canada Inc., Markham, ON, Canada). The two solvents that were used for the elution step were 10% formic acid (A) and 100% methanol (B). The volume of the injection was 10 μL, at a flow rate of 1.0 mL/min. The anthocyanins from the purple maize flour extract were identified based on the retention time of the standard compounds and the data reported in the literature.

2.6. DPPH Radical Scavenging Analysis

A volume of 0.1 mmol solution of methanolic DPPH solution was prepared. The initial absorbance of the DPPH in ethanol was measured at 515 nm and did not change throughout the period of assay. Aliquots (0.1 mL) of each sample were added to 2.9 mL of DPPH solution. Discolorations were measured at 515 nm after incubation for 60 min at 25 °C in the dark. Working solutions of Trolox (range from 0 to 10 mmol) were used for calibration. The antioxidant capacity was calculated from the linear calibration curve expressed as mmol Trolox/g DW [15].

2.7. Thermal Treatment

Aliquots of 2.0 mL purple maize flour anthocyanins extract were put into screw-cap test tubes already equilibrated in a digital block heater (Stuart) at temperatures ranging from 80 to 180 °C, with an increase of 10 °C. At regular time intervals (2, 5, 7, 10, 20, 30, and 40 min), samples were removed from the bath and rapidly cooled by plunging into an ice water bath. The analysis was conducted immediately.

2.8. Kinetic Parameters

Data for the change in ANC content and antioxidant activity over time were fitted to a first-order degradation kinetics model. The kinetic rate constant (k) of thermal degradation, decimal reduction time (D), z-values, and activation energy (E_a) were defined as described by Peron et al. [16].

2.9. Static In Vitro Digestion

The simulated in vitro digestion model was prepared using a modified method described by Lee et al. [17], involving gastric and intestinal fractions. The purple maize flour anthocyanin extracts (equivalent to 10 mg C3G/g DW) were dissolved in simulated gastric juice (containing 100 mL HCl and 100 mg pepsin, pH 2.0) and digested sequentially at 37 °C using a water bath for 2 h. Aliquots of 2 mL were collected, centrifuged, filtered through a 0.22 μm filter, and analyzed for TAC and antioxidant activity at every 30 min. After 2 h, 5 mL of the digested sample was mixed with 10 mL of simulated intestinal juice (100 mL $NaHCO_3$ and 200 mg pancreatin, pH 7.0) and digested for another 2 h. The samples were prepared in triplicate ($n = 3$).

2.10. Single-Molecule-Level Investigations on Anthocyanin Behavior at Thermal Treatment

Molecular modeling techniques were further employed to simulate the heat-induced behavior of main anthocyanins from purple maize, as indicated by the HPLC analysis. The Hyperchem 8.0 software (Hypercube Inc., Gainesville, FL, USA) was used to prepare and optimize the molecular models of cyanidin-3-*O*-glucoside and its acylated form cyanidin-3-*O*-(6″-malonylglucoside). The geometry optimization of these molecules was performed using in sequence the steepest descent and conjugate gradient algorithms. The optimized molecules, obtained when getting potential energy gradients lower than 0.001 kcal/Å·mol, were further used as models for investigating the thermal-dependent self-association and intramolecular copigmentation reactions. In order to study the effect of the thermal treatment on the anthocyanins' behavior, six different starting models, consisting of anthocyanins of the same type, were prepared so as to favor all types of self-associations, in agreement with poses proposed by Castaneda-Ovando et al. [18]. The starting models were obtained by merging two anthocyanin molecules of the same type, so as to get the following three complexes: complex 1—having the oxygen of benzopyrylium of one anthocyanin molecule interfaced with the phenyl aromatic ring of the second anthocyanin molecule; complex 2—obtained by placing the oxygen atoms of the phenyl aromatic ring of one anthocyanin molecule nearby one of the hydroxyls of benzopyrylium of the second anthocyanin molecule; complex 3—obtained by interfacing the hydroxyls of benzopyrylium of one anthocyanin molecule with the sugar moiety of the second anthocyanin molecule [19]. The orientation and the distance of about 1.5 Å between functional groups of the two molecules of each complex were chosen so as to allow possible hydrogen bonding. Molecular dynamics steps were further employed to heat and equilibrate each of the six starting models at 25 and 180 °C for 100 ps. The equilibrated models were characterized in terms of atomic-level details important for predicting eventual heat-dependent self-association or co-pigmentation events.

2.11. Statistical Analysis

All analyses were conducted in at least three independent replicates, and the results are reported as mean ± standard deviation. Statistical comparisons were made by one-way analysis of variance

(ANOVA). Differences were considered to be significant when the *p* values were <0.05. The parameters of kinetic models and Arrhenius equation were estimated by linear regression. Mathematical models were selected by comparing correlations coefficients.

3. Results

3.1. Characterization of the Extract

It is well known that the concentration of ANCs may vary among foods produced by a given plant species due to different external and internal factors, such as genetic and agronomic factors, intensity and type of light, temperature, processing, and storage, as explained by de Pascual-Teresa et al. [20]. In our study, the purple maize flour extract showed a total monomeric anthocyanin content (TAC) content of 520.42 ± 23.88 mg C3G/g DW. Our results are similar with those reported by Li et al. [13], who obtained a TAC from purple corn extract of 4933.1 ± 43.4 mg C3G/kg dry corn, whereas Saikaew et al. [21] showed a TAC of the untreated purple kernels of 101.76 ± 1.16 mg CGE/100 g DW, respectively.

The ANC composition of purple maize flour was determined by HPLC at 520 nm. The HPLC-DAD profile of the purple maize flour extract displayed the presence of six main compounds such as cyanidin-3-*O*-glucoside, pelargonidin-3-*O*-glucoside, peonidin-3-*O*-glucoside, cyanidin-3-*O*-(6″-malonylglucoside), pelargonidin-3-*O*-(6″-malonylglucoside), and peonidin-3-*O*-(6″-malonylglucoside) (Figure 1). The two major compounds were cyanidin-3-*O*-glucoside and its acylated form cyanidin-3-*O*-(6″-malonylglucoside), with a concentration that represented 47% of the total anthocyanin content for cyanidin-3-*O*-glucoside, while the concentration of cyanidin-3-*O*-(6″-malonylglucoside) accounted for 18% of the total anthocyanins content. Nonetheless, the major anthocyanin, cyanidin-3-*O*-glucoside, registered a concentration of 9.85 mg/g DW. The compound that registered the lowest content was the acylated form of pelargonidin-3-*O*-glucoside. Li et al. [13] suggested also that cyanidin-3-*O*-glucoside concentration was the highest in whole corn (1135.4 ± 12.9 mg/kg corn). Among the acylated ANCs, these authors suggested that the concentration of cyanidin-3-*O*-(6″-malonylglucoside) was higher than that of pelargonidin-3-*O*-(6″-malonylglucoside) and peonidin-3-*O*-(6″-malonylglucoside).

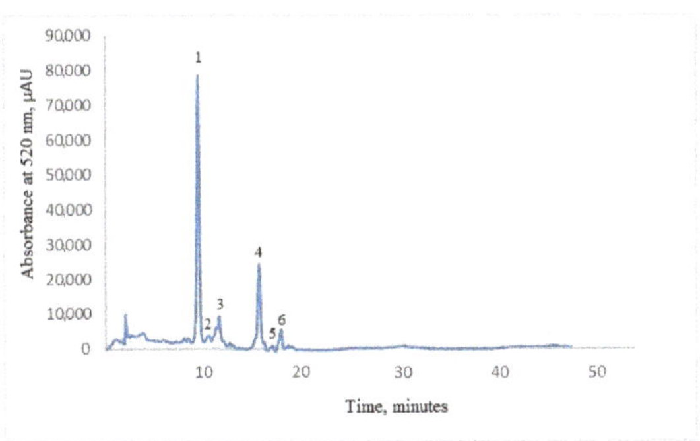

Figure 1. Chromatographic profile of the purple maize flour extract: Peak 1—cyanidin-3-*O*-glucoside; Peak 2—pelargonidin-3-*O*-glucoside; Peak 3—peonidin-3-*O*-glucoside; Peak 4—cyanidin-3-*O*-(6″-malonylglucoside); Peak 5—pelargonidin-3-*O*-(6″-malonylglucoside), and Peak 6—peonidin-3-*O*-(6″-malonylglucoside).

Moreover, cyanidin-based ANCs were present in greater abundance than pelargonidin- and peonidin-based anthocyanins in purple maize. Yang and Zhai [15] identified three kinds of anthocyanins extracted from the seed and cob of purple corn, accounting for 75.7%, 8.3%, and 16.0%, respectively, of the total amount of all the anthocyanins, and their retention times were 13.5, 15.7, and 16.6 min, respectively. The three major anthocyanins were cyaniding-3-*O*-glucoside, pelargonidin-3-*O*-glucoside, and peonidin-3-*O*-glucoside.

3.2. Degradation Kinetics of Anthocyanins from Purple Maize Flour Extract

No significant heat-induced changes were found in total anthocyanins in the temperature range of 80 to 110 °C (data not shown). The degradation of ANCs between 120 and 180 °C followed a first-order model, as shown in Figure 2a. Our results are in good agreement with studies that reported the use of the first-order kinetic model for the thermal degradation of anthocyanins [22,23].

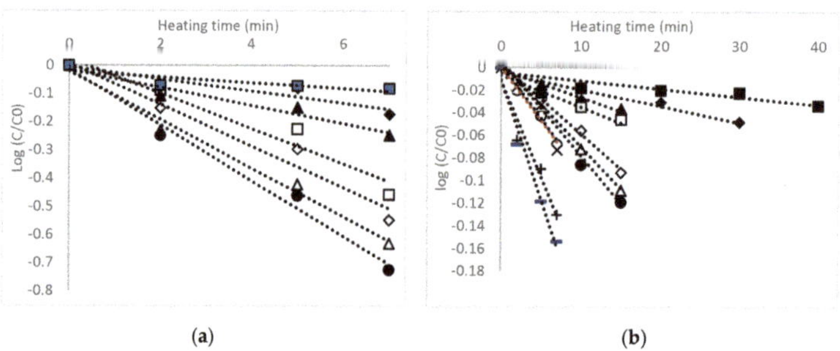

(a)   (b)

Figure 2. First-order kinetic model for the thermal degradation of total monomeric anthocyanin content (a) and antioxidant activity (b) in the purple maize flour extract. Temperature: ■ 80 °C, ♦ 90 °C, ▲ 100 °C, □ 110 °C, ◊ 120 °C, △ 130 °C, ● 140 °C, ○ 150 °C, ×160 °C, +170 °C, −180 °C.

Hou et al. [22] studied the thermal stability of the four anthocyanins (cyanidin-3-*O*-glucoside, peonidin-3-*O*-glucoside, cyanidin-3,5-*O*-diglucoside, cyanidin-3-*O*-rutinoside) in black rice extract at selected temperatures (80, 90, and 100 °C) in the pH range of 1.0–6.0, suggesting a first-order reaction kinetics with respect to temperature. In the temperature range studied, the kinetic rate constants (k) (Table 1) for the degradation of ANCs increases with increasing temperature, with the greatest difference observed at 180 °C. Consequently, the half-life of ANCs was shorter at the higher temperature of 180 °C. Furthermore, the k values for the ANCs extracted from purple maize flour showed clear differences when subjected to thermal degradation at different temperatures (Table 1).

Table 1. Kinetic parameters for the thermal degradation of total monomeric anthocyanins in the purple maize flour extract.

Temperature (°C)	$k \times 10^{-2}$ (min^{-1})	D Value (min)	$t_{1/2}$ (min)
120	2.28 ± 0.024 *	101.01 ± 2.66	30.40 ± 1.24
130	4.99 ± 0.035	46.08 ± 3.72	13.86 ± 1.98
140	7.39 ± 0.125	31.15 ± 0.61	9.37 ± 0.97
150	14.37 ± 0.17	16.02 ± 0.39	4.82 ± 0.67
160	16.95 ± 0.23	13.58 ± 0.19	4.08 ± 0.45
170	19.29 ± 0.12	11.93 ± 0.25	3.59 ± 0.22
180	22.68 ± 0.20	10.15 ± 0.14	3.05 ± 0.12
E_a 55.75 ± 6.83 kJ/mol (0.93)		z_T 61.72 ± 2.28 °C (0.90)	

* Standard deviation. k—kinetic rate constant; D—decimal reduction time, z_T—the temperature-increase necessary to induce a 10-fold reduction in D, decimal reduction time; $t_{1/2}$—time required for 50% reduction in the original content at the same temperature, E_a, activation energy.

Bolea et al. [24] studied the thermal degradation of anthocyanins from different milled fractions of black rice and reported values for k ranging from 17.43×10^{-2} min^{-1} at 60 °C to 22.42×10^{-2} min^{-1} at 100 °C for the first fraction flour.

Sui and Zhou [23] studied the thermal stability of cyanidin-3-*O*-glucoside in an aqueous system at temperatures ranging from 100 to 165 °C, followed by the investigation of the impact of thermal treatment on the antioxidant capacity. These authors suggested a significantly lower value for the degradation rate constant of cyanidin-3-*O*-glucoside at 132.5 °C of 0.0047 s^{-1}, whereas Harbourne et al. [25] reported values of 0.0028 s^{-1} for the total monomeric anthocyanins degradation at 140 °C in blackcurrant juice. The *D* values showed clear differences regarding the thermal sensitivity at different temperatures. For example, the decimal reduction time at 120 °C showed values of 101.01 ± 2.66 min, whereas an approximately 10 times decrease was observed by increasing the temperature to 180 °C (10.15 ± 0.14 min).

The *z*-value obtained confirms that the thermal resistance factors for food attributes, such as color given by related bioactives, is greater than that for spores or vegetative cells (*z* = 5–12 °C). Therefore, the rates of destruction of color are very much less temperature sensitive [26]. It is well known that the thermal degradation of ANCs starts with opening of the central ring followed by hydrolysis of the molecule, establishing colorless products. Hou et al. [22] pointed out that the stability of anthocyanins may increase via intramolecular copigmentation, especially in grain extracts, characterized by high anthocyanin content, with mixtures of different compounds, serving as copigments for intramolecular association with anthocyanins. However, these authors suggested that sugars and their degradation products tend to accelerate the degradation of anthocyanins. The Maillard reaction rate increases with the temperature, and the pigment–copigment complexes become less stable; therefore, the hydration prevails over copigmentation explaining the overall stability of the copigment on anthocyanin stability during heating [27].

The E_a value calculated for the degradation of anthocyanins in the purple maize flour extract was 55.76 ± 6.83 kJ mol^{-1}. Peron et al. [16] suggested values of 99.77 ± 0.87 kJ mol^{-1} in the juçara extract and 93.62 ± 0.44 kJ mol^{-1} in the "Italia" grape extract, whereas Heldman [28] previously reported that the activation energy for anthocyanin degradation ranged from 35 to 125 kJ mol^{-1}.

Molecular modeling investigations were further employed to predict the events responsible for the thermal-dependent behavior of the main anthocyanins from purple maize. In agreement with the HPLC analysis, the cyanidine-3-*O*-glucoside and cyanidin-3-*O*-(6″-malonylglucoside) molecules were used for the single-molecule-level analysis. Three different starting complexes including two molecules of cyanidine-3-*O*-glucoside or cyanidin-3-*O*-(6″-malonylglucoside) were heated at 25 and 180 °C.

The simulations performed on the cyanidine-3-*O*-glucoside indicated that, when heating the complexes where the intermolecular interactions between functional groups of the flavyliums are favored (complex 1 and complex 2), the aglycons got twisted with the phenyl ring out of the plane defined by the benzopyrylium. Moreover, in the case of complex 2, the sugar moiety bended toward the aglycon, forming an angle between the two planes of the sugar moiety and of the phenyl from the flavylium structure of about 90 °C. A similar observation was reported by Dumitraşcu et al. [19] when simulating the behavior of the main anthocyanins from Cornelian cherries at temperatures of 100 and 150 °C. A different thermal behavior of the cyanidine-3-*O*-glucoside molecules was noticed when the intermolecular contacts between the sugar and the aglycon were favored in the starting complex. A close look at complex 3 revealed that at 25 °C a hydrogen bond (Hb) of 2.73 Å connected the hydroxyl groups of benzopyrylium and of C6 from the glucose moiety on the same anthocyanin molecule. Unlike complexes 1 and 2, in which heating at 180 °C resulted in the increase of the distance between the two cyanidine-3-glucoside molecules, in the case of complex 3 a tendency of aligning the benzopyrylium moieties of the two anthocyanin molecules of the complex was noticed at an intermolecular distance of ~3.9 Å. In line with these observations, the interaction energy (E) values increased with the temperature increase, suggesting lower affinity between cyanidine-3-glucoside molecules from complex 1 (*E* increased from −15.74 kcal/mol at 25 °C to −0.01 kcal/mol at 180 °C) and complex 2 (*E* increased from −16.89 kcal/mol at 25 °C to −0.02 kcal/mol at 180 °C). On the other hand, no significant variation of *E* was noticed when heating the complex 3 (*E* of −12.90 and −12.49 kcal/mol at 25 and 180 °C, respectively), suggesting that the intermolecular affinity due to the hydroxyls of benzopyrylium and of the sugar moiety is preserved even at high temperature.

Detailed analysis of the complexes made up of cyanidin-3-*O*-(6″-malonylglucoside) indicated the increase of the interaction energy with the temperature in all studied cases. The two cyanidin-

3-*O*-(6″-malonylglucoside) molecules of the complex 1 equilibrated at 25 °C, established one Hb of 2.82 Å between the malonyl group on one anthocyanin and the phenyl of the second anthocyanin molecule. The attractive forces acting between the molecules are strong enough to keep them close together, even when the kinetic energy increases because of the thermal agitation of the atoms. In fact, the Hb identified at 25 °C is preserved at 180 °C (length of Hb of 2.46 Å). In a similar manner, a Hb of 3.15 Å was observed to connect the malonyl group on one anthocyanin by the hydroxyl group from C7 of the benzopyrylium in complex 2, but the temperature increase resulted in spacing out the two molecules. An interesting intramolecular copigmentation event was observed at 180 °C, consisting in glucose molecules twisting in respect to the initial model. If at 25 °C the malonyl tends to be aligned with the benzopyrylium, at 180 °C, it approaches the aromatic ring, which is carbon–carbon bound to the benzopyrylium (O9_O71 interatomic distance of 3.49 Å). The results of the molecular modeling investigations are in agreement with Nayak et al. [29] and Dumitraşcu et al. [19], indicating that the intramolecular copigmentation and intermolecular self-association might concur with anthocyanin degradation at high temperature and reduction of their color.

3.3. Degradation Kinetics of Antioxidant Activity from Purple Maize Flour Extract

The DPPH radical scavenging method was used in this study to evaluate the antioxidant activity of the heat-treated extract. In the un-treated state, the purple maize flour extract showed an antioxidant activity of 85.72 ± 0.73 mmol/g DW. Peron et al. [16] obtained a half maximal effective concentration (EC_{50}) value for antioxidant activity of 53.9 µg/mL in juçara extract. After thermal degradation of the anthocyanin extracts at 120 °C for 30 min, there was a reduction in the expected antioxidant activity by 20%. When heating at higher temperature of 180 °C for 7 min, the reduction in antioxidant activity was of approximatively 30%. Therefore, even with this reduction and considering the intense applied time–temperature combination, it can be appreciated that the antioxidant potential of the extracts was maintained in the whole temperature range studied. The loss in antioxidants from cooked corn can be attributed to synergistic combinations or interactions of several types of chemical reactions, diffusion of water soluble compounds, and the formation or breakdown of them, as explained by Harakotr et al. [30].

Sui and Zhou [23] suggested that the thermal treatments had little impact on the overall antioxidant capacity of the anthocyanin solutions. These authors reported values for DPPH radical scavenging activity for untreated anthocyanins solutions of 0.202 mg Trolox/mL and a similar value of 0.212 mg Trolox/mL after heating at 160 °C for 30 min. The authors concluded that it is possible that the loss of antioxidant capacity through the degradation of anthocyanins is compensated by the phenolic yield.

The thermal degradation of the antioxidant activity of the purple maize flour extract also followed a first-order kinetic model (Figure 2b). Therefore, the k values varied from $0.16 \pm 0.01 \times 10^{-3}$ min^{-1} at 80 °C and significantly increased to $4.85 \pm 0.29 \times 10^{-2}$ min^{-1} at 180 °C (Table 2). Mercali et al. [31] also suggested that anthocyanin thermal degradation fitted a first-order reaction model with the rate constants ranging from 5.9 to 19.7×10^{-3} min^{-1}.

The $t_{1/2}$ values (Table 2) for antioxidant thermal degradation varied from 429.96 ± 13.45 min (about 7.2 h) at 80 °C to 14.26 ± 1.82 min (about 0.23 h) at 180 °C. Significant thermal differences may be observed in Table 2 when analyzing D values, demonstrating clear differences in thermal sensitivity at different temperatures. By increasing the temperature by 10 °C from 80 to 90 °C, the D value was 2 times lower, while increasing to 120 °C caused a reduction of almost 10 times. Wang and Zu [9] studied the degradation kinetics of anthocyanins in blackberry juice and concentrate and reported $t_{1/2}$ values varying from 16.7 to 2.9 h for 8.9° Brix samples at 60, 70, 80, and 90 °C, whereas Cemeroğlu et al. [32] reported that $t_{1/2}$ values for anthocyanins degradation were 54.3, 22.5, and 8.1 h in sour cherry juice at 60, 70, and 80 °C, respectively.

Table 2. Kinetic parameters for the thermal degradation of antioxidant activity of the purple maize flour extract.

Temperature (°C)	$k \times 10^{-2}$ (min^{-1})	D Value (min)	$t_{1/2}$ (min)
80	0.16 ± 0.01 *	1428.57 ± 25.67	429.96 ± 13.45
90	0.34 ± 0.02	666.66 ± 17.59	200.65 ± 11.26
100	0.57 ± 0.10	400 ± 19.35	120.39 ± 16.78
110	0.64 ± 0.09	357.14 ± 8.89	107.49 ± 11.27
120	1.42 ± 0.28	161.29 ± 7.66	48.54 ± 2.24
130	1.70 ± 0.26	135.13 ± 3.82	40.67 ± 2.87
140	1.93 ± 0.25	119.04 ± 4.61	35.83 ± 2.98
150	2.18 ± 0.27	105.26 ± 2.39	31.68 ± 1.97
160	2.23 ± 0.13	103.09 ± 2.19	31.02 ± 2.67
170	3.91 ± 0.18	58.82 ± 1.12	17.70 ± 1.43
180	4.85 ± 0.29	47.39 ± 1.14	14.26 ± 1.82
E_a 41.12 ± 3.00 kJ/mol (0.95)		z_T 75.75 ± 2.87 °C (0.93)	

* Standard deviation.

To determine the effect of temperature on the parameters studied, the k values were fitted to an Arrhenius-type equation. The E_a value estimated for the degradation of antioxidant activity in the purple maize flour extract was 41.12 ± 3.00 kJ mol^{-1}, suggesting that a higher energy is needed to thermally degrade the antioxidant activity, an aspect that can be explained by the different thermostability of the compounds from the extract responsible for the antioxidant activity.

Bolea et al. [24] suggested higher values for antioxidant activity after thermal degradation of phytochemicals from black rice flour extracts, with values ranging from 1.33×10^{-2} min^{-1} at 60 °C to 2.18×10^{-2} min^{-1} at 100 °C. Wang and Zu [9] suggested E_a value of 58.95 kJ/mol for anthocyanin degradation during heating the 8.9° Brix blackberry juice, whereas Mercali et al. [31] suggested value of 74.8 kJ/mol for degradation kinetics of monomeric anthocyanins in acerola pulp during thermal treatment by ohmic and conventional heating. As expected, the z-value denotes that the thermal resistance of the compounds responsible for the antioxidant activity (75.75 ± 2.87 °C) is greater than that for spores or vegetative cells, suggesting that the rates of thermal destruction of bioactives are very much less temperature sensitive.

3.4. *In Vitro Digestibility of ANCs in the Purple Maize Flour Extract as Affected by Thermal Treatment*

The unheated ANCs extracts (equivalent to 10 mg C3G/mL) were digested sequentially in the gastric and intestinal simulated juices. Prior to digestion step, the extracts were heat treated at 80 °C for 40 min, and at 120 °C and 180 °C for 7 min. The TAC and antioxidant activity in the different fractions were quantified at every 30 min. During in simulated gastric digestion, the untreated TAC content showed no changes in the first 60 min of reaction, with a significant decrease of approximatively 21% after 120 min (Figure 3a).

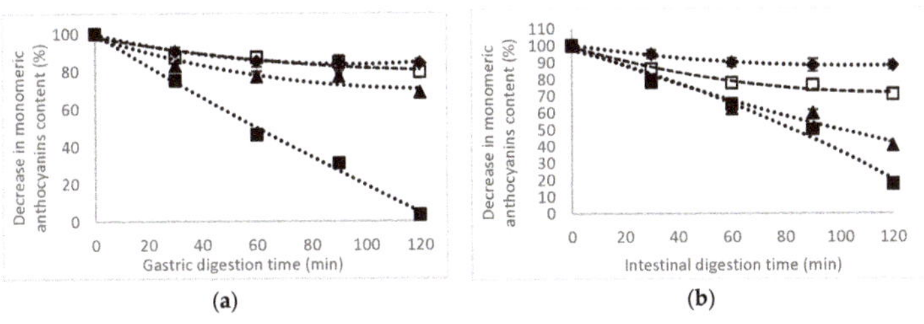

Figure 3. The decrease in total monomeric anthocyanins in the purple maize flour extract in gastric (**a**) and intestinal (**b**) simulated in vitro digestion as affected by thermal treatment. Temperature: ◆ 25 °C, □ 80 °C, ▲ 120 °C, ■ 180 °C.

The extract heated at 80 °C showed a high stability of ANCs at gastric digestion, with changes in ANCs during 120 min of digestion of approximatively 16%. A sequential digestion pattern was observed in the case of extract heated at 120 °C (Figure 3a), with a maximum decrease registered after 120 min of gastric digestion of 32%. When heating the extract at the higher temperature of 180 °C, the digestion ended quickly after 60 min, and low amounts of ANCs were detected after 120 min (Figure 3a). McGhie and Walton [33] suggested that the acidic conditions of the simulated gastric fluid contributed to the stability of ANCs, whereas Talavera et al. [34] reported that anthocyanin was absorbed via the stomach in rats. Thus, it can be appreciated that the high stability of the anthocyanins after gastric digestion may be very important because it suggests that in vivo, circulating metabolites may be present such as the anthocyanin metabolites found in the gastric fluid, as suggested by Kim et al. [35].

After simulated intestinal digestion, the TAC of the purple maize extract at 25 °C was slowly decreased from 5% after 30 min of digestion to a maximum of 12% after 120 min, when compared to that observed after simulated gastric digestion (Figure 3b). Heat treatment increased the degradation rate of ANCs in intestinal juice, with a maximum levels registered after 120 min of digestion of approximatively 60% after a heat treatment at 80 °C, 30% after heating at 120 °C, and 83% after a preliminary heating at 180 °C, respectively.

Regarding the mechanism of ANC metabolism, Stevens and Maier [36] reported that the process involved the opening of the intramolecular heterocyclic flavylium ring under alkaline conditions in the intestinal fraction. It is well known that ANCs are typically stable at an acidic pH but unstable at an alkaline pH. Moreover, the pH stability of anthocyanins depends on their chemical structures [30]. The methoxyl groups on the B-ring of anthocyanins seem to enhance the stability of anthocyanins at an alkaline pH. For example, it has been reported that malvidin-3-*O*-glucoside, which has methoxyl groups on the B-ring, exhibited higher stability than cyanidin-3-*O*-glucoside across the alkaline pH range [37].

3.5. In Vitro Digestibility of Antioxidant Activity as Affected by Thermal Treatment

The DPPH radical scavenging activity results of untreated extracts in simulated gastric and intestinal in vitro digestion model are presented in Figure 4.

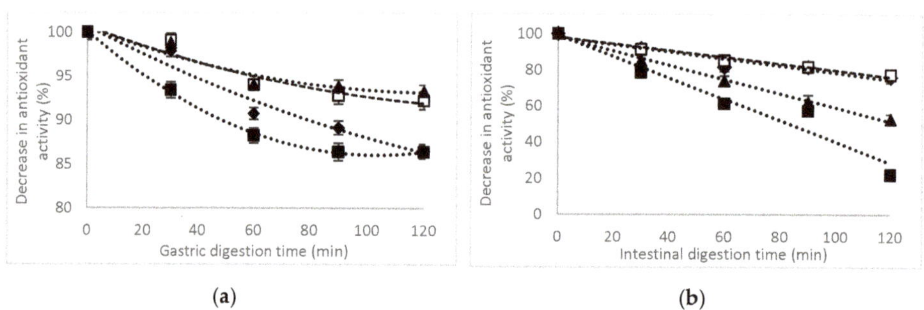

Figure 4. The antioxidant activity of the purple maize flour extract in gastric (**a**) and intestinal (**b**) simulated in vitro digestion as affected by thermal treatment. Temperature: ♦ 25 °C, □ 80 °C, ▲ 120 °C, ■ 180 °C.

The DPPH radical scavenging activity decreased for the untreated extract from the gastric to the intestinal fraction from approximatively 14% to 24%, respectively, after 120 min of digestion for each phase. In the gastric simulated digestion fraction, the heat treatment had no significant ($p > 0.05$) effect on DPPH radical scavenging activity, thus registering decreasing variations of approximatively 11% after heating at 80 °C, 7% after a heat treatment at 120 °C, and of 14% by heating at 180 °C, after 120 min of gastric digestion (Figure 4a). Significant decreases ($p < 0.05$) were found in intestinal simulated juice, up to 78% for the extract heated at 180 °C, whereas for the extracts with no heating or heated at 80 °C, a slow decrease up to 24% was found (Figure 4b). These changes in antioxidant activity are highly correlated with the loss of anthocyanins in gastric and intestinal fraction.

4. Conclusions

Purple maize extract has high anthocyanin content and high antioxidant activity, which may increase its popularity, as a significant source of bioactives. The chromatographic profile of the purple maize flour extract displayed the presence of six main compounds, namely, cyanidin-3-*O*-glucoside, pelargonidin-3-*O*-glucoside, peonidin-3-*O*-glucoside, cyanidin-3-*O*-(6″-malonylglucoside), pelargonidin-3-*O*-(6″-malonylglucoside), and peonidin-3-*O*-(6″-malonylglucoside), with the two major compounds being cyanidin-3-*O*-glucoside and its acylated form cyanidin-3-*O*-(6″-malonylglucoside). The thermostability of anthocyanins in extracts at temperatures between 80 and 120 °C was suggested, thus highlighting the opportunity to use these extracts in different processes involving high temperature–short time parameters. At temperatures ranging from 120 °C to 180 °C, both anthocyanins and antioxidant activity degraded following a first-order kinetic model. The z-values confirmed that the thermal resistance factors for food attributes are greater than those for spores or vegetative cells. Molecular modeling tests, performed on starting bimolecular complexes in which all types of self-association between main anthocyanins from purple corn are favored, indicated that, depending on the starting models, both intramolecular copigmentation and intermolecular self-association events are possible at 180 °C, therefore explaining the experimental results on the degradation of anthocyanins at high temperature. The thermal treatment affected the anthocyanin and antioxidant stability during the in vitro digestion. Different patterns in anthocyanins and antioxidant activity were found, both in gastric and intestinal digestion, highlighting a good stability when preliminarily heated up to 120 °C.

Author Contributions: Conceptualization, N.S.; methodology, I.A.; software, I.A.; validation, N.S., I.A. and G.R.; formal analysis, M.S. and Ş.A.M.; investigation, E.E.; resources, G.E.B. and G.R.; writing—original draft preparation, N.S. and I.A.; writing—review and editing, N.S. and I.A.; project administration, N.S.; funding acquisition, G.E.B. All authors have read and agreed to the published version of the manuscript.

Funding: This research received no external funding.

Acknowledgments: This work was supported by the project "Excellence, performance and competitiveness in the Research, Development, and Innovation activities at "Dunarea de Jos" University of Galati", acronym "EXPERT", financed by the Romanian Ministry of Research and Innovation in the framework of Program 1-Development of the national research and development system, Sub-program 1.2-Institutional Performance-Projects for financing excellence in Research, Development, and Innovation, contract no. 14PFE/17.10.2018. The Integrated Center for Research, Expertise, and Technological Transfer in Food Industry is acknowledged for providing technical support.

Conflicts of Interest: The authors declare no conflict of interest.

References

1. Spence, C. On the psychological impact of food colour. *Flavour* **2015**, *4*, 21. [CrossRef]
2. Stevens, L.J.; Kuczek, T.; Burgess, J.R.; Stochelski, M.A.; Arnold, L.E.; Galland, L. Mechanisms of behavioral, atopic, and other reactions to artificial food colors in children. *Nutr. Rev.* **2013**, *71*, 268–281. [CrossRef]
3. Somavat, P.; Kumar, D.; Singh, V. Techno-economic feasibility analysis of blue and purple corn processing for anthocyanin extraction and ethanol production using modified dry grind process. *Ind. Crops Prod.* **2018**, *115*, 78–87. [CrossRef]
4. Assous, M.T.M.; Abdel-Hady, M.M.; Medany, G.M. Evaluation of red pigment extracted from purple carrots and its utilization as antioxidant and natural food colorants. *Ann. Agric. Sci.* **2014**, *59*, 1–7. [CrossRef]
5. Fernandes, I.; Faria, A.; Calhau, C.; de Freitas, V.; Mateus, N. Bioavailability of anthocyanins and derivatives. *J. Funct. Food* **2014**, *7*, 54–66. [CrossRef]
6. Esposito, D.; Damsud, T.; Wilson, M.; Grace, M.H.; Strauch, R.; Li, X.; Lila, M.A.; Komarnytsky, S. Black currant anthocyanins attenuate weight gain and improve glucose metabolism in diet-induced obese mice with intact, but not disrupted, gut microbiome. *J. Agric. Food Chem.* **2015**, *63*, 6172–6180. [CrossRef]

7.   He, F.; Liang, N.; Mu, L.; Pan, Q.; Wang, J.; Reeves, M.J.; Duan, C. Anthocyanins and their variation in red wines I. Monomeric Anthocyanins and their color expression. *Molecules* **2012**, *17*, 1571–1601. [CrossRef] [PubMed]

8.   Flores, G.; del Castillo, M.L.R.; Costabile, A.; Klee, A.; Bigetti Guergoletto, K.; Gibson, G.R. *In vitro* fermentation of anthocyanins encapsulated with cyclodextrins: Release, metabolism and influence on gut microbiota growth. *J. Funct. Food* **2015**, *16*, 50–57. [CrossRef]

9.   Wang, W.-D.; Xu, S.-Y. Degradation kinetics of anthocyanins in blackberry juice and concentrate. *J. Food Eng.* **2007**, *82*, 271–275. [CrossRef]

10.   Patras, A.; Brunton, N.P.; O'Donnell, C.; Tiwari, B.K. Effect of thermal processing on anthocyanin stability in foods; mechanisms and kinetics of degradation. *Trends Food Sci. Technol.* **2010**, *21*, 3–11. [CrossRef]

11.   Costa, H.C.B.; Silva, D.O.; Vieira, L.G.M. Physical properties of açai-berry pulp and kinetics study of its anthocyanin thermal degradation. *J. Food Eng.* **2018**, *239*, 104–113. [CrossRef]

12.   Abdel-Aal, E.-S.M.; Young, J.C.; Rabalski, I. Anthocyanin composition in black, blue, pink, purple, and red cereal grains. *J. Agric. Food Chem.* **2006**, *54*, 4696–4704. [CrossRef]

13.   Li, Q.; Somavat, P.; Singh, V.; Chatham, L.; Gonzalez de Mejia, E. A comparative study of anthocyanin distribution in purple and blue corn coproducts from three conventional fractionation processes. *Food Chem.* **2017**, *231*, 332–339. [CrossRef] [PubMed]

14.   Giusti, M.M.; Worsltad, R.E. Characterization and measurement of anthocyanins by UV–visible spectroscopy. In *Current Protocols in Food Analytical Chemistry New York*; Wiley: Hoboken, NJ, USA, 2001.

15.   Yang, Z.; Zhai, W. Identification and antioxidant activity of anthocyanins extracted from the seed and cob of purple corn (*Zea mays* L.). *Innov. Food Sci. Emerg. Technol.* **2010**, *11*, 169–176. [CrossRef]

16.   Peron, D.V.; Fraga, S.; Antelo, F. Thermal degradation kinetics of anthocyanins extracted from juçara (*Euterpe edulis* Martius) and "Italia" grapes (*Vitis vinifera* L.), and the effect of heating on the antioxidant capacity. *Food Chem.* **2017**, *232*, 836–840. [CrossRef] [PubMed]

17.   Lee, S.-J.; Lee, S.Y.; Hur, S.J. Effect of Escherichia coli and Lactobacillus casei on luteolin found in simulated human digestion system. *J. Food Nutr. Res.* **2015**, *3*, 311–316. [CrossRef]

18.   Castaneda-Ovando, A.; de Lourdes Pacheco-Hernández, M.; Páez-Hernández, M.E.; Rodríguez, J.A.; Galán-Vidal, C.A. Chemical studies of anthocyanins: A review. *Food Chem.* **2009**, *113*, 859–871. [CrossRef]

19.   Dumitraşcu, L.; Enachi, E.; Stănciuc, N.; Aprodu, I. Optimization of ultrasound assisted extraction of phenolic compounds from cornelian cherry fruits using response surface methodology. *CyTA-J. Food* **2019**, *17*, 814–823. [CrossRef]

20.   Pascual-Teresa, S.; Santos-Buelga, C.; Rivas-Gonzalo, J.C. LC-MS analysis of anthocyanin from purple corn cob. *J. Sci. Food Agric.* **2002**, *82*, 1003–1006. [CrossRef]

21.   Saikaew, K.; Lertrat, K.; Meenune, M.; Tangwongchai, R. Effect of high-pressure processing on colour, phytochemical contents and antioxidant activities of purple waxy corn (*Zea mays* L. var. *ceratina*) kernels. *Food Chem.* **2018**, *243*, 328–337. [CrossRef]

22.   Hou, Z.; Qin, P.; Zhang, Y.; Cui, S.; Ren, G. Identification of anthocyanins isolated from black rice (*Oryza sativa* L.) and their degradation kinetics. *Food Res. Int.* **2013**, *50*, 691–697. [CrossRef]

23.   Sui, X.; Zhou, W. Monte Carlo modelling of non-isothermal degradation of two cyanidin-based anthocyanins in aqueous system at high temperatures and its impact on antioxidant capacities. *Food Chem.* **2014**, *148*, 342–350. [CrossRef]

24.   Bolea, C.A.; Grigore-Gurgu, L.; Aprodu, I.; Vizireanu, C.; Stănciuc, N. Process-Structure-Function in Association with the Main Bioactive of Black Rice Flour Sieving Fractions. *Foods* **2019**, *8*, 131. [CrossRef]

25.   Harbourne, N.; Jacquier, J.C.; Morgan, D.J.; Lyng, J.G. Determination of the degradation kinetics of anthocyanins in a model juice system using isothermal and non-isothermal methods. *Food Chem.* **2008**, *111*, 204–208. [CrossRef]

26.   Holdsworth, S.D. Optimisation of thermal processing A review. *J. Food Eng.* **1985**, *4*, 89–116. [CrossRef]

27.   Gradinaru, G.; Biliaderis, C.G.; Kallithraka, S.; Kefalas, P.; Garcia-Viguera, C. Thermal stability of *Hibiscus sabdariffa* L. anthocyanins in solution and in solid state: Effects of copigmentation and glass transition. *Food Chem.* **2003**, *83*, 423–436. [CrossRef]

28.   Heldman, D.R. Food preservation process design. In *Advances in Food Process Engineering Research and Applications*; Springer: New York, NY, USA, 2011; pp. 489–497.

29. Nayak, B.; Berrios, J.D.J.; Powers, J.R.; Tang, J. Thermal degradation of anthocyanins from purple potato (cv. Purple Majesty) and impact on antioxidant capacity. *J. Agric. Food Chem.* **2011**, *59*, 11040–11049. [CrossRef]

30. Harakotr, B.; Suriharn, B.; Tangwongchai, R.; Scott, M.P.; Lertrat, K. Anthocyanin, phenolics and antioxidant activity changes in purple waxycorn as affected by traditional cooking. *Food Chem.* **2014**, *164*, 510–517. [CrossRef] [PubMed]

31. Mercali, G.D.; Jaeschke, D.P.; Tessaro, I.C.; Ferreira Marczak, L.D. Degradation kinetics of anthocyanins in acerola pulp: Comparison between ohmic and conventional heat treatment. *Food Chem.* **2013**, *136*, 853–857. [CrossRef]

32. Cemeroğlu, B.; Velioğlu, S.; Isik, S. Degradation kinetics of anthocyanins in sour cherry juice and concentrate. *J. Food Sci.* **1994**, *59*, 1216–1218. [CrossRef]

33. McGhie, T.K.; Walton, M.C. The bioavailability and absorption of anthocyanins: Towards a better understanding. *Mol. Nutr. Food Res.* **2007**, *51*, 702–713. [CrossRef]

34. Talavera, S.; Felgines, C.; Texier, O.; Besson, C.; Lamaison, J.-L.; Rémésy, C. Anthocyanins are efficiently absorbed from the stomach in anesthetized rats. *J. Nutr.* **2003**, *133*, 4178–4182. [CrossRef] [PubMed]

35. Kim, I.; Moon, J.K.; Hur, S.J.; Lee, J. Structural changes in mulberry (*Morus Microphylla*. Buckl) and chokeberry (*Aronia melanocarpa*) anthocyanins during simulated in vitro human digestion. *Food Chem.* **2020**, *318*, 126449. [CrossRef]

36. Stevens, J.F.; Maier, C.S. The chemistry of gut microbial metabolism of polyphenols. *Phytochem. Rev.* **2016**, *15*, 425–444. [CrossRef]

37. Loypimai, P.; Moongngarm, A.; Chottanom, P. Thermal and pH degradation kinetics of anthocyanins in natural food colorant prepared from black rice bran. *J. Food Sci. Technol.* **2016**, *53*, 461–470. [CrossRef]

Publisher's Note: MDPI stays neutral with regard to jurisdictional claims in published maps and institutional affiliations.

Article

Revalorization of Coffee Husk: Modeling and Optimizing the Green Sustainable Extraction of Phenolic Compounds

Miguel Rebollo-Hernanz [1,2], **Silvia Cañas** [1,2], **Diego Taladrid** [2], **Vanesa Benítez** [1,2], **Begoña Bartolomé** [2], **Yolanda Aguilera** [1,2,*] and **María A. Martín-Cabrejas** [1,2]

[1] Department of Agricultural Chemistry and Food Science, Universidad Autónoma de Madrid, 28049 Madrid, Spain; miguel.rebollo@uam.es (M.R.-H.); silvia.cannas@uam.es (S.C.); vanesa.benitez@uam.es (V.B.); maria.martin@uam.es (M.A.M.-C.)

[2] Institute of Food Science Research, CIAL (UAM-CSIC), 28049 Madrid, Spain; d.taladrid@csic.es (D.T.); b.bartolome@csic.es (B.B.)

* Correspondence: yolanda.aguilera@uam.es

Abstract: This study aimed to model and optimize a green sustainable extraction method of phenolic compounds from the coffee husk. Response surface methodology (RSM) and artificial neural networks (ANNs) were used to model the impact of extraction variables (temperature, time, acidity, and solid-to-liquid ratio) on the recovery of phenolic compounds. All responses were fitted to the RSM and ANN model, which revealed high estimation capabilities. The main factors affecting phenolic extraction were temperature, followed by solid-to-liquid ratio, and acidity. The optimal extraction conditions were 100 °C, 90 min, 0% citric acid, and 0.02 g coffee husk mL^{-1}. Under these conditions, experimental values for total phenolic compounds, flavonoids, flavanols, proanthocyanidins, phenolic acids, *o*-diphenols, and in vitro antioxidant capacity matched with predicted ones, therefore, validating the model. The presence of chlorogenic, protocatechuic, caffeic, and gallic acids and kaemferol-3-*O*-galactoside was confirmed by UPLC-ESI-MS/MS. The phenolic aqueous extracts from the coffee husk could be used as sustainable food ingredients and nutraceutical products.

Keywords: coffee by-products; phenolic compounds; antioxidant capacity; response surface methodology; artificial neural networks

Citation: Rebollo-Hernanz, M.; Cañas, S.; Taladrid, D.; Benítez, V.; Bartolomé, B.; Aguilera, Y.; Martín-Cabrejas, M.A. Revalorization of Coffee Husk: Modeling and Optimizing the Green Sustainable Extraction of Phenolic Compounds. *Foods* **2021**, *10*, 653. https://doi.org/10.3390/foods10030653

Academic Editor: Antonio Cilla

Received: 1 March 2021
Accepted: 16 March 2021
Published: 19 March 2021

Publisher's Note: MDPI stays neutral with regard to jurisdictional claims in published maps and institutional affiliations.

1. Introduction

The United Nations Sustainable Development Goals promote the warranting of sustainable consumption and production patterns such as achieving the efficient use of natural resources and reducing waste generation through prevention, reduction, recycling, and reusing [1]. Likewise, the Food and Agriculture Organization (FAO) is focused on re-designing the global food system to be more productive, environmentally sustainable, and able to deliver healthy and nutritious foods to the population [2]. In this respect, the reduction of the food industry wastes and by-products is a clear need. The development of ground-breaking strategies for revalorizing food by-products via their conversion into food-grade novel ingredients is a critical challenge in the food chain [3].

The processing of the coffee cherries into coffee beans is very complex and produces a variety of residues. Mature fruits are collected and transformed by basically two methods: dry or wet processing [4]. Within the dry processing, the coffee cherries are sun-dried. The coffee husk is separated during the following de-hulling step. The coffee husk comprises the skin, pulp, mucilage, parchment, and parts of the silverskin. In the wet processing, conversely, water is used to separate ripened and unripped coffee cherries. After this separation, the coffee pulp and skin are removed using a pulper. The coffee parchment remains attached to the coffee bean [5]. Therefore, wet processing produces pulp and parchment, whereas dry processing produces just the coffee husk, which contains both parts (Figure 1). The disposal of coffee husk without treatment causes environmental problems in the producing countries, owing to the high content of caffeine and phenolic compounds composing it [6]. The traditional use of the coffee husk is the preparation of the "Cascara beverage", traditionally consumed in Yemen and Ethiopia [7]. Current uses for the coffee husk include mushroom cultivation [8], lignin extraction [9], and recovery of biomolecules from the lignin alkali hydrolysate [10], bioethanol production [11], biosorbents preparation [12], and composting [13]. Although these strategies for valorizing the coffee husk contribute to the coffee industry's sustainability, the production of high-value-added food ingredients is a more desirable approach to promoting a more productive, environmentally sustainable food system.

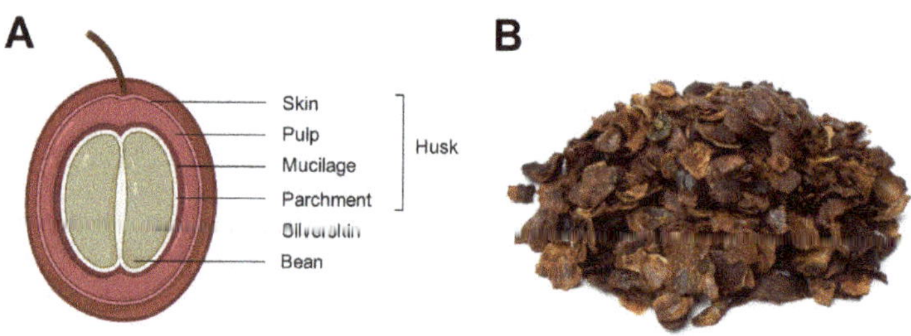

Figure 1. Coffee cherry anatomy (**A**) from the outside to the inside (skin, pulp, mucilage, parchment (comprising the coffee husk), silverskin, and coffee bean) and appearance of dried coffee husk once separated from the coffee bean (**B**).

The coffee husk is a source of phenolic compounds and caffeine. Chlorogenic, protocatechuic, and gallic acids are the main phenolic compounds comprised in it and responsible for the coffee husk antioxidant properties [7]. Extracting these compounds could be a strategy to valorize the coffee husk and develop food ingredients with health-promoting properties. Coffee antioxidants are associated with preventing chronic diseases such as obesity, diabetes, and other cardiometabolic diseases [14]. Limited research has evaluated the extraction of these phytochemicals from the coffee husk [15]. Green extraction methods are needed to ensure the sustainability of the system. Even though there is research pointing out the use of ultrasound- and microwave-assisted extractions and pressurized liquid extraction [16], heat-assisted extraction (HAE) is still the most common extraction technique in the food industry [17]. Consequently, there is a need to model, optimize, and validate the green HAE of phenolic compounds from the coffee husk.

Response surface methodology (RSM) is one of the chemometric techniques often employed to optimize procedures in food production and analysis [18]. The parameters influencing the extraction are modeled and optimized using RSM, principally to reduce energy and, therefore, extraction costs [19]. Machine learning algorithms may also be employed for these purposes. Artificial neural networks (ANNs) have gained interest in modeling and optimization processes. This methodology allows for the study of the relationships between the extraction and response variables, i.e., the extraction yield, employing fewer experimental measurements [20]. ANNs are computational models based on the structure and functions of the nervous system and the brain and have extraordinary learning and predictive abilities [21]. Hence, these mathematical and computational tools can be used to enhance the effectiveness of the extraction process, making it more economically and environmentally sustainable.

We hypothesized that the modification of extraction parameters would increase the recovery yield of phenolic compounds from the coffee husk, allowing the establishment of a green sustainable extraction method. Hence, this study aimed to model the process conditions to maximize the sustainable aqueous extraction of phenolic compounds from the coffee husk using response surface methodology and artificial neural networks, optimize it, and comprehensively characterize the obtained extracts using UPLC-MS/MS analysis. Multivariate statistics were used to gain insight into the effects of extraction conditions on the phenolic composition and its relationship with the in vitro antioxidant capacity of the obtained aqueous phenolic extracts.

2. Materials and Methods

2.1. Material and Sample Preparation

The coffee husk, mechanically separated from the sun-dried cherries of the Arabica species variety Caturra, was supplied by "Las Morenitas" (Nicaragua). The milling of the coffee husk was carried out in a pilot-scale ball mill over three days, then sieved with a pilot-scale sieve, selecting the fraction with a particle size of <250 µm. Milled coffee husk was stored in closed and sealed plastic bags and preserved in dark and dry conditions to avoid oxidation until further extraction and analysis.

2.2. Experimental Design

2.2.1. Response Surface Methodology (RSM)

Box–Behnken, being a spherical RSM, consists of a central point and several middle points on the edges of a cube superimposed on the sphere, which requires fewer experiments than other statistical designs. We employed a four-factor, three-level Box–Behnken design coupled to RSM to find the optimal extraction conditions to achieve the highest extraction of phenolic compounds from the coffee husk. The experimental conditions for the aqueous extraction of phytochemicals from the coffee husk are presented in Table 1. The statistical design comprised 27 experimental runs with three levels ($-1, 0, 1$) for each of the variables: temperature (°C) (X_1), time (min) (X_2), acidity as the percentage of citric acid in water (%) (X_3), and solid-to-liquid ratio (S/L, g mL^{-1}) (X_4). Those parameters were selected according to previous studies found in the literature and tested on preliminary experiments to guarantee they exerted an influence on the extraction of phenolic compounds from coffee parchment [17]. The impact of extraction temperature was investigated in the range from 30 to 100 °C, time from 5 to 90 min, S/L ratio, 0.02–0.05 g mL^{-1}, and acidity, 0–2% citric acid. The variables were coded according to the following equation:

$$X = \frac{x_i - x_0}{\Delta x} \tag{1}$$

where X is the coded value; x_i, the corresponding actual value; x_0, the real value at the center of the domain; and Δx, the increment of x_i corresponding to a variation of 1 unit of x. The response variables were fitted to the following second-order polynomial model equation, which described the relationship between the responses and the independent variables.

$$Y = \beta_0 + \sum_{i=1}^{k} \beta_i X_i + \sum_{i=1}^{k} \beta_{ii} X_{ii}^2 + \sum_{i}^{k-1} \sum_{j}^{k} \beta_{ij} X_{ij} \tag{2}$$

where Y was the response variables; X_i and X_j were independent coded variables; β_0 was the constant coefficient; β_i was the linear coefficient; β_{ii} was the quadratic coefficient, and β_{ij} was the cross-product coefficients.

Table 1. Experimental conditions (independent variables) and their corresponding responses values (phenolic compounds content) according to the Box–Behnken design. S/L ratio: solid-to-liquid ratio; TPC: total phenolic compounds; TF: total flavonoids; TFL: total flavanols; PAC: total proanthocyanidin; TPA: total phenolic acid; TOD: total *ortho*-diphenols; AC: antioxidant capacity.

Run	Independent Variables				Responses						
	Temperature (X_1, °C)	Time (X_2, Min)	Acidity (X_3, %)	S/L Ratio (X_4, g mL^{-1})	TPC (mg g^{-1})	TF (mg g^{-1})	TFL (mg g^{-1})	PAC (mg g^{-1})	TPA (mg g^{-1})	TOD (mg g^{-1})	AC (mg g^{-1})
1	30 (−1)	5 (−1)	1 (0)	0.035 (0)	4.19 ± 0.12	5.52 ± 0.27	0.74 ± 0.14	2.48 ± 0.25	1.28 ± 0.08	0.86 ± 0.11	11.95 ± 0.08
2	100 (1)	5 (−1)	1 (0)	0.035 (0)	5.45 ± 0.12	7.94 ± 0.25	0.84 ± 0.07	2.63 ± 0.22	2.29 ± 0.11	1.23 ± 0.06	15.61 ± 0.05
3	30 (−1)	90 (1)	1 (0)	0.035 (0)	4.41 ± 0.09	6.30 ± 0.24	0.73 ± 0.08	2.34 ± 0.09	1.69 ± 0.18	1.03 ± 0.10	11.90 ± 0.09
4	100 (1)	90 (1)	1 (0)	0.035 (0)	5.44 ± 0.10	7.34 ± 0.18	0.96 ± 0.08	2.74 ± 0.13	2.42 ± 0.06	1.28 ± 0.06	15.37 ± 0.07
5	65 (0)	47.5 (0)	0 (−1)	0.02 (1)	4.47 ± 0.13	8.60 ± 0.23	1.09 ± 0.12	2.99 ± 0.27	2.98 ± 0.16	1.56 ± 0.10	17.43 ± 0.05
6	65 (0)	47.5 (0)	2 (1)	0.02 (1)	4.56 ± 0.13	6.38 ± 0.34	0.71 ± 0.03	3.48 ± 0.11	1.55 ± 0.07	1.13 ± 0.05	10.32 ± 0.05
7	65 (0)	47.5 (0)	0 (−1)	0.05 (−1)	3.47 ± 0.15	6.99 ± 0.30	0.75 ± 0.05	2.23 ± 0.26	1.95 ± 0.07	1.10 ± 0.05	13.19 ± 0.05
8	65 (0)	47.5 (0)	2 (1)	0.05 (−1)	4.07 ± 0.10	5.97 ± 0.28	0.70 ± 0.05	2.45 ± 0.30	1.43 ± 0.06	1.00 ± 0.06	11.33 ± 0.09
9	65 (0)	47.5 (0)	1 (0)	0.035 (0)	3.70 ± 0.16	6.81 ± 0.30	0.65 ± 0.09	2.51 ± 0.18	1.77 ± 0.08	0.97 ± 0.05	12.17 ± 0.09
10	30 (−1)	47.5 (0)	1 (0)	0.02 (1)	3.94 ± 0.15	6.69 ± 0.34	0.67 ± 0.05	2.41 ± 0.15	1.37 ± 0.12	1.04 ± 0.06	11.09 ± 0.08
11	100 (1)	47.5 (0)	1 (0)	0.02 (1)	5.93 ± 0.15	10.07 ± 0.24	1.10 ± 0.10	3.04 ± 0.19	2.88 ± 0.10	1.56 ± 0.05	18.67 ± 0.08
12	30 (−1)	47.5 (0)	1 (0)	0.05 (−1)	3.64 ± 0.12	6.17 ± 0.25	0.66 ± 0.09	2.16 ± 0.09	1.69 ± 0.15	0.92 ± 0.08	11.28 ± 0.08
13	100 (1)	47.5 (0)	1 (0)	0.05 (−1)	4.07 ± 0.17	7.32 ± 0.28	0.78 ± 0.06	2.22 ± 0.27	2.19 ± 0.08	1.06 ± 0.05	13.47 ± 0.06
14	65 (0)	5 (−1)	0 (−1)	0.035 (0)	4.30 ± 0.21	7.09 ± 0.34	0.76 ± 0.06	2.51 ± 0.20	2.14 ± 0.13	1.26 ± 0.04	16.20 ± 0.08
15	65 (0)	90 (1)	0 (−1)	0.035 (0)	4.27 ± 0.25	7.60 ± 0.27	0.97 ± 0.09	2.55 ± 0.13	2.74 ± 0.10	1.58 ± 0.07	17.33 ± 0.08
16	65 (0)	5 (−1)	2 (1)	0.035 (0)	4.50 ± 0.23	6.89 ± 0.35	0.63 ± 0.14	2.47 ± 0.16	1.54 ± 0.06	1.11 ± 0.06	12.52 ± 0.07
17	65 (0)	90 (1)	2 (1)	0.035 (0)	4.18 ± 0.10	5.55 ± 0.12	0.75 ± 0.09	2.67 ± 0.25	1.53 ± 0.18	1.06 ± 0.07	10.44 ± 0.06
18	65 (0)	47.5 (0)	1 (0)	0.035 (0)	3.68 ± 0.10	6.00 ± 0.21	0.69 ± 0.14	2.52 ± 0.16	1.74 ± 0.11	0.94 ± 0.06	12.04 ± 0.09
19	30 (−1)	47.5 (0)	0 (−1)	0.035 (0)	4.22 ± 0.10	7.06 ± 0.27	0.79 ± 0.10	2.49 ± 0.18	2.23 ± 0.06	1.26 ± 0.07	15.55 ± 0.04
20	100 (1)	47.5 (0)	0 (−1)	0.035 (0)	4.88 ± 0.14	10.10 ± 0.27	1.26 ± 0.09	2.89 ± 0.19	3.93 ± 0.15	1.85 ± 0.07	18.77 ± 0.09
21	30 (−1)	47.5 (0)	2 (1)	0.035 (0)	3.28 ± 0.12	6.54 ± 0.19	0.60 ± 0.06	2.22 ± 0.12	1.28 ± 0.08	0.95 ± 0.04	9.90 ± 0.06
22	100 (1)	47.5 (0)	2 (1)	0.035 (0)	5.00 ± 0.11	8.38 ± 0.21	0.85 ± 0.06	2.88 ± 0.29	2.53 ± 0.08	1.29 ± 0.07	14.12 ± 0.06
23	65 (0)	5 (−1)	1 (0)	0.02 (1)	4.40 ± 0.12	7.47 ± 0.18	0.76 ± 0.10	2.59 ± 0.22	2.09 ± 0.19	1.03 ± 0.04	13.99 ± 0.07
24	65 (0)	90 (1)	1 (0)	0.02 (1)	5.52 ± 0.08	8.15 ± 0.22	0.82 ± 0.09	2.66 ± 0.14	1.89 ± 0.05	1.18 ± 0.05	13.09 ± 0.07
25	65 (0)	5 (−1)	1 (0)	0.05 (−1)	3.89 ± 0.12	7.10 ± 0.18	0.65 ± 0.09	2.25 ± 0.13	1.58 ± 0.07	0.97 ± 0.04	12.76 ± 0.06
26	65 (0)	90 (1)	1 (0)	0.05 (−1)	3.58 ± 0.15	6.67 ± 0.20	0.51 ± 0.14	2.08 ± 0.35	1.19 ± 0.10	0.72 ± 0.04	10.00 ± 0.09
27	65 (0)	47.5 (0)	1 (0)	0.035 (0)	3.69 ± 0.07	6.79 ± 0.17	0.69 ± 0.11	2.59 ± 0.14	1.82 ± 0.07	0.94 ± 0.05	12.45 ± 0.08

Based on the analysis of variance (ANOVA), the regression coefficients of individual linear, interaction, and quadratic terms were determined. The numerical magnitude of the standardized model coefficients evidenced their significance in the obtained model. Among standardized coefficients, the larger values are more effective. Plots depicting response surface 3D plots were constructed for all the response variables (Figure 1A).

The polynomial equation's fitness to the responses was assessed through the coefficient of determination (R^2). The significance of all the terms within the polynomial equation was analyzed statistically by analyzing the F-value at $p < 0.05$. Equations were created, selecting the significant ($p < 0.05$) non-standardized coefficients (including non-significant terms if needed to ensure that the model was hierarchical), and their statistical parameters (F-value and R^2) were determined again.

2.2.2. Artificial Neural Networks (ANNs)

A multilayer perceptron (MLP)-based feed-forward ANN was applied for modeling the extraction of phenolic compounds from the coffee husk. MATLAB version R2020a was used to model the data using ANNs. The experimental data was constructed using the regression-based network approach. The Broyden–Fletcher–Goldfarb–Shanno (BFGS) quasi-Newton back-propagation (TRAINBFG) method was selected since it is an efficient training function because it performs non-smooth optimizations and smaller networks [22]. The gradient descent method (LEARNGDM) as the adaptive learning function was used to minimize the mean squared error (MSE) between the network output and the actual error rate [23]. The hyperbolic tangent sigmoid transfer function (TANSIG) and linear transfer function (PURELIN) were used to calculate a layer's output from its net input [24]. All these functions were used to train the neural network and built the best ANN. Multiple feed-forward neural networks were trained and, subsequently, tested by determining the number of neurons in the hidden layer to select an optimized ANN topology, with the lowest root mean square error (RMSE) and highest R^2 values. However, the number of epochs (or cycles through the whole training dataset) was restricted to a minimum to avoid over-fitting while establishing an optimal topology. Increased epoch numbers may cause model over-fitting issues. The network architecture (Figure 2B) consisted of an input layer with four neurons (temperature (T), time, (t), acidity, and S/L ratio), one hidden layer with ten neurons, and an output layer with one neuron, which represented each of the response variables (total phenolic compounds, TPC, in Figure 2B).

2.3. Comparison of the Prediction Ability of RSM and ANN

Several statistical parameters, including the coefficient of determination (R^2), the root mean square error (RMSE), and the absolute average deviation (AAD), were calculated for the comparison of the estimation capabilities of RSM and ANN, according to the following equations.

$$R^2 = 1 - \frac{\sum_{i=1}^{n}\left(Y_{pre} - Y_{exp}\right)^2}{\sum_{i=1}^{n}\left(Y_m - Y_{exp}\right)^2} \tag{3}$$

$$RMSE = \sqrt{\frac{\sum_{i=1}^{n}\left(Y_{pre} - Y_{exp}\right)^2}{n}} \tag{4}$$

$$ADD\ (\%) = \left(\frac{\sum_{i=1}^{n}\left(\left|Y_{pre} - Y_{exp}\right|/Y_{pre}\right)^2}{n}\right)\cdot 100 \tag{5}$$

where Y_{pre} is the predicted response variable (by either RSM or ANN), Y_{exp} is the observed response variable, Y_m is the average response variable, and n is the number of experiments.

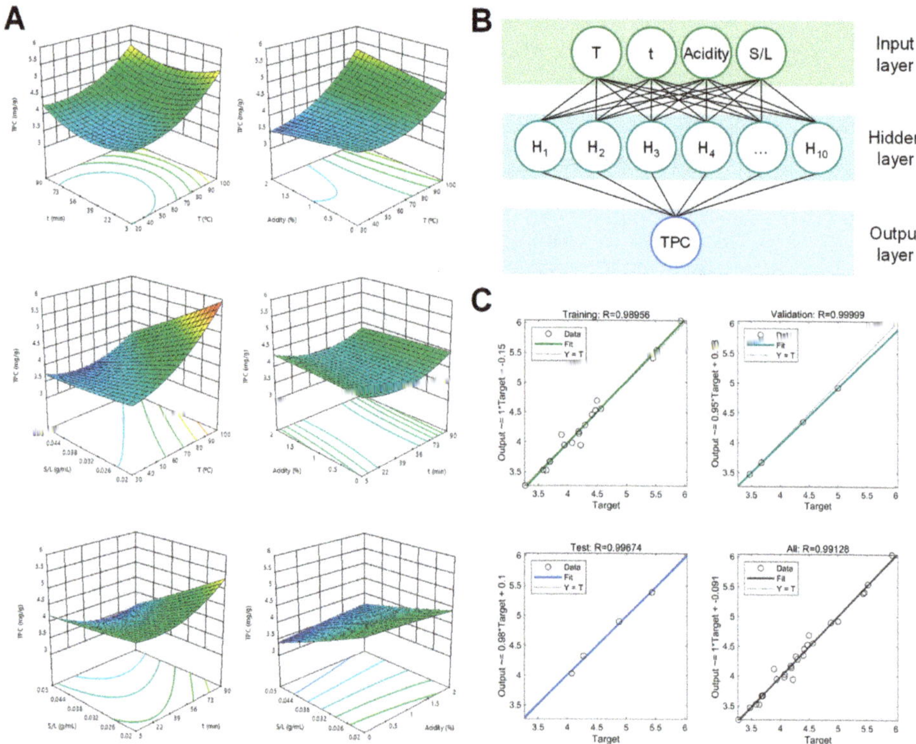

Figure 2. Representative 3D plots illustrating the behavior of total phenolic compounds (TPC) extraction (**A**). Responses (mg g^{-1}) are graphed against two paired variables: T (temperature in °C), acidity (% citric acid), t (time in min) and S/L (solid:liquid ratio in g mL^{-1}); the topology of the multilayer feed-forward neural network for TPC (**B**), and scatter plot between the experimental and predicted yield by artificial neural networks (ANNs) for training, validation, testing, and overall data fitting for TPC (**C**).

2.4. Validation of the Model

The extraction conditions were optimized for the maximum yield of phenolic compounds (total phenolic compounds (TPC), total flavonoids (TF), total flavanols (TFL), total proanthocyanidins (PAC), total phenolic acid (TPA), and total *ortho*-diphenols (TOD)) and the antioxidant capacity (AC) by employing RSM. Then, the responses were determined under the optimal and suboptimal extraction conditions. Finally, the experimental values were compared with predicted values (from RSM and ANN) based on the coefficient of variation, CV (%), to determine the model's validity. The UPLC-ESI-MS/MS profiles of phenolic compounds were also determined at the optimized conditions.

2.5. Heat-Assisted Extraction (HAE)

The HAE was carried out in closed vessels in a temperature-controlled water bath with continuous stirring. According to the experimental design, the milled coffee husk was extracted at various temperatures, times, acidity values, and S/L ratios as described in Table 1. Once HAE was finished, the solubilized phytochemicals were separated by centrifugation (4000× *g*, 4 °C, 15 min), and the supernatants were freeze-dried. The samples were resuspended in Milli-Q water (10 mL) after neutralization and preserved at −20 °C until analysis.

2.6. Organic Solvent Extraction of Free and Bound Phenolic Compound Fractions

Free and bound phenolic fractions from the coffee husk were extracted as described by Rebollo-Hernanz et al. [25] Phenolic compounds from the coffee husk were recovered using a conventional organic method to compare these conditions to those of the optimized methodology using just water as the extracting agent. Here, phenolic compounds' total content was calculated as the sum of the free and bound phenolic fraction.

2.6.1. Extraction of Free Phenolic Compounds

Milled coffee husk (1.0 g) was macerated for 30 min using methanol-HCl (0.1%)/H_2O (80:20, v/v) (50 mL) in an ultrasonic bath. After that, samples were maintained under stirring for 16 h at 40 °C. The samples were centrifuged (4000× g, 4 °C, 15 min), and the supernatants were collected. This process was repeated two times. All the methanolic fractions were combined and evaporated under vacuum. The free phenolic compound fractions were redissolved in methanol (10 mL) and were preserved at −20 °C until analysis.

2.6.2. Extraction of Bound Phenolic Compounds

The insoluble residues from the free phenolic compound extraction were hydrolyzed using 4 mol L^{-1} NaOH (20 mL) under an atmosphere of N_2 under continuous shaking (1 h, 25 °C). The hydrolysates were acidified with 8 mol L^{-1} HCl until reaching pH 2. Then, the samples were centrifuged (4000× g, 4 °C, 15 min), and bound phenolic compounds were extracted from the aqueous alkali phase. Liquid:liquid extraction with diethyl ether:ethyl acetate (50:50, v/v) was repeated three times, and the three organic phases were mixed. Organic fractions were dried under vacuum, redissolved in methanol (10 mL), and preserved at −20 °C until analysis.

2.7. Determination of Phenolic Compounds
2.7.1. Total Phenolic Compounds (TPC)

Total phenolic compounds were quantified using the Folin–Ciocalteu colorimetric method, following the protocol of Singleton, Orthofer, and Lamuela-Raventós [26] adapted to the micromethod format. Samples (10 μL) were mixed with the Folin–Ciocalteu reagent (diluted 1:14, v/v in Milli-Q water) (150 μL). After incubating for 3 min, 20% Na_2CO_3 (50 μL) was added to each well, and the mixture was homogenized. Plates were incubated for 2 h at room temperature. The absorbance was read at 750 nm using a microplate reader incubation. Calibration curves were prepared using solutions of gallic acid, and results were expressed as mg of gallic acid equivalents per gram (mg GAE g^{-1}) of dry coffee husk.

2.7.2. Total Flavonoids (TF)

The content of total flavonoids was determined using the aluminum chloride method adjusted to the micromethod format [27]. Briefly, samples and standards (100 μL) were mixed with 5% Na_2NO_2 (30 μL) and incubated for 5 min at 20 °C. Subsequently, 10% $AlCl_3$ (30 μL) was added. The mixture was further homogenized and incubated for 6 min. Then, 2 mol L^{-1} NaOH (100 μL) was added, and the solution was finally homogenized. The absorbance was recorded at 510 nm. The content of total flavonoids was estimated with a quercetin calibration curve, and the results were expressed as mg of quercetin equivalents per gram (mg QE g^{-1}) of dry coffee husk.

2.7.3. Total Flavanols (TFL)

The content of total flavanols was assessed by the vanillin method adapted [28]. Samples (10 μL) were added to each well, and 8.4 mol L^{-1} vanillin 1% HCl (50 μL) and 37% HCl (250 μL) were added and let to react (15 min, 20 °C). The absorbance was measured at 500 nm, and the concentration of total flavanols was calculated using a standard curve of catechin. The results were expressed as mg of catechin equivalents per gram (mg CE g^{-1}) of dry coffee husk.

2.7.4. Total Proanthocyanidins (PAC)

The content of total proanthocyanidin was determined using a modification of the Bate-Smith method [29]. Briefly, 10 µL of each extract and 1 mL of 0.54 mmol L^{-1} $FeSO_4$ in butanol/HCl (50:50) were incubated at 90 °C for 1 h. After cooling, the absorbance was measured at 550 nm against an unheated blank prepared in the same way than each sample. Cyanidin chloride was used as a standard to construct the calibration curve. Results were expressed as mg of cyanidin chloride equivalents per gram of dry coffee husk (mg CCE g^{-1}).

2.7.5. Total Phenolic Acids (TPA)

The content of total phenolic acids was measured following the method described by Vukic et al. [30]. Samples or standards (10 µL) were mixed with Milli-Q water (50 µL). Then, a Na_2MoO_4 solution and 0.1 mol L^{-1} HCl (50 µL) were combined with the diluted sample, and then NaOH was added (100 µL, 0.1 mol L^{-1}). The absorbance was measured at 490 nm and the content of total phenolic acids was estimated using a calibration curve of caffeic acid. The results were expressed as mg of caffeic acid equivalents per gram (mg CAE g^{-1}) of dry coffee husk.

2.7.6. Total ortho-Diphenols (TOD)

The content of *ortho*-diphenols was estimated according to the method described by Granato et al. [31]. Samples (50 µL) were mixed with 0.05 g mL^{-1} $Na_2MoO_4 \cdot 2H_2O$. The absorbance was measured at 370 nm after a 25 min incubation, and the *o*-diphenols content was calculated using a calibration curve of caffeic acid. The results were expressed as mg of caffeic acid equivalents per gram (mg CAE g^{-1}) of dry coffee husk.

2.7.7. Assessment of In Vitro Antioxidant Capacity (AC)

The coffee husk phenolic extracts' in vitro antioxidant capacity was assessed using the $ABTS^{\bullet+}$ assay, as previously described [32]. $ABTS^{\bullet+}$ radical cations were generated by mixing ABTS solution with $K_2S_2O_8$ (dark and room temperature for 12–16 h, under continuous shaking before use). The $ABTS^{\bullet+}$ assay solution was prepared by dilution in PBS (5 mmol L^{-1}, pH 7.4) to reach an absorbance of 0.70 ± 0.02 at 734 nm. The samples and standards (30 µL) were mixed with the 270 µL of $ABTS^{\bullet+}$ assay solution, and the absorbance of the samples at 734 nm was read after a 10 min incubation. Calibration curves were prepared using standard solutions of Trolox, and the results were expressed as mg Trolox equivalents per gram (mg TE g^{-1}) of dry coffee husk.

2.8. UPLC-ESI-MS/MS Analysis of Phenolic Compounds

The targeted phenolic compounds were analyzed using UPLC-ESI–MS/MS according to a method previously described [33]. Extracts were suspended in water, filtered (0.22 µm), and the internal standard 4-hydroxybenzoic-2,3,5,6-d4 acid solution (Sigma–Aldrich, St. Louis, MO, USA) was added to the samples in a proportion of 1:5 (*v/v*). Data were collected under the multiple reaction monitoring mode for the quantification, tracking the specific transition of parent and product ions for each compound. The electrospray ionization was operated in negative mode. All phenolics were quantified using the calibration curves of their corresponding standards. Injections were carried out in triplicate (*n* = 3).

2.9. Statistical Analysis

Statistical analysis of the experimental results was performed using the statistical programs Design Expert 11, MATLAB version R2020a, and SPSS 24.0. All data are presented as the mean $\pm$ standard deviation (SD) of at least three independent experiments (*n* = 3), where each experiment had a minimum of three replicates for each sample. For comparisons among extraction conditions, data were analyzed by one-way analysis of variance (ANOVA) and the post hoc Tukey test. Differences were considered significant at *p* < 0.05. The statistical design, RSM model, and optimization were calculated with

Design Expert. ANN models were constructed, tested, and validated using MATLAB. The chemometric analysis was carried out to describe the phenolic extracts better. Principal component analysis (PCA) was used to classify samples according to their phenolic composition. Partial least squares analysis (PLSA) was used to rank the spectrophotometric and chromatographic parameters according to their importance (variable importance in projection (VIP) scores) on the variability among extracts. An agglomerative hierarchical cluster analysis coupled to a heatmap was generated to depict the variability among extracts. Principal components regression (PCR) and principal least squares regression (PLS-R) were constructed to evaluate the influence of individual phenolic compounds on the in vitro antioxidant capacity. Pearson's linear correlations were calculated to assess the association between spectrophotometric techniques and results and chromatographic methods, using the concentration of phenolic compounds obtained.

3. Results and Discussion

3.1. Fitting of the RSM and ANN Models

The experimental results for each of the 27 conditions from the Box–Behnken design are presented in Table 1. The RSM fitting for each response variable (TPC, TF, TFL, PAC, TPA, TOD, and AC) was produced using second-order polynomial equations. Response surface 3D plots were generated for each response variable. Figure 2A depicts the behavior of TPC when modifying extraction variables as a representative response. The graphs were generated by plotting the response (TPC) against two independent variables while keeping the other independent variables at a fixed level (in its intermediate value).

The non-significant terms ($p > 0.05$) were not considered to improve the models' fitting and prediction. Both complete (RSM) and simplified models, including just significant ($p < 0.05$) terms (RSM $_{ST}$), statistical parameters measuring the predictive ability of models are presented in Table 2. RSM models exhibited R^2 values between 0.8919 and 0.9744, demonstrating a high linear correlation between experimental and predicted values. The RSM $_{ST}$ models exhibited lower R^2 values (0.7101–0.9597); nonetheless, the model's mathematical fitting continues to show a strong correlation between experimental–predicted values. The lower RMSE and ADD are, the better is the fit between experimental and predicted values. Thus, it was observed that the complete RSM models exhibited lower RMSE and ADD than the RSM $_{ST}$ models for all the response variables.

The ANN was used to predict non-linear associations between the extraction parameters (X_1, X_2, X_3, and X_4) and the response variables (TPC, TF, TFL, PAC, TPA, TOD, and AC). The experimental values used to create the RSM model were also employed to build the ANN model: 70% (19 points) for network training, 15% (4 points) for validation, and the remaining 15% (4 points) for network testing (Figure 2C). The output responses were calculated by passing the weighted sum of input variables to each neuron via an activation function represented by the ANN architecture's hidden layer. The interconnected weights were randomly initialized and adjusted to minimize residual errors between the target and the models' actual outputs (Figure 2B). The optimal number of neurons in the hidden layer was identified through a systematic trial-and-error method using the TPC input. According to this principle, the best results were acquired with feed-forward network topologies, with three layers: input, output, and one hidden layer, with ten neurons, trained with the back-propagation algorithm. These architectures were then used for all the response variables. The correlation coefficient between experimental response variables and the ANN's predicted values was higher than 0.9 for training, validation, testing, and overall fitting for all variables. As an example, Figure 2C depicts the scatter plots for TPC modeling. Table 2 shows the high R^2 and low RMSE and ADD obtained from the ANN models. R^2 values ranged from 0.9802–0.9950. RMSE and ADD values were lower than those of RMS and RSM $_{ST}$, between 0.02–0.24 and 0.19–2.79, respectively. Therefore, it was proved that ANNs are a complex optimization and simulation computational method that displays great potential due to their robust prediction and estimation abilities [34].

Table 2. Comparison of optimization and prediction capabilities of response surface methodology (RSM) and ANN for the extraction of total phenolic compounds (TPC), total flavonoids (TF), total flavanols (TFL), total proanthocyanidins (PAC), total phenolic acids (TPA), total *o*-diphenols (TOD), and the in vitro antioxidant capacity (AC). R^2: the coefficient of determination; AAD: absolute average deviation.

Response	Modeling Method	R^2	RMSE	AAD (%)
TPC	RSM	0.9402	0.16	3.02
	RSM $_{ST}$	0.9110	0.20	3.58
	ANN	0.9802	0.09	1.48
TF	RSM	0.8919	0.36	4.20
	RSM $_{ST}$	0.7101	0.63	6.95
	ANN	0.9950	0.08	0.57
TFL	RSM	0.9222	0.05	0.50
	RSM $_{ST}$	0.8639	0.06	0.65
	ANN	0.9882	0.02	0.19
PAC	RSM	0.9413	0.06	1.97
	RSM $_{ST}$	0.8747	0.09	2.92
	ANN	0.9879	0.03	0.78
TPA	RSM	0.9355	0.16	6.61
	RSM $_{ST}$	0.8901	0.20	8.40
	ANN	0.9860	0.07	2.79
TOD	RSM	0.9627	0.05	3.47
	RSM $_{ST}$	0.9489	0.06	4.19
	ANN	0.9882	0.03	1.59
AC	RSM	0.9744	0.41	2.47
	RSM $_{ST}$	0.9597	0.51	3.16
	ANN	0.9912	0.24	1.52

RSM $_{ST}$: RSM simplified models including only significant ($p < 0.05$) terms.

3.2. Effect of HAE Parameters on the Different Response Variables

RSM regression equations, extraction variables contributions, and statistical parameters (ANOVA) are presented in Table 3. All the response variables adjusted to second-order polynomial equations explained the variation in the different responses as a function of the extraction parameters. The *p*-values were used to evaluate the significance of each coefficient. Low *p*-values, below 0.05, 0.01, and 0.001, indicated that the model terms were significant, highly significant, and remarkably significant, respectively, and *p*-values greater than 0.05 indicate that the model terms were not significant [35]. Temperature (X_1) and S/L ratio (X_4) significantly ($p < 0.01$) contributed to all response variables. Acidity (X_3) significantly ($p < 0.01$) influenced TF, TFL, TPA, TOD, and AC. The impact of time (X_2) was just significant for AC ($p < 0.05$). The quadratic influence of extraction parameters was restricted to TPC (temperature and time), TF (temperature), TFL (temperature and acidity), PAC (S/L ratio), TPA, TOD, and AC (temperature and acidity). Similarly, the interactive effects of the variables were limited to temperature–acidity in TPC; the temperature–S/L ratio in TPC, TFL, PAC, TPA, TOD, and AC; time–acidity in TOD and AC; the time–S/L ratio in TPC and TOD; and acidity–S/L ratio in TFL, PAC, TOD, and AC. The quadratic (5.1–28.6%) and interactive (6.2–17.1%) effects exhibited a low contribution to the models. Contrariwise, the linear effect accounted for 61.3–78.6% of the contribution on the extraction.

Table 3. Regression coefficient (β), contribution, coefficient of determination (R^2 and Adj. R^2), and F-test value of the predicted second-order polynomial models for the phenolic compounds and antioxidant capacity.

	TPC		TF		TFL		PAC		TPA		TOD		AC	
	β	Contrib. (%)	β	Contrib. (%)	β	Contrib. (%)	β	Contrib. (%)	β	Contrib. (%)	β	Contrib. (%)	β	Contrib. (%)
Constant (X_0)	3.689 ***		6.769 ***		0.676 ***		2.541 ***		1.775 ***		0.950 ***		12.221 ***	
Linear		61.3 ***		69.7 ***		65.6 ***		69.5 ***		73.1 ***		61.3 ***		78.6 ***
X_1	0.592 ***	35.1 ***	1.073 ***	41.8 ***	0.133 ***	29.2 ***	0.191 ***	27.4 ***	0.558 ***	36.4 ***	0.185 ***	24.5 ***	2.028 ***	28.0 ***
X_2	0.056	0.3	−0.034	0.0	0.030	1.5	0.010	0.1	0.045	0.2	0.032	0.7	−0.408 *	1.1 *
X_3	−0.001	0.0	−0.645 **	15.1 **	−0.114 ***	21.5 ***	−0.042	1.3	−0.509 ***	30.4 ***	−0.172 ***	21.2 ***	−2.487 ***	42.1 ***
X_4	−0.508 ***	25.9 ***	−0.595 ***	12.9 ***	−0.090 ***	13.4 ***	−0.232 ***	40.7 ***	−0.229 **	6.1 **	−0.144 ***	14.9 ***	−1.045 **	7.4 ***
Quadratic		28.6 ***		8.1 *		17.3 *		5.1		11.5 **		27.2 ***		9.6 ***
X_1^2	0.559 ***	13.9 ***	0.550 *	4.9 *	0.114 **	9.7 **	0.025	0.2	0.320 **	5.3 **	0.149 ***	7.1 ***	1.255 ***	4.7 ***
X_2^2	0.517 ***	11.9 ***	−0.179	0.5	0.007	0.0	−0.041	0.6	−0.105	0.6	0.020	0.1	0.439	0.6
X_3^2	0.153	1.0	0.263	1.1	0.100 **	7.4 **	0.066	1.4	0.324 **	5.5 **	0.251 ***	20.0 ***	1.183* **	4.2 ***
X_4^2	0.196	1.7	0.317	1.6	0.018	0.2	−0.093 *	2.9 *	−0.055	0.2	0.016	0.1	−0.127	0.0
Interaction		12.5 **		10.9		11.2 *		17.1 **		6.2		9.2 *		10.1 **
X_{12}	−0.060	0.1	−0.344	1.4	0.032	0.6	0.066	1.1	0.070	0.2	−0.033	0.3	−0.048	0.0
X_{13}	0.267 *	2.4 *	−0.298	1.1	−0.057	1.8	0.065	1.1	−0.113	0.5	−0.063	1.0	0.249	0.1
X_{14}	−0.390 **	5.1 **	−0.559	3.8	−0.079 *	3.4 *	−0.145 **	5.3 **	−0.254 *	2.5 *	−0.094 *	2.1 *	−1.345 ***	4.1 ***
X_{23}	−0.072	0.2	−0.460	2.6	−0.021	0.2	0.042	0.5	−0.153	0.9	−0.089 *	1.9 *	−0.799 *	1.4 *
X_{24}	−0.357 *	4.3 *	−0.280	1.0	−0.051	1.4	−0.060	0.9	−0.047	0.1	−0.099 *	2.3 *	−0.465	0.5
X_{34}	0.126	0.5	0.301	1.1	0.082*	3.7 *	0.182 **	8.3 **	0.228	2.0	0.082 *	1.6 *	1.312 **	3.9 **
Model		94.0 ***		89.2 ***		92.2 ***		94.1 ***		93.6 ***		96.3 ***		97.4 ***
R^2	0.9402		0.8919		0.9222		0.9413		0.9355		0.9627		0.9744	
Adj. R^2	0.8705		0.8296		0.8314		0.8728		0.8603		0.9192		0.9446	
F value (model)	13.48 ***		7.07 ***		10.16 ***		13.75 ***		12.43 ***		22.13 ***		32.67 ***	

X_1: extraction temperature (°C), X_2: extraction time (min), X_3: acidity (% citric acid), X_4: solid-to-liquid ratio (g mL^{-1}), R^2: Coefficient of determination. Level of significance: * $p < 0.05$, ** $p < 0.01$, *** $p < 0.001$.

121

The TPC varied from 3.28 to 5.93 mg g^{-1} (Table 1). The model, explaining 94.0% of the variation, was mainly influenced by temperature (35.1%) and the S/L ratio (25.9%) (Table 3). Generally, a high extraction temperature is associated with an increase in the solubility of phenolic compounds from the matrix [36]. A decrease in the S/L ratio enhances the extraction of phenolic compounds from plant matrices by reducing the saturation effects due to the concentration of phenolic compounds [37]. The TF ranged from 5.52–10.10 mg g^{-1} (Table 1). Similar to TPC, temperature exhibited the highest contribution (41.8%), whereas acidity and S/L ratio were the following variables contributing to the variability of the model (15.1% and 12.9%, respectively), which responded to 89.2% of the total variability (Table 3). Previous studies demonstrated a positive impact of this parameter on flavonoid extraction [38]. The TFL oscillated from 0.51 to 1.26 mg g^{-1} (Table 1). The model was mainly contributed by temperature (29.2%), acidity (21.5%), and S/L ratio (13.4%), as observed for TF (Table 3). Recently, Silva et al. [15] observed that higher TFL content was obtained from coffee husk using HAE at 60 °C than with ultrasound-assisted extraction alt 35 °C. Therefore, the extraction temperature showed a key role in the phenolic recovery. The PAC fluctuated less than other responses (2.08–3.48 mg g^{-1}) (Table 1). In contrast to other responses, the main factor affecting PAC extraction was the S/L ratio (40.7%) (Table 3). Temperature showed a remarkably significant ($p < 0.001$) effect (27.4%). Procyanidins extraction entails higher difficulty due to the lower polarity in aqueous solvents [39]. As observed, higher water volumes and high temperatures would be needed to recover the maximum PAC yield. The TPA reached 1.19–3.93 mg g^{-1} (Table 1). This extraction model was primarily influenced by temperature and acidity (36.4% and 30.4%, respectively) (Table 3). Likewise, the TOD varied from 0.72 to 1.85 mg g^{-1} (Table 1). Contrary to the other responses, TOD was not only highly influenced by temperature and acidity (24.5 and 21.2, respectively) linearly, but also a remarkably significant ($p < 0.001$) effect was observed quadratically by acidity (20.0%) (Table 3). High acidity can degrade chlorogenic acid. This compound is unstable under these conditions; thus, lower acidity is preferred to extract when chlorogenic acid is present [40]. Finally, the AC of the extracts from the coffee husk oscillated from 9.90 to 18.77 mg g^{-1} (Table 1). Acidity was the main extraction variable affecting AC (42.1%), followed by temperature (28.0%) (Table 3). This model explained 97.4% of the total viability, being the best one among the studied response variables. As previously mentioned, the effect of acidity/low pH needs to be taken into account in extracts with a high concentration of chlorogenic acid, such as coffee, and its by-products [41]. In summary, the positive impact of temperature and S/L ratio was evidenced for all the responses (TPC, TF, TFL, PAC, TPA, TOD, and AC). Temperature increases the water diffusivity, and the lower S/L ratio favors the mass transfer. Moreover, the negative effect of acidity was proved in most of them, attributable to the degradation of chlorogenic acids, highly distributed in all the coffee cherry tissues. Time did not affect the extraction significantly ($p < 0.05$).

All models exhibited remarkably significant fitting ($p < 0.001$), F-value (7.07–32.67). R^2 values exhibited a very strong correlation, being close to the unity and similar to the Adj. R^2 values. To increase the significance of the models', the non-significant linear, quadratic, and interactive terms were excluded from the model, and the mathematical model was recalculated, resulting in the polynomial equations shown in Table 4. R^2 values diminished since these models did not include all the variability of the extraction parameters. Nonetheless, they were much more significant ($p < 0.0001$), showing F-values from 15.77 to 44.96. Hence, the obtained models are presented as an approach for predicting the real behavior of the extraction of phenolic compounds from the coffee husk when modifying the studied parameters (temperature, time, acidity, and S/L ratio).

Table 4. Non-coded equations and their statistical parameters for the extraction of total phenolic compounds (TPC), total flavonoids (TF), total flavanols (TFL), total proanthocyanidins (PAC), total phenolic acids (TPA), total *o*-diphenols (TOD), and the in vitro antioxidant capacity (AC).

Non-Coded Equation	R^2	*F*-Value	*p*-Value
$Y_{TPC} = 4.0 - 1.5 \times 10^{-2}\,x_1 - 1.7 \times 10^{-3}x_2 - 5.0 \times 10^{-1}x_3 + 4.1 \times 10^{+1}x_4 + 3.8 \times 10^{-4}x_1{}^2 + 2.4 \times 10^{-4}x_2{}^2 + 7.6 \times 10^{-3}x_{13} - 7.4 \times 10^{-1}x_{14} - 5.6 \times 10^{-1}x_{24}$	0.9110	19.33	<0.0001
$Y_{TF} = 8.6 - 1.9 \times 10^{-2}x_1 - 6.4 \times 10^{-1}\,x_3 - 4.0 \times 10^{+1}\,x_4 + 3.8 \times 10^{-4}\,x_1{}^2$	0.7414	15.77	<0.0001
$Y_{TFL} = 1.1 - 2.4 \times 10^{-3}x_1 - 4.9 \times 10^{-1}x_3 - 1.7x_4 + 8.8 \times 10^{-5}x_1{}^2 + 9.4 \times 10^{-2}x_3{}^2 - 1.5 \times 10^{-1}x_{14} + 5.4x_{34}$	0.8639	17.23	<0.0001
$Y_{PAC} = 2.0 + 1.5 \times 10^{-2}x_1 - 4.7 \times 10^{-1}x_3 + 2.2 \times 10^{+1}x_4 - 4.6 \times 10^{+2}x_4{}^2 - 2.8 \times 10^{-1}x_{14} + 1.2 \times 10^{+1}x_{34}$	0.8747	23.27	<0.0001
$Y_{TPA} = 2.2 - 5.3 \times 10^{-3}x_1 - 1.2x_3 + 1.6 \times 10^{+1}x_4 + 2.9 \times 10^{-4}x_1{}^2 + 3.6 \times 10^{-1}x_3{}^2 - 4.8 \times 10^{-1}x_{14}$	0.8901	26.99	<0.0001
$Y_{TOD} = 1.3 - 3.3 \times 10^{-3}x_1 + 8.3 \times 10^{-3}x_2 - 7.5 \times 10^{-1}x_3 + 4.0x_4 + 1.1 \times 10^{-4}x_1{}^2 + 2.4 \times 10^{-1}x_3{}^2 - 1.8 \times 10^{-1}x_{14} - 2.1 \times 10^{-3}x_{23} - 1.6 \times 10^{-1}x_{24} + 5.4x_{34}$	0.9489	29.73	<0.0001
$Y_{AC} = 1.6 \times 10^{+1} + 2.3 \times 10^{-2}x_1 + 9.2 \times 10^{-3}x_2 - 6.9x_3 + 9.4x_4 + 9.6 \times 10^{-4}x_1{}^2 + 1.1x_3{}^2 - 2.6x_{14} - 1.9 \times 10^{-2}x_{23} + 8.7 \times 10^{+1}x_{34}$	0.9597	44.96	<0.0001

3.3. Evaluation and Experimental Validation of Optimal Conditions

Maximizing the desirability of all the responses (TPC, TF, TFL, PAC, TPA, TOD, and AC) conduced to two optimal extraction conditions varying just on the extraction time (100 °C, 0% acid, 0.02 g mL^{-1}, and 90 and 5 min, respectively). These conditions predicted the maximum yield of phenolic compounds and antioxidant capacity (among the conditions established). Their responses were evaluated experimentally and validated.

RSM and ANN-predicted values and experimental results obtained after extraction under optimal and suboptimal conditions are shown in Table 5. The extraction at optimal conditions yielded an extract with a high content of phenolic compounds and antioxidant capacity. The experimental results did not differ from the predicted values from RSM (0.0–5.5%) and ANN (2.3–11.8%). At suboptimal conditions, the difference was slower for the RSM model (0.0–1.6%) and the ANN model (0.2–9.8%). Therefore, both models, showing low CV (%) values, could be validated. The optimal and suboptimal conditions generated extract with significantly different concentrations of phenolic compounds and antioxidant capacity. However, the reduction in time could be of interest to the industry. A considerable time reduction (18-fold) would mean a reduction in energy consumption (the extracting agent, here water, has to reach 100 °C and be maintained for 5 or 90 min). Thus, selecting the suboptimal conditions as the most appropriate for the food industry would result in a higher extraction method sustainability. At these optimal and suboptimal conditions, the need for milling the sample was assessed: extractions were carried using non-milled or raw samples, and the same analyses were performed. No differences ($p > 0.05$) were found for most of the response variables at optimal conditions. On the contrary, all response variables from the raw coffee husk exhibited significant differences ($p < 0.05$) with the milled one. Milling increased the extraction yield by 20–74%. We have previously evidenced that milling increases the extraction of total phenolic compounds from coffee parchment (one of the teguments composing the coffee husk) [42]. A smaller particle size favors mass transfer from the coffee pulp to the extracting water.

Table 5. Validation of predicted values at optimal conditions of the aqueous extraction and comparison with organic solvent extraction of the phenolic compounds from the coffee husk.

Response	Optimal Conditions Aqueous Extraction								Organic Solvent Extraction		
	100 °C, 90 min, 0% acid, 0.02 g mL^{-1}				100 °C, 5 min, 0% acid, 0.02 g mL^{-1}				MeOH:H$_2$O	NaOH-AcEt	Σ
	Predicted (CV, %)		Experimental		Predicted (CV, %)		Experimental		*Free Phenolics*	*Bound Phenolics*	*Total Phenolics*
	RSM	ANN	Milled	Raw	RSM	ANN	Milled	Raw			
TPC (mg g^{-1})	6.56 (3.5)	5.83 (11.8)	6.89 ± 0.13 [d]	6.31 ± 0.46 [d]	5.70 (1.4)	4.97 (8.2)	5.59 ± 0.06 [c]	3.94 ± 0.13 [b]	15.99 ± 1.16 [e]	2.39 ± 0.29 [a]	18.38
TF (mg g^{-1})	11.29 (1.2)	9.82 (11.0)	11.48 ± 0.05 [d]	11.01 ± 0.41 [d]	10.56 (0.8)	9.43 (7.2)	10.44 ± 0.08 [c]	7.08 ± 0.05 [b]	26.82 ± 2.65 [e]	4.72 ± 0.47 [a]	31.55
TFL (mg g^{-1})	1.60 (5.5)	1.61 (5.1)	1.73 ± 0.07 [d]	1.67 ± 0.06 [d]	1.33 (1.6)	1.18 (9.8)	1.36 ± 0.04 [c]	0.78 ± 0.01 [b]	1.63 ± 0.30 [d]	0.20 ± 0.06 [a]	1.83
PAC (mg g^{-1})	3.42 (0.0)	3.14 (5.9)	3.42 ± 0.05 [d]	3.24 ± 0.09 [c]	3.23 (0.2)	3.06 (3.7)	3.22 ± 0.12 [c]	2.68 ± 0.03 [b]	3.91 ± 0.37 [e]	1.32 ± 0.08 [a]	5.23
TPA (mg g^{-1})	4.87 (0.5)	5.01 (2.5)	4.84 ± 0.48 [d]	4.12 ± 0.29 [c]	3.90 (0.8)	3.76 (3.4)	3.94 ± 0.18 [c]	23.64 ± 0.08 [b]	6.88 ± 0.23 [e]	0.98 ± 0.06 [a]	7.86
TOD (mg g^{-1})	2.25 (1.6)	1.97 (7.9)	2.20 ± 0.06 [e]	2.04 ± 0.03 [d]	1.87 (0.0)	1.88 (0.2)	1.87 ± 0.08 [c]	1.28 ± 0.06 [b]	5.56 ± 0.30 [f]	0.91 ± 0.11 [a]	6.47
AC (mg g^{-1})	23.64 (0.7)	22.69 (2.3)	23.42 ± 0.15 [e]	22.36 ± 0.24 [d]	21.12 (0.8)	19.35 (7.0)	21.36 ± 0.66 [d]	14.99 ± 0.18 [b]	22.79 ± 1.54 [d e]	5.31 ± 0.42 [a]	28.10

Results are reported as mean ± SD (n = 3). Mean values followed by different superscript letters significantly differ (among columns) when subjected to ANOVA analysis and Tukey multiple range *post hoc* test ($p < 0.05$).

Compared with the organic solvent extraction, HAE extraction at optimal and suboptimal conditions yielded 40–70% of TPC, TF, and TOD but extracted 87% of PAC and 100% of TFL and AC. Therefore, while reducing the concentration of phenolic acids, flavanols would be exerting a higher antioxidant capacity, resulting in 100% antioxidant capacity maintenance. The bound phenolic fraction was also studied. This fraction, bound to the coffee husk cell walls, accounted for 11–25% of the total phenolics and antioxidant capacity. From these results, it is evidenced that the residue resulted from the aqueous extraction would still contain a high concentration of free and bound phenolic compounds, and consequently, antioxidant capacity. This residue could be used as a source of antioxidant dietary fiber, as we have recently proposed [43], but also further treated to separate the phenolic compounds associated with dietary fiber [44,45]. The literature gathers scarce information about the extraction of bioactive compounds from the coffee husk.

The concentration of TPC varies among the diverse extraction conditions used by different authors. Silva et al. [15] extracted phenolic compounds with ethanol and water: ethanol mixtures. Ruesgas-Ramon at al. [46] used deep eutectic solvents, whereas Andrade et al. [47] employed supercritical fluids to extract phenolic compounds from the coffee husk. The aqueous extract presented concentrations similar to those obtained with eutectic solvents but much lower than those obtained with ethanol and supercritical carbon dioxide. Torres-Valenzuela et al. [48] extracted high contents of caffeine, chlorogenic acid, and protocatechuic acid using supramolecular solvents from the coffee pulp. In general, the phytochemical load in the coffee husk depends on the coffee variety, coffee cherry processing, and the plants' stress and climatic and soil conditions [41,49].

3.4. UPLC-ESI-MS/MS Phenolic Compound Profile and Chemometric Analysis

The UPLC-ESI-MS/MS analysis of the phenolic compounds profile from the different coffee husk extracts (Table 6) rendered a better comprehension of the composition of the extracts and the extraction behavior. Representative chromatograms of the optima condition HAE extract, free, and bound phenolic extracts are illustrated in Figure 3A. The main phenolic compounds composing the aqueous extracts was chlorogenic acid (670–906 μg g^{-1}), followed by protocatechuic acid (55–128 μg g^{-1}), kaempferol-3-O-galactoside (12–32 μg g^{-1}), and gallic acid (9–23 μg g^{-1}). The reduction of the extraction time (from 90 to 5 min) significantly ($p < 0.05$) reduced the concentration of all phenolics compounds, but syringic, p-coumaric, and ferulic acids, which were not found or their concentration was reduced in the optimal conditions of extraction (100 °C, 90 min, 0% acid, 0.02 g mL^{-1}). Moreover, the effect of milling was also significant ($p < 0.05$). (+)-Catechin and procyanidin B1 were just released from the coffee husk matrix in the milled samples. Likewise, vanillic and 3,4-dihydroxyphenylacetic acids, (−)-epicatechin, and procyanidin B2 were primarily present in the aqueous extract from the optimal conditions and in a much lower concentration when reducing time or skipping the milling step. The free phenolic compounds fraction (extracted with methanol) contained the highest concentration of chlorogenic acid (1428 μg g^{-1}), kaempferol-3-O-galactoside (40 μg g^{-1}), (+)-catechin (30 μg g^{-1}), and (−)-epicatechin (25 μg g^{-1}). On the other hand, the protocatechuic acid concentration was lower. The high temperature used in HAE may be liberating protocatechuic acid from the bound phenolic fraction [50]. The bound phenolic compounds fraction (extracted after an alkali hydrolysis) was mainly composed of caffeic and protocatechuic acids (90 and 43 μg g^{-1}, respectively), with caffeic acid's concentration being 3.4-fold higher than in the free fraction.

Table 6. UPLC-ESI-MS/MS phenolic compounds profile of the coffee husk extracts obtained by the aqueous extraction using optimal conditions and the organic solvent extraction of the free and bound phenolic fractions.

Compound (μg g^{-1})	R_t (min)	Mass Spectral Data		Optimal Conditions Aqueous Extraction				Organic Solvent Extraction		
		[M − H]$^-$ (*m/z*)	MS^2 (*m/z*)	100 °C, 0% acid, 0.02 g mL^{-1}				MeOH:H$_2$O	NaOH-AcEt	Σ
				90 Min		5 Min		*Free Phenolics*	*Bound Phenolics*	*Total Phenolics*
				Milled	Raw	Milled	Raw			
Hydroxybenzoic acids										
Gallic acid	1.73	169	125	22.79 ± 1.45 [d]	10.94 ± 0.81 [b]	12.11 ± 0.89 [c]	8.51 ± 0.10 [a]	23.15 ± 0.30 [d]	-	23.15
Protocatechuic acid	3.34	153	109	127.96 ± 6.86 [f]	83.53 ± 5.37 [d]	66.20 ± 8.73 [c]	55.08 ± 2.1 [b]	99.94 ± 1.33 [e]	42.82 ± 8.59 [a]	142.76
4-hydroxybenzoic acid	4.43	137	93	3.51 ± 0.34 [c]	2.47 ± 0.19 [b]	3.19 ± 0.36 [c]	1.56 ± 0.07 [a]	4.30 ± 0.07 [d]	3.75 ± 0.51 [c d]	8.06
Vanillic acid	5.43	167	152	6.00 ± 2.52 [a]	-	-	-	9.55 ± 0.20 [b]	-	9.55
Syringic acid	5.96	197	182	-	-	0.20 ± 0.03 [a]	-	1.66 ± 0.10 [b]	0.20 ± 0.01 [a]	1.86
Salicylic acid	8.96	137	93	0.81 ± 0.03 [b]	0.90 ± 0.08 [b]	0.22 ± 0.04 [a]	-	1.19 ± 0.02 [c]	0.25 ± 0.03 [a]	1.43
Hydroxycinnamic acids										
Chlorogenic acid	5.38	353	191	905.67 ± 18.50 [e]	747.17 ± 36.10 [c]	840.04 ± 30.79 [d]	669.54 ± 40.06 [b]	1428.40 ± 25.80 [f]	0.67 ± 0.08 [a]	1429.07
Caffeic acid	5.48	179	135	15.16 ± 0.51 [b]	10.51 ± 1.07 [a]	14.00 ± 1.50 [b]	9.35 ± 0.11 [a]	26.21 ± 0.79 [c]	89.93 ± 9.73 [d]	116.14
p-coumaric acid	6.81	163	119	2.27 ± 0.04 [b]	1.39 ± 0.11 [a]	4.22 ± 0.14 [d]	1.37 ± 0.06 [a]	2.94 ± 0.04 [c]	8.21 ± 1.31 [e]	11.15
Ferulic acid	7.81	193	134	-	-	4.25 ± 0.36 [b]	-	1.74 ± 0.08 [a]	5.08 ± 0.67 [c]	6.81
Phenylacetic acids										
3,4-dihydroxyphenylacetic acid	4.18	167	123	1.46 ± 0.51 [b]	1.04 ± 0.04 [a]	-	-	-	-	-
Flavan-3-ols: monomers										
(+)-catechin	5.22	289	245	0.45 ± 0.05 [b]	-	0.31 ± 0.03 [a]	-	30.31 ± 0.28 [d]	0.58 ± 0.08 [c]	30.89
(−)-epicatechin	6.27	289	245	4.71 ± 0.51 [c]	1.45 ± 0.06 [b]	-	-	25.07 ± 0.23 [d]	0.64 ± 0.05 [a]	25.71
Flavan-3-ols: dimers										
Procyanidin B1	4.90	577	289	5.84 ± 0.69	-	-	-	-	-	-
Procyanidin B2	5.93	577	289	3.03 ± 0.46 [c]	0.93 ± 0.17 [b]	0.76 ± 0.09 [a]	0.77 ± 0.08	-	-	-
Flavonols										
Quercetin-3-*O*-glucoside	8.34	463	301	15.03 ± 0.96 [d]	8.89 ± 1.17 [b]	10.45 ± 0.73 [c]	10.39 ± 0.44	18.16 ± 0.42 [e]	0.66 ± 0.03 [a]	18.82
Quercetin-3-*O*-galactoside	8.65	463	301	14.33 ± 0.14 [d]	8.79 ± 0.42 [b]	10.20 ± 1.91 [c]	10.25 ± 0.26	18.17 ± 0.41 [e]	0.67 ± 0.05 [a]	18.84
Kaempferol-3-*O*-galactoside	9.46	447	284	32.12 ± 0.94 [e]	18.27 ± 0.83 [d]	11.50 ± 0.24 [b]	14.13 ± 1.46	40.32 ± 0.85 [f]	1.05 ± 0.11 [a]	41.37

Results are reported as mean ± SD ($n = 3$). Mean values followed by different superscript letters significantly differ (among columns) when subjected to ANOVA analysis and Tukey's multiple range *post hoc* test ($p < 0.05$).

126

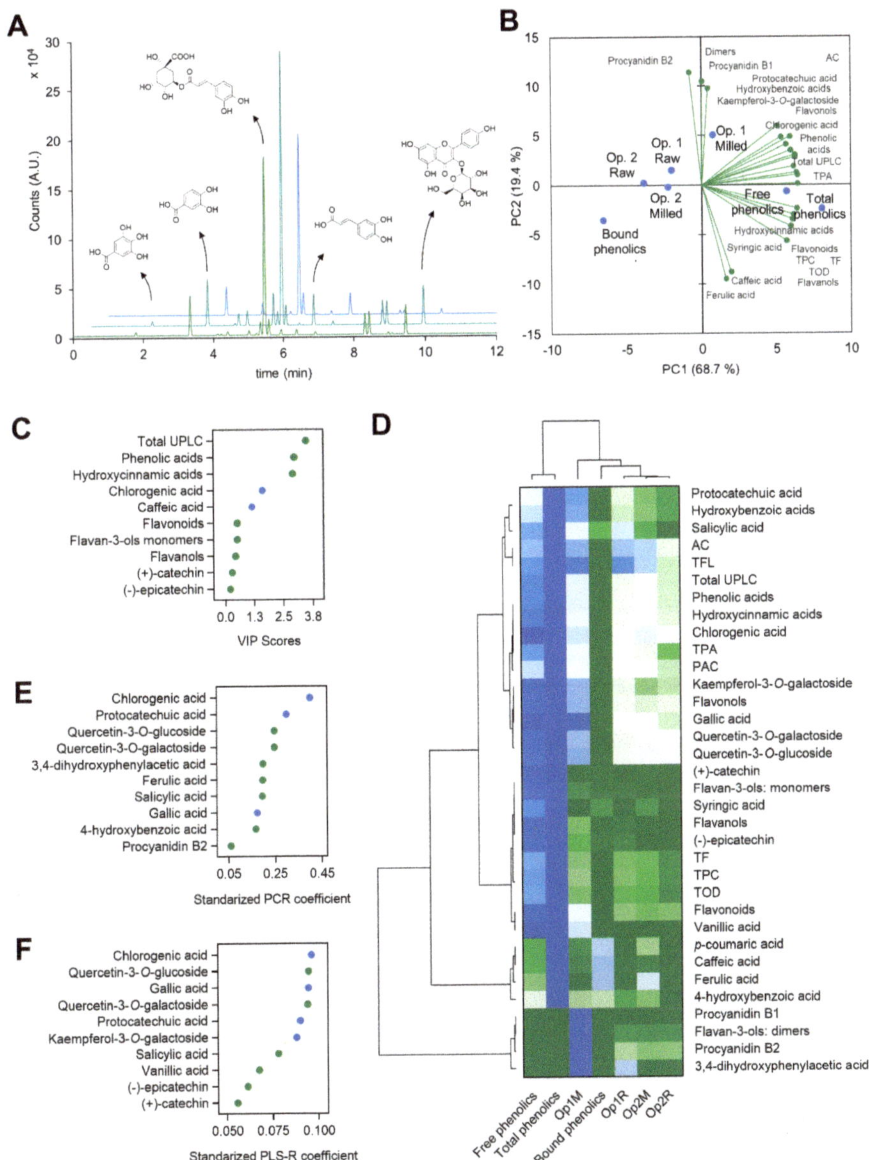

Figure 3. Superimposed chromatograms of the extract at optimal conditions, free, and bound phenolics and the chemical structures of gallic, protocatechuic, chlorogenic, caffeic acids and kaempferol-3-*O*-galactoside, major phenolic compounds found in the coffee husk (**A**), biplot (scores of samples and load factors of each variable) of the principal component analysis (PCA) (**B**), Variable importance in projection (VIP) scores from partial least squares analysis (PLSA) (**C**), agglomerative hierarchical cluster analysis coupled to heatmap (from the lowest (■) to the highest (■) value for each parameter) (**D**) showing the associations among the measured parameters and classifying phenolic extracts from coffee husk according to them, and the ten most significant coefficients from principal components regression, PCR (**E**) and principal least squares regression, PLS-R (**F**). Circles in different colors indicate minor phenolic or phenolic family, green (●), major phenolic, blue (●).

PCA (Figure 3B) revealed the intrinsic grouping among samples. PCA extracted five factors or principal components (PCs) to explain the phytochemical variability among the aqueous and organic extracts from the coffee husk. The two first PCs (the ones graphed) explained 88.1% of the variability; PC1 and PC2 represented 68.7 and 19.4% of the whole variability. The PC1 exhibited a positive influence on all the in vitro determinations (TPC, TF, PAC, TPA, TOD, and AC) but TFL, and most compounds measured by UPLC-MS/MS, excluding caffeic, ferulic, p-coumaric, and 3,4-dihydroxyphenylacetic acids, and the flavan-3-ols dimes, procyanidins B1 and B2, which were correlated with PC2. The PCA's clustering grouped the optimum condition (Op.1 Milled) with the sample of free phenolic compounds. In turn, the three other aqueous extraction conditions (100 °C, 0% acid, 0.02 g mL^{-1}; 90 min, using raw coffee pulp, Op.1 Raw; and 5 min using both raw (Op.2 Raw) and milled (Op.2 Milled) coffee pulp) were grouped together, between the free and bound phenolic extracts. Total phenolics were depicted on the right edge of the graph. Therefore, the extracts were classified from left to right according to the total phenolic content.

Figure 3C represents the VIP scores from the PLS analysis. Total phenolics measured by UPLC (total UPLC), phenolic, and hydroxycinnamic acids were the three most variable parameters among the samples. Chlorogenic and caffeic acids were the individual phenolic compounds exhibiting the highest variation, and therefore, showing the most significant impact on sample classification. As the heatmap coupled to the dendrogram of hierarchical clustering shows (Figure 3D), the extract at optimum condition (Op.1) was depic-ted separately, between free and total phenolic, which were considered similar, and bound phenolics and the other three aqueous extracts (Op.1 Raw and Op.2 Milled and Raw). The differences in the extracts' phenolic composition and antioxidant capacity define the extraction at optimal conditions (100 °C, 90 min, 0% citric acid, and 0.02 g mL^{-1} S/L ratio) as the best extraction, being the most similar to the conventional extraction of free phenolics. A comprehensive analysis was carried out to find 18 compounds, including hydroxybenzoic, hydroxycinnamic, phenylacetic acids, monomeric and dimeric flavan-3-ols, and flavonols. Previous studies have been focused on the main phenolic compounds (chlorogenic, protocatechuic, gallic, and caffeic acids) and the caffeine content [15,46,47].

Phenolic compounds from the coffee husk have been demonstrated to possess antioxidant potential, as revealed in a previous study [51]. Identifying the compounds responsible for these properties arouses great interest. Extensive research has focused on the separation and purification of active biomolecules from food and natural products [52,53]. The isolation and purification of the phytochemicals extracted following the proposed green extraction method could strengthen their biological activity to be then used as food ingredients or nutraceutical products. The ten most significant coefficients from principal component regression (PCR) and partial least squares regression (PLS-R) are depicted in Figure 3E,F. From the PCR coefficients, chlorogenic and protocatechuic acids were the main phenolic compounds responsible for the in vitro antioxidant capacity. According to the PLS-R coefficients, gallic acid and quercetin-3-O-glucoside, and 3-O-galactoside were also significant contributors to the antioxidant properties of the extracts from the coffee husk. These phenolic compounds have been formerly associated with potent antioxidant properties in vitro and in vivo [54–57].

Pearson's correlations were studied to analyze the relationship among in vitro parameters and the phenolic compounds quantified chromatographically. The obtained associations were illustrated in a heatmap (Figure 4).

The concentration of numerous phenolic compounds correlated with the in vitro assays results. The concentration of chlorogenic acid in the extracts from the coffee husk strongly correlated with the content of TPC and TPA ($r = 0.8973$, $p < 0.01$ and $r = 0.9619$, $p < 0.001$, respectively). Protocatechuic acid also showed a strong association with TPA ($r = 0.8745$, $p < 0.01$). Kaempferol-3-O-galactoside significantly correlated with TPC ($r = 0.9008$, $p < 0.001$) and TF ($r = 0.8864$, $p < 0.001$). The sum of flavonoids exhibited strong association with TF ($r = 0.9584$, $p < 0.001$). Furthermore, the sum of the concentration of all individual phenolic compounds (total UPLC) presented a significant ($p < 0.01$)

correlation (r = 0.8333–0.9842) with all the in vitro methods. Consequently, the use of these spectrophotometric techniques to screen the best extraction conditions could be validated, as indicated by Granato et al. [58]. In vitro methods are consistent during screening steps, as long as more specific and comprehensive techniques are used for phytochemical profile analysis.

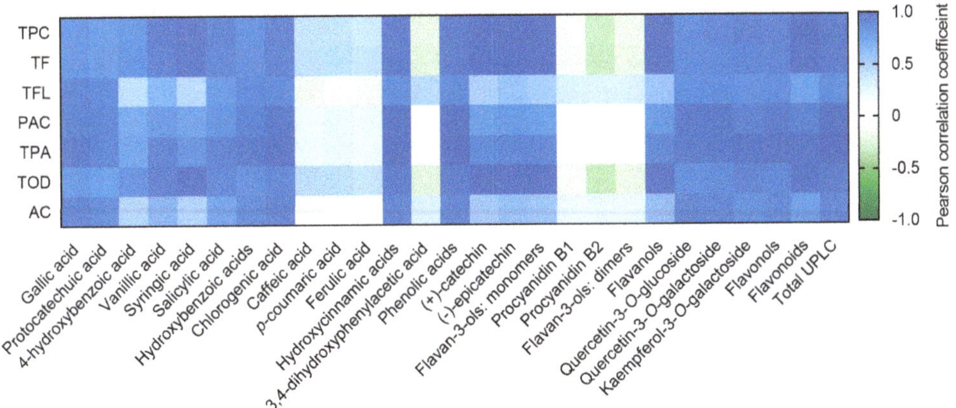

Figure 4. Heatmap depicting the Pearson correlation coefficients from the associations among in vitro determinations of phenolic families and the different phenolic compounds quantified using UPLC-ESI-MS/MS.

This study presents the most comprehensive analysis of phenolic compounds composing the coffee husk. Here, we present the aqueous soluble phenolic compounds extracted with HAE and the complete phenolic profile from free and bound phenolic compounds' fractions. The protective effects of these compounds against oxidative stress and the development of chronic diseases have been widely reported in the literature [51,59]. Chlorogenic acid, the major phenolic compound found in the coffee husk, displayed the highest PCR and PLS-R coefficients, and positively correlated with the in vitro AC (r = 0.8977, p < 0.01). This compound has been demonstrated to be an excellent radical scavenger following different antioxidant mechanisms and activating cellular antioxidant response (Nrf2-ARE signaling pathways) [54,60]. Additionally, chlorogenic acid presents other biological properties, including the modulations of glucose and lipid metabolism [61], promotion of adipocyte browning [62], and prevention of inflammation [63]. These health-promoting properties elicit the use of the coffee husk as a sustainable source of chlorogenic acid, among other phytochemicals. Thus, using these aqueous extracts from the coffee husk as healthy ingredients could be a great strategy in valorizing coffee by-products and producing novel sustainable products. Although the present work is limited to a variety of coffee husk, the sustainable conditions established could be applied in the extraction of other coffee varieties and even to the extraction of phenolic compounds from the coffee pulp (a by-product comparable to the coffee husk but obtained through the wet processing).

4. Conclusions

A green sustainable extraction method for recovering the high-value phenolic compounds from the coffee husk was modeled and validated. The modification of the extraction variables (temperature, time, acidity, and S/L ratio) lead to an improved extraction of phenolic compounds and in vitro antioxidant capacity. The use of RSM and ANN permitted one to model and optimize the aqueous extraction of total phenolic compounds, flavonoids, and flavanols proanthocyanidins, phenolic acids, and o-diphenols, and a high in vitro antioxidant capacity. Thus, the optimal conditions (100 °C, 90 min, 0% citric acid, and 0.02 g mL^{-1} S/L ratio), producing phenolic-rich extracts from the coffee husk using water as the only extracting agent were established. The presence of chlorogenic, protocatechuic, caffeic, and gallic acids and several flavonols, with kaemferol-3-*O*-galactoside being the primary one, was confirmed by the UPLC-ESI-MS/MS results. Chemometric techniques defined chlorogenic acid as the main antioxidant compound. Likewise, multivariate analysis permitted us to validate spectrophotometric techniques for screening the best extraction methods since they showed strong correlations with the chromatographic results. This green extraction may revalorize the coffee husk, a by-product generated globally and of outstanding chemical and biological interest, as a new food ingredient with potential antioxidant and health-promoting properties.

Author Contributions: Conceptualization, M.R.-H., Y.A. and M.A.M.-C.; methodology, M.R.-H. and B.B.; formal analysis, M.R.-H.; investigation, M.R.-H., S.C., D.T. and V.B.; writing—original draft preparation, M.R.-H.; writing—review and editing, M.R.-H., S.C., D.T., B.B., Y.A. and M.A.M.-C.; visualization, M.R.-H.; supervision, B.B., Y.A. and M.A.M.-C.; project administration, M.A.M.-C.; funding acquisition, M.A.M.-C. All authors have read and agreed to the published version of the manuscript.

Funding: This research was funded by the Spanish Ministry of Science and Innovation, SUS-COFFEE (grant number AGL2014–57239-R), COCARDIOLAC (grant number RTI 2018-097504-B-I00) projects, and the Community of Madrid and UAM Agreement (2019–2023). M. Rebollo-Hernanz thanks to the FPU program of the Ministry of Universities for his predoctoral fellowship (grant number FPU15/04238).

Institutional Review Board Statement: Not applicable.

Informed Consent Statement: Not applicable.

Data Availability Statement: The data presented in this study are available from the corresponding author on reasonable request.

Conflicts of Interest: The authors declare no conflict of interest.

References

1. United Nations Sustainable Consumption and Production. Available online: https://www.un.org/sustainabledevelopment/sustainable-consumption-production/ (accessed on 17 February 2021).
2. FAO. *Sustainable Food Systems Concept and Framework. What Is a Sustainable Food System? Why Take a Food Systems Approach? Changing Food Systems*; FAO: Rome, Italy, 2018; pp. 1–8.
3. Torres-León, C.; Ramírez-Guzman, N.; Londoño-Hernandez, L.; Martinez-Medina, G.A.; Díaz-Herrera, R.; Navarro-Macias, V.; Alvarez-Pérez, O.B.; Picazo, B.; Villarreal-Vázquez, M.; Ascacio-Valdes, J.; et al. Food Waste and Byproducts: An Opportunity to Minimize Malnutrition and Hunger in Developing Countries. *Front. Sustain. Food Syst.* **2018**, *2*, 52. [CrossRef]
4. de Sousa e Silva, J.; Moreli, A.P.; Donzeles, S.M.L.; Soares, S.F.; Vitor, D.G. Harvesting, Drying and Storage of Coffee. In *Food Engineering Series*; Springer: Berlin/Heidelberg, Germany, 2021; pp. 1–64. [CrossRef]
5. Klingel, T.; Kremer, J.I.; Gottstein, V.; Rajcic de Rezende, T.; Schwarz, S.; Lachenmeier, D.W. A Review of Coffee By-Products Including Leaf, Flower, Cherry, Husk, Silver Skin, and Spent Grounds as Novel Foods within the European Union. *Foods* **2020**, *9*, 665. [CrossRef] [PubMed]
6. Kumar, S.S.; Swapna, T.S.; Sabu, A. Coffee Husk: A Potential Agro-Industrial Residue for Bioprocess. In *Waste to Wealth. Energy, Environment, and Sustainability*; Singhania, R., Agarwal, R., Kumar, R., Sukumaran, R., Eds.; Springer: Singapore, 2018; pp. 97–109. [CrossRef]
7. Heeger, A.; Kosińska-Cagnazzo, A.; Cantergiani, E.; Andlauer, W. Bioactives of coffee cherry pulp and its utilisation for production of Cascara beverage. *Food Chem.* **2017**, *221*, 969–975. [CrossRef] [PubMed]
8. Da Silva, M.C.S.; Naozuka, J.; Da Luz, J.M.R.; De Assunão, L.S.; Oliveira, P.V.; Vanetti, M.C.D.; Bazzolli, D.M.S.; Kasuya, M.C.M. Enrichment of *Pleurotus ostreatus* mushrooms with selenium in coffee husks. *Food Chem.* **2012**, *131*, 558–563. [CrossRef]
9. de Carvalho Oliveira, F.; Srinivas, K.; Helms, G.L.; Isern, N.G.; Cort, J.R.; Gonçalves, A.R.; Ahring, B.K. Characterization of coffee (*Coffea arabica*) husk lignin and degradation products obtained after oxygen and alkali addition. *Bioresour. Technol.* **2018**, *257*, 172–180. [CrossRef] [PubMed]
10. Sarrouh, B.; de Souza, R.O.A.; da Silva Florindo, R.H.; Lofrano, R.C.Z.; de Oliveira, A.M. Extraction and Identification of Biomolecules from Lignin Alkaline Hydrolysate from Coffee Husk. *Waste Biomass Valorization* **2021**, *12*, 787–794. [CrossRef]
11. Morales-Martínez, J.L.; Aguilar-Uscanga, M.G.; Bolaños-Reynoso, E.; López-Zamora, L. Optimization of Chemical Pretreatments Using Response Surface Methodology for Second-Generation Ethanol Production from Coffee Husk Waste. *Bioenergy Res.* **2020**, 1–13. [CrossRef]
12. Alhogbi, B.G. Potential of coffee husk biomass waste for the adsorption of Pb(II) ion from aqueous solutions. *Sustain. Chem. Pharm.* **2017**, *6*, 21–25. [CrossRef]

13. Dadi, D.; Daba, G.; Beyene, A.; Luis, P.; Van der Bruggen, B. Composting and co-composting of coffee husk and pulp with source-separated municipal solid waste: A breakthrough in valorization of coffee waste. *Int. J. Recycl. Org. Waste Agric.* **2019**, *8*, 263–277. [CrossRef]

14. del Castillo, M.D.; Iriondo-DeHond, A.; Fernandez-Gomez, B.; Martinez-Saez, N.; Rebollo-Hernanz, M.; Martín-Cabrejas, M.A.; Farah, A. Coffee Antioxidants in Chronic Diseases. In *Coffee: Consumption and Health Implications*; Farah, A., Ed.; Royal Society of Chemistry: Cambridge, UK, 2019; pp. 20–56. [CrossRef]

15. Silva, M.d.O.; Honfoga, J.N.B.; de Medeiros, L.L.; Madruga, M.S.; Bezerra, T.K.A. Obtaining Bioactive Compounds from the Coffee Husk (*Coffea arabica* L.) Using Different Extraction Methods. *Molecules* **2020**, *26*, 46. [CrossRef]

16. Ameer, K.; Shahbaz, H.M.; Kwon, J.-H. Green Extraction Methods for Polyphenols from Plant Matrices and Their Byproducts: A Review. *Compr. Rev. Food Sci. Food Saf.* **2017**, *16*, 295–315. [CrossRef]

17. Aguilera, Y.; Rebollo-Hernanz, M.; Cañas, S.; Taladrid, D.; Martín-Cabrejas, M.A. Response surface methodology to optimise the heat-assisted aqueous extraction of phenolic compounds from coffee parchment and their comprehensive analysis. *Food Funct.* **2019**, *10*, 4739–4750. [CrossRef]

18. Ferreira, S.L.C.; Silva Junior, M.M.; Felix, C.S.A.; da Silva, D.L.F.; Santos, A.S.; Santos Neto, J.H.; de Souza, C.T.; Cruz Junior, R.A.; Souza, A.S. Multivariate optimization techniques in food analysis—A review. *Food Chem.* **2019**, *273*, 3–8. [CrossRef] [PubMed]

19. Yolmeh, M.; Jafari, S.M. Applications of Response Surface Methodology in the Food Industry Processes. *Food Bioprocess Technol.* **2017**, *10*, 413–433. [CrossRef]

20. Khadir, M.T. Artificial neural networks for food processes: A survey. In *Artificial Neural Networks in Food Processing*; De Gruyter: Berlin, Germany, 2021; pp. 27–50. [CrossRef]

21. Gonzalez-Fernandez, I.; Iglesias-Otero, M.A.; Esteki, M.; Moldes, O.A.; Mejuto, J.C.; Simal-Gandara, J. A critical review on the use of artificial neural networks in olive oil production, characterization and authentication. *Crit. Rev. Food Sci. Nutr.* **2019**, *59*, 1913–1926. [CrossRef] [PubMed]

22. Kass, R.E.; Dennis, J.E.; Schnabel, R.B. *Numerical Methods for Unconstrained Optimization and Nonlinear Equations*; SIAM: Philadelphia, PA, USA, 1985; Volume 80, ISBN 0898713641. [CrossRef]

23. Priyadarshini, R. Functional Analysis of Artificial Neural Network for Dataset Classification. *Spec. Issue IJCCT* **2010**, *1*, 3–5. [CrossRef]

24. Vogl, T.P.; Mangis, J.K.; Rigler, A.K.; Zink, W.T.; Alkon, D.L. Accelerating the convergence of the back-propagation method. *Biol. Cybern.* **1988**, *59*, 257–263. [CrossRef]

25. Rebollo-Hernanz, M.; Aguilera, Y.; Herrera, T.; Cayuelas, L.T.; Dueñas, M.; Rodríguez-Rodríguez, P.; Ramiro-Cortijo, D.; Arribas, S.M.; Martín-Cabrejas, M.A. Bioavailability of melatonin from lentil sprouts and its role in the plasmatic antioxidant status in rats. *Foods* **2020**, *9*, 330. [CrossRef] [PubMed]

26. Singleton, V.L.; Orthofer, R.; Lamuela-Raventós, R.M. Analysis of total phenols and other oxidation substrates and antioxidants by means of folin-ciocalteu reagent. *Methods Enzymol.* **1999**, *299*, 152–178. [CrossRef]

27. Rebollo-Hernanz, M.; Fernández-Gómez, B.; Herrero, M.; Aguilera, Y.; Martín-Cabrejas, M.A.; Uribarri, J.; Del Castillo, M.D. Inhibition of the Maillard reaction by phytochemicals composing an aqueous coffee silverskin extract via a mixed mechanism of action. *Foods* **2019**, *8*, 438. [CrossRef] [PubMed]

28. Swain, T.; Hillis, W.E. The phenolic constituents of *Prunus domestica*. I.—The quantitative analysis of phenolic constituents. *J. Sci. Food Agric.* **1959**, *10*, 63–68. [CrossRef]

29. Ribéreau-Gayón, P.; Stonestreet, E. Dógase des tanins du vin rouges et détermination du leur structure. *Chem. Anal.* **1966**, *48*, 188–196.

30. Vukic, M.D.; Vukovic, N.L.; Djelic, G.T.; Obradovic, A.; Kacaniova, M.M.; Markovic, S.; Popović, S.; Baskić, D. Phytochemical analysis, antioxidant, antibacterial and cytotoxic activity of different plant organs of *Eryngium serbicum* L. *Ind. Crops Prod.* **2018**, *115*, 88–97. [CrossRef]

31. Granato, D.; Margraf, T.; Brotzakis, I.; Capuano, E.; van Ruth, S.M. Characterization of Conventional, Biodynamic, and Organic Purple Grape Juices by Chemical Markers, Antioxidant Capacity, and Instrumental Taste Profile. *J. Food Sci.* **2015**, *80*, C55–C65. [CrossRef]

32. Benítez, V.; Rebollo-Hernanz, M.; Aguilera, Y.; Bejerano, S.; Cañas, S.; Martín-Cabrejas, M.A. Extruded coffee parchment shows enhanced antioxidant, hypoglycaemic, and hypolipidemic properties by releasing phenolic compounds from the fibre matrix. *Food Funct.* **2021**, *12*, 1097. [CrossRef]

33. Sánchez-Patán, F.; Monagas, M.; Moreno-Arribas, M.V.; Bartolomé, B. Determination of Microbial Phenolic Acids in Human Faeces by UPLC-ESI-TQ MS. *J. Agric. Food Chem.* **2011**, *59*, 2241–2247. [CrossRef]

34. Shanmuganathan, S. Artificial neural network modelling: An introduction. In *Studies in Computational Intelligence*; Springer: New York, NY, USA, 2016; Volume 628, pp. 1–14. [CrossRef]

35. Dahmoune, F.; Remini, H.; Dairi, S.; Aoun, O.; Moussi, K.; Bouaoudia-Madi, N.; Adjeroud, N.; Kadri, N.; Lefsih, K.; Boughani, L.; et al. Ultrasound assisted extraction of phenolic compounds from *P. lentiscus* L. leaves: Comparative study of artificial neural network (ANN) versus degree of experiment for prediction ability of phenolic compounds recovery. *Ind. Crops Prod.* **2015**, *77*, 251–261. [CrossRef]

36. Roselló-Soto, E.; Martí-Quijal, F.J.; Cilla, A.; Munekata, P.E.S.; Lorenzo, J.M.; Remize, F.; Barba, F.J. Influence of temperature, solvent and pH on the selective extraction of phenolic compounds from tiger nuts by-products: Triple-TOF-LC-MS-MS characterization. *Molecules* **2019**, *24*, 797. [CrossRef]

37. Andres, A.I.; Petron, M.J.; Lopez, A.M.; Timon, M.L. Optimization of Extraction Conditions to Improve Phenolic Content and In Vitro Antioxidant Activity in Craft Brewers' Spent Grain Using Response Surface Methodology (RSM). *Foods* **2020**, *9*, 1398. [CrossRef] [PubMed]

38. Mokrani, A.; Madani, K. Effect of solvent, time and temperature on the extraction of phenolic compounds and antioxidant capacity of peach (*Prunus persica* L.) fruit. *Sep. Purif. Technol.* **2016**, *162*, 68–76. [CrossRef]

39. Wong-Paz, J.E.; Guyot, S.; Aguilar-Zárate, P.; Muñiz-Márquez, D.B.; Contreras-Esquivel, J.C.; Aguilar, C.N. Structural characterization of native and oxidized procyanidins (condensed tannins) from coffee pulp (*Coffea arabica*) using phloroglucinolysis and thioglycolysis-HPLC-ESI-MS. *Food Chem.* **2021**, *340*, 127830. [CrossRef]

40. Narita, Y.; Inouye, K. Degradation kinetics of chlorogenic acid at various pH values and effects of ascorbic acid and epigallocatechin gallate on its stability under alkaline conditions. *J. Agric. Food Chem.* **2013**, *61*, 966–972. [CrossRef] [PubMed]

41. Janissen, B.; Huynh, T. Chemical composition and value-adding applications of coffee industry by-products: A review. *Resour. Conserv. Recycl.* **2018**, *128*, 110–117. [CrossRef]

42. Benítez, V.; Rebollo-Hernanz, M.; Hernanz, S.; Chantres, S.; Aguilera, Y.; Martin-Cabrejas, M.A. Coffee parchment as a new dietary fiber ingredient: Functional and physiological characterization. *Food Res. Int.* **2019**, *122*, 105–113. [CrossRef] [PubMed]

43. Cañas, S.; Rebollo-Hernanz, M.; Cano-Muñoz, P.; Aguilera, Y.; Benítez, V.; Braojos, C.; Gila-Díaz, A.; Rodríguez-Rodríguez, P.; Cobeta, I.M.; López De Pablo, Á.L.; et al. Critical Evaluation of Coffee Pulp as an Innovative Antioxidant Dietary Fiber Ingredient: Nutritional Value, Functional Properties and Acute and Sub-Chronic Toxicity. *Proceedings* **2021**, *70*, 7623. [CrossRef]

44. Saura-Calixto, F. Dietary Fiber as a Carrier of Dietary Antioxidants: An Essential Physiological Function. *J. Agric. Food Chem.* **2011**, *59*, 43–49. [CrossRef]

45. Saura-Calixto, F. The Story of the Introduction of Non-extractable Polyphenols into Polyphenol Research: Origin, Development and Perspectives. In *Non-Extractable Polyphenols and Carotenoids: Importance in Human Nutrition and Health*; Royal Society of Chemistry: London, UK, 2018; Volume 2018, pp. 1–16. ISBN 9781782627081. [CrossRef]

46. Ruesgas-Ramón, M.; Suárez-Quiroz, M.L.; González-Ríos, O.; Baréa, B.; Cazals, G.; Figueroa-Espinoza, M.C.; Durand, E. Biomolecules extraction from coffee and cocoa by- and co-products using deep eutectic solvents. *J. Sci. Food Agric.* **2020**, *100*, 81–91. [CrossRef]

47. Andrade, K.S.; Gonçalvez, R.T.; Maraschin, M.; Ribeiro-do-Valle, R.M.; Martínez, J.; Ferreira, S.R.S. Supercritical fluid extraction from spent coffee grounds and coffee husks: Antioxidant activity and effect of operational variables on extract composition. *Talanta* **2012**, *88*, 544–552. [CrossRef]

48. Torres-Valenzuela, L.S.; Ballesteros-Gómez, A.; Rubio, S. Supramolecular solvent extraction of bioactives from coffee cherry pulp. *J. Food Eng.* **2020**, *278*, 109933. [CrossRef]

49. Geremu, M.; Tola, Y.B.; Sualeh, A. Extraction and determination of total polyphenols and antioxidant capacity of red coffee (*Coffea arabica* L.) pulp of wet processing plants. *Chem. Biol. Technol. Agric.* **2016**, *3*, 1–6. [CrossRef]

50. Călinoiu, L.; Vodnar, D. Thermal Processing for the Release of Phenolic Compounds from Wheat and Oat Bran. *Biomolecules* **2019**, *10*, 21. [CrossRef]

51. Rebollo-Hernanz, M.; Zhang, Q.; Aguilera, Y.; Martín-Cabrejas, M.A.; Gonzalez de Mejia, E. Phenolic compounds from coffee by-products modulate adipogenesis-related inflammation, mitochondrial dysfunction, and insulin resistance in adipocytes, via insulin/PI3K/AKT signaling pathways. *Food Chem. Toxicol.* **2019**, *132*, 110672. [CrossRef]

52. Chemat, F.; Abert Vian, M.; Fabiano-Tixier, A.S.; Nutrizio, M.; Režek Jambrak, A.; Munekata, P.E.S.; Lorenzo, J.M.; Barba, F.J.; Binello, A.; Cravotto, G. A review of sustainable and intensified techniques for extraction of food and natural products. *Green Chem.* **2020**, *22*, 2325–2353. [CrossRef]

53. Rebollo-Hernanz, M.; Cañas, S.; Taladrid, D.; Bartolomé, B.; Aguilera, Y.; Martin-Cabrejas, M.A. Extraction of phenolic compounds from cocoa shell: Modeling using response surface methodology and artificial neural networks. *Sep. Purif. Technol.* **2021**, *266*.

54. Sato, Y.; Itagaki, S.; Kurokawa, T.; Ogura, J.; Kobayashi, M.; Hirano, T.; Sugawara, M.; Iseki, K. In vitro and in vivo antioxidant properties of chlorogenic acid and caffeic acid. *Int. J. Pharm.* **2011**, *403*, 136–138. [CrossRef] [PubMed]

55. Rebollo-Hernanz, M.; Zhang, Q.; Aguilera, Y.; Martín-Cabrejas, M.A.; de Mejia, E.G. Cocoa Shell Aqueous Phenolic Extract Preserves Mitochondrial Function and Insulin Sensitivity by Attenuating Inflammation between Macrophages and Adipocytes In Vitro. *Mol. Nutr. Food Res.* **2019**, *63*, 1801413. [CrossRef] [PubMed]

56. Badhani, B.; Sharma, N.; Kakkar, R. Gallic acid: A versatile antioxidant with promising therapeutic and industrial applications. *RSC Adv.* **2015**, *5*, 27540–27557. [CrossRef]

57. Lesjak, M.; Beara, I.; Simin, N.; Pintać, D.; Majkić, T.; Bekvalac, K.; Orčić, D.; Mimica-Dukić, N. Antioxidant and anti-inflammatory activities of quercetin and its derivatives. *J. Funct. Foods* **2018**, *40*, 68–75. [CrossRef]

58. Granato, D.; Shahidi, F.; Wrolstad, R.; Kilmartin, P.; Melton, L.D.; Hidalgo, F.J.; Miyashita, K.; van Camp, J.; Alasalvar, C.; Ismail, A.B.; et al. Antioxidant activity, total phenolics and flavonoids contents: Should we ban *in vitro* screening methods? *Food Chem.* **2018**, *264*, 471–475. [CrossRef]

59. Rebollo-Hernanz, M.; Zhang, Q.; Aguilera, Y.; Martín-Cabrejas, M.A.; de Mejia, E.G. Relationship of the phytochemicals from coffee and cocoa by-products with their potential to modulate biomarkers of metabolic syndrome in vitro. *Antioxidants* **2019**, *8*, 279. [CrossRef]

60. Yao, J.; Peng, S.; Xu, J.; Fang, J. Reversing ROS-mediated neurotoxicity by chlorogenic acid involves its direct antioxidant activity and activation of Nrf2-ARE signaling pathway. *BioFactors* **2019**, *45*, biof.1507. [CrossRef] [PubMed]

61. Rebollo-Hernanz, M.; Aguilera, Y.; Martin-Cabrejas, M.A.; de Mejia, E.G. Regulation of lipid and glucose metabolism in hepatocytes by phytochemicals from coffee by-products and prevention of non-alcoholic fatty liver disease in vitro. *Proceedings* **2020**, *61*, 6996. [CrossRef]

62. Sudhakar, M.; Sasikumar, S.J.; Silambanan, S.; Natarajan, D.; Ramakrishnan, R.; Nair, A.J.; Kiran, M.S. Chlorogenic acid promotes development of brown adipocyte-like phenotype in 3T3-L1 adipocytes. *J. Funct. Foods* **2020**, *74*, 104161. [CrossRef]

63. Hwang, S.J.; Kim, Y.-W.; Park, Y.; Lee, H.-J.; Kim, K.-W. Anti-inflammatory effects of chlorogenic acid in lipopolysaccharide-stimulated RAW 264.7 cells. *Inflamm. Res.* **2014**, *63*, 81–90. [CrossRef] [PubMed]

Article

The Effect of *Ilex × meserveae* S. Y. Hu Extract and Its Fractions on Renal Morphology in Rats Fed with Normal and High-Cholesterol Diet

Piotr Kuropka [1], Anna Zwyrzykowska-Wodzińska [2], Robert Kupczyński [2], Maciej Włodarczyk [3], Antoni Szumny [4] and Renata M. Nowaczyk [1,*]

[1] Department of Biostructure and Animal Physiology, Division of Histology and Embryology, Wroclaw University of Environmental and Life Sciences, Norwida 25, 50-375 Wroclaw, Poland; piotr.kuropka@upwr.edu.pl

[2] Department of Environment Hygiene and Animal Welfare, Wroclaw University of Environmental and Life Sciences, Chelmońskiego 38C, 51-631 Wroclaw, Poland; anna.zwyrzykowska@upwr.edu.pl (A.Z.-W.); robert.kupczyński@upwr.edu.pl (R.K.)

[3] Department of Pharmacognosy and Herbal Medicines, Faculty of Pharmacy, Wroclaw Medical University, Borowska 211a, 50-556 Wroclaw, Poland; maciej.wlodarczyk@umed.wroc.pl

[4] Department of Chemistry, Wroclaw University of Environmental and Life Sciences, Norwida 25, 50-375 Wroclaw, Poland; antoni.szumny@upwr.edu.pl

* Correspondence: renata.nowaczyk@upwr.edu.pl

Citation: Kuropka, P.; Zwyrzykowska-Wodzińska, A.; Kupczyński, R.; Włodarczyk, M.; Szumny, A.; Nowaczyk, R.M. The Effect of *Ilex × meserveae* S. Y. Hu Extract and Its Fractions on Renal Morphology in Rats Fed with Normal and High-Cholesterol Diet. *Foods* **2021**, *10*, 818. https://doi.org/10.3390/foods10040818

Academic Editors: Yolanda Aguilera and Vanesa Benítez García

Received: 16 March 2021
Accepted: 6 April 2021
Published: 9 April 2021

Publisher's Note: MDPI stays neutral with regard to jurisdictional claims in published maps and institutional affiliations.

Abstract: Therapeutic properties of *Ilex* species are widely used in natural medicine. *Ilex × meserveae* may become a potential substitute for *Ilex paraguariensis* (Yerba Mate). As a part of the preliminary safety verification of this European *Ilex* hybrid vs. Yerba Mate, an eight-week study concerning the impact of regular administration of leaves of both species on kidneys was conducted. The standard water infusion and three dominant fractions of *Ilex × meserveae* leaves' constituents (polyphenols, saponins and less polar terpenoids) were separately tried on 96 male Wistar rats divided into 8-member groups. Animals were divided into two basic nutritional groups: the first one was rats fed standard feed and the second on was rats fed with high-cholesterol diet (20 g of cholesterol per kg of standard feed). Postmortem morphometric evaluation of stained kidney samples concerned the filtration barrier elements, which are crucial in proper diuresis. The results showed that saponins present in the hydroalcoholic dry extract (administered in a dose of 10 mg/kg of body weight/day) as well as in water infusions (1:20) from *Ilex × meserveae* and *Ilex paraguariensis* do not demonstrate nephrotoxicity but conversely, have a protective role on kidney status in animals fed with a normal diet and in a high-cholesterol diet.

Keywords: *Ilex × meserveae*; Yerba Mate; kidney filtration barrier; high-cholesterol diet; saponins; terpenoids; polyphenols

1. Introduction

Ilex L. species, especially *Ilex paraguariensis* A. St.-Hil., are frequently used in traditional medicine. The brew (infusion) of *Ilex paraguariensis* leaves, known as Yerba Mate, was ritually used by Native South Americans before the colonizers' arrival. This beverage's consumption is recently being expanded to many North America, Asia, and European countries. In trade, Yerba Mate, in its ground form, contains mainly leaves, together with fragments of young branches, flowers, and peduncles. Simultaneously, *I. paraguariensis* extracts are distributed as an additive in various products, e.g., cosmetics and food supplements as well as functional foods [1,2]. The research confirmed that the therapeutic effect of *Ilex* spp. concern arthritis, diabetes, immune diseases, hemorrhoids, headaches, liver dysfunction, and obesity [3]. Many studies have shown that obesity and related diseases are significant health problems [4,5]. In this plant's case, the research indicated

135

that *I. paraguariensis* water extract lowered total cholesterol and low-density lipoprotein in people with high levels of serum lipid, and thus is promiscuous to treat obesity [5,6], while hyperlipidemia can be a risk factor for the progression of renal diseases and changes of glomerular structure and thickness of basement membrane [7]. It is noteworthy that cholesterol modulates the bilayer structure parameters of biological membranes, such as thickness, compressibility, water penetration. However, high cholesterol and triglyceride plasma levels have been demonstrated to be important risk factors for the progression of kidney disease and high total cholesterol or reduced HDL (high-density lipoprotein) cholesterol can decrease glomerular filtration rate [8].

In terms of phytochemical research, *I. paraguariensis* has probably been the subject of the most intensive investigations among all *Ilex* spp. [3]. It should be noted that the content of active biological compounds present in *Ilex* species depends on extraction methods, phenotype, environmental variability, as well as harvest time [9]. Generally, *I. paraguariensis* extracts contain polyphenols (including flavonoids, tannins, chlorogenic acid, and its derivatives), purine alkaloids (methylxanthines; like caffeine or theobromine), vitamins (A, B, C, and E) as well as some triterpene saponins (derived mainly from ursolic acid) [10]. The main phenolic compounds in *I. paraguariensis* are associated with caffeoylquinic esters [11]. Our preliminary study [12] confirmed that *Ilex* species other than *I. paraguariensis* also contain a high amount of polyphenolic fraction (rich in rutin, quinic acid, and its caffeoyl esters), triterpenes, as well as their glycosides (saponins). The similarity of phytoconstituents in European *Ilex* × *meserveae* S. Y. Hu "Blue Angel" and South American *I. paraguariensis* allows us to consider *I. meserveae* "Blue Angel" as a promising source of bioactive compounds [12]. Of the above, caffeoylquinic esters were generally found to be interesting remedies for lowering blood cholesterol [5,6].

The *Ilex* secondary metabolites are mainly eliminated via the kidney route and impact the diuresis level due to dynamic changes in the glomerular filtration barrier (GFB) structure. The glomerular basement membrane provides stability for the filtration process and constitutes the GFB [13]. According to Jarad and Miner [14], the glomerular capillary wall consists of three layers: the glomerular basement membrane and the fenestrated endothelium, with its glycocalyx; the podocyte with interfoot processes and the fenestral diaphragm; and the glomerular basement membrane. These structures have been considered as the significant determinants of glomerular permeability with functional importance of two additional layers: the endothelial surface layer and the subpodocyte space; which all of the above structures have highly restrictive dimensions and contribute to the hydraulic resistance and ultrafiltration characteristics of the glomerulus [13,15]. Disturbed glomerular filtration barrier functions play a crucial role in developing many kidney diseases, including proteinuria [14,16].

Generally, it has been observed that drinking the infusion of many leaf extracts increases urination in animal models. There are only rare reports about the influence of *I. paraguariensis* beverages [17] on the urinary system, while no report on *I. meserveae* was found. However, the medical properties, including the impact on the high-cholesterol diet and also the side effects of *Ilex* spp., expected of *Ilex paraguariensis*, have not been studied in detail yet. Therefore, based upon our earlier study [12] on detailed characterization of polar, semipolar (polyphenolic, saponin), and less-polar (terpenoid) fractions from plants belonging to the *Ilex* genus, we decided to comparatively investigate the effects of *I. paraguariensis* and *I. meserveae* extract on kidney structure in animals fed with a normal and high-cholesterol diet. As factors modulating the impact of the adverse effects of cholesterol were used the extracts of *I. paraguariensis* (Yerba Mate) and *I. meserveae*, and fractions of the last one: polyphenols, terpenoids, and saponins.

2. Materials and Methods

2.1. Reagents and Plant Materials

The following solvents were used for UHPLC-MS (ultra-high-performance liquid chromatography coupled with mass spectrometry detector): acetonitrile (MS-purity; Sigma-

Aldrich, St. Louis, MO, USA), water (LC-gradient; Merck, Kenilworth, NJ, USA), and formic acid (p.p.a. (pure per analysis), 98–100%, Merck). Analytically pure methanol (Chempur, Karlsruhe, Germany) and distilled water were used to extract *Ilex* leaves. Ballast substances were precipitated with lead (II) acetate (p.p.a, Chempur). The octadecyl bed for the SPE (solid-phase extraction) process was from J. T. Baker. Suitable solvents and reagents used for histological examinations were bought from Archem, Ommen, The Netherlands.

The leaves of *Ilex* × *meserveae* S. Y. Hu "Blue Angel" (*Aquifoliaceae*, voucher Il.6/06.2016) were obtained from a nursery (Grodziszów, Poland) and authenticated by Professor Przemysław Bąbelewski. Immediately after harvesting, the leaves were frozen (−20 °C) and lyophilized to avoid enzymatic degradation of metabolites (20 h in 0.25 mBar, followed by 4 h in 0.025 mBar; Alpha 1–4 LDplus, Martin Christ). Directly after the lyophilization process, the leaves were reduced to a powder and extracted. Commercial Yerba Mate leaves (*Ilex paraguariensis* A. St.-Hil.), used as well-recognized standards in a saponin identification protocol, were purchased from two independent distributors (vouchers Il.1a/06.2016 and Il.1b/06.2016). Voucher specimens were deposited at the Department of Horticulture, The Faculty of Life Sciences and Technology, Wroclaw University of Environmental and Life Sciences, Wroclaw, Poland.

2.2. Plant Extracts

2.2.1. Water Extracts

Infusion of *I. paraguariensis and I. meserveae* were prepared day by day (across the animal experiment period) in the same manner by adding 50.0 g of dried and ground leaves to 1L of boiled water (80 °C), left for 20 min, then filtered. The typical process efficiency for *I. meserveae* was 1.4% (calc. on DM (dry mass)). Suspected active compounds in this extract were: simple sugars, polysaccharides, amino acids and other acids (including phenolic acids), small polar glycosides, polar polyphenols, saponins, and terpenoids listed in Appendix A (Tables A1–A3).

2.2.2. Polyphenols

Extracts from *I. meserveae* dried leaves were prepared with 80% methanol as the solvent, according to the method presented by Zwyrzykowska et al. [12]. The resulting fine powder was sufficiently soluble in water at experimental concentration. Process yield was 1.3% (calc. on DM). Detailed HPLC composition was published in [12].

2.2.3. Saponins

Extracts from *I. meserveae* dried leaves were prepared and analyzed using a modified methodology developed by Włodaczyk et al. [18]. Briefly, the powdered plant material was cold-macerated with 70% methanol; the polar compounds (polysaccharides, peptides, phenolics) were precipitated and removed. Simultaneously, the supernatant was diluted and cleaned by solid-phase extraction (SPE) on an RP-18 bed to obtain a refined saponin fraction. This fraction was concentrated to dryness by consecutive vacuum evaporation and lyophilization. The resulting finely powdered dry extract fortified with saponins was sufficiently soluble in water at experimental concentrations. The yield of the saponin-enriched fraction was 2.5% (calc. on starting leaf DM). Suspected active compounds in this extract were mainly triterpenoid glycosides (saponins). Detailed isolation protocol and UHPLC-MS data are presented in Appendix A (Tables A2 and A3).

2.2.4. Terpenoids

According to a protocol published in Polish Patent applications P.437122 and P.437123, the terpenoid fraction was obtained. The isolation details are presented in Appendix A.1. Terpenoid fraction was dissolved in sunflower oil and added to animal feed. At the same time, the extract was monitored by gas chromatography with mass spectrometry (GC-MS). The yield of *I. meserveae* lipophilic fraction was 1.8% (calc. on starting DM). The detailed composition of the terpenoid fraction is presented in Appendix A (Table A1).

2.3. Animals, Housing, and Diets

All experimental procedures used in this study were approved by the II Local Ethics Committee in Wroclaw, Poland (permission No. 94/2015). The study was carried out on 96 male Wistar rats, aged 8 weeks, weight 250–280 g, kept in standardized environmental conditions (12/12 h light/dark cycle, the temperature approximately 22 °C, humidity about 55%). Every single cage counted two individuals. The animals were divided into two basic nutritional groups. The first main group (assigned with Roman numerals) was rats fed a standard feed and the second main group (assigned with Roman numerals followed by the letter a) was rats fed with a high-cholesterol diet (20 g of cholesterol per kg of standard feed). Each of the two main groups was divided into six subgroups consisted of 8 animals each. As factors modulating the impact of the negative effects of cholesterol were used the extracts of *I. paraguariensis* (Yerba Mate; **II, IIa**), *I. meserveae* (**III, IIIa**) and *I. meserveae* fractions: polyphenols (**IV, IVa**), terpenoids (**V, Va**) and saponins (**VI, VIa**). The subgroups **I** and **Ia** were the control groups (received no additional herbal extracts), while the other subgroups differed in the type of extracts added (data presented in Table 1). The doses of examined herbal extracts have been selected based on the de Resende et al. [19] study with our modification.

Table 1. Research groups of experimental animals. Control groups: **I, Ia**, and other groups with diet modified by the addition of *Ilex* extracts (**II–VI**) or both *Ilex* extracts and cholesterol (**IIa–VIa**).

Group	Diet Type (8 Animals in Each Diet Group)
I	rats fed with a standard diet
Ia	rats fed as a group **I** but with the addition of 20 g of cholesterol per kilogram of diet
II	rats receiving instead of drinking water the water extract of *I. paraguariensis* (each day, freshly infused extract was prepared by extraction of 50 g of leaves with 1L boiled water; every two animals had free access to 250 mL of this sole source of drink per day)
IIa	rats fed as group **II** but with the addition of 20 g of cholesterol per kilogram of diet
III	rats receiving instead of drinking water the water extract of *I. meserveae* "Blue Angel" (each day, freshly infused extract was prepared by extraction of 50 g of leaves with 1L boiled water; every two animals had free access to 250 mL of this sole source of drink per day)
IIIa	rats fed as group **III** but with the addition of 20 g of cholesterol per kilogram of diet
IV	rats receiving additionally polyphenol fraction from *I. meserveae* "Blue Angel" (each day, the dry extract was freshly solubilized in water in a dose of 10 mg/kg BW; every two animals had free access to 250 mL of this sole source of drink per day)
IVa	rats fed as group **IV** but with the addition of 20 g of cholesterol per kilogram of diet
V	rats receiving additionally terpenoid fraction from *I. meserveae* "Blue Angel" (each day, 200 mL of oil was mixed with terpenoids (in a dose 10 mg/kg BW) and 1 kg of feed and left overnight; every two animals had free access to diet and drinking water, supplied ad libitum as in group **I**)
Va	rats fed as group **V** but with the addition of 20 g of cholesterol per kilogram of diet
VI	rats receiving additionally saponin fraction from *I. meserveae* "Blue Angel" (each day, the dry extract was freshly solubilized in water in a dose of 10 mg/kg BW; every two animals had free access to 250 mL of this sole source of drink per day)
VIa	rats fed as group **VI** but with the addition of 20 g of cholesterol per kilogram of diet

During the eight weeks of the experiment period, rats were fed ad libitum with a standard pelleted feed (composition: dry mass—906.04 (g/kg); energy content—19.78 (MJ/kg); total protein—17.42 (% DM); crude fat—2.13 (% DM); crude fiber—9.45 (% DM); Hybrid-pellet, Animalab, Poland). All animals had free access to drinking fluids, either water or *Ilex* infusion or *Ilex* fraction drink (250 mL/24 h/cage). Food and liquid ingestion and body weight (BW) were monitored daily throughout the experiment. At the end of the study, the animals were anesthetized with isoflurane and sacrificed by abdominal aorta exsanguinations. The kidneys were examined by the pathologist macroscopically in situ, based on the position, color, shape, size, and consistency of the organs.

2.4. Specimen Processing and Staining

The kidney samples were taken during the necropsy and immediately fixed in 10% neutral buffered formalin for three days, then washed in tap water for 24 h, dehydrated in a graded alcohol series, cleared in xylene, and finally embedded in paraffin. The 5 µm thick sections were routinely stained with hematoxylin and eosin (H&E, Sigma-Aldrich) and Alcian blue (Sigma-Aldrich) according to our modification. Histopathological observations were performed using a Nikon Eclipse 80i light microscope.

2.5. Statistical Analysis

Morphometric studies were carried out using the Nis-Elements Ar software (Nikon). A minimum of ten measurements of the glomerulus, the vascular loop, and the basement membrane's thickness in the vascular loop was performed from each individual of the experimental group. The average was calculated based upon ten representative areas from the kidney cortex in each sample. All data were presented as mean $\pm$ SD. Statistical analysis was made using one-way analysis of variance (ANOVA) and performed using Statistica 6.0 (StatSoft), taking $p < 0.05$ as significant.

3. Results

3.1. Morphological Studies

Macroscopically, the kidneys' structure, with a clear border between cortex and medulla, was well preserved in all experimental groups. The kidney capsule was not strongly anchored to the organ parenchyma. However, the kidneys' edema and blood congestion were observed in some individuals of the group's **II, IIa, IIIa, IVa, Va,** and **VIa**. The edema was mainly associated with increased diuresis.

3.2. Hematoxylin and Eosin Staining

No significant changes were observed in the control group—**Ia** (Figure 1a). It should be noted that urogenesis and the transfer of urine through the tubules were slowed down. Proximal and distal tubules had enlarged light and were lined with a regular cuboidal epithelium (Figure 1a,b). In the case of the high-cholesterol control group—**Ib**, the renal glomeruli were congested, but there was no high urine content in the capsule (Figure 1b).

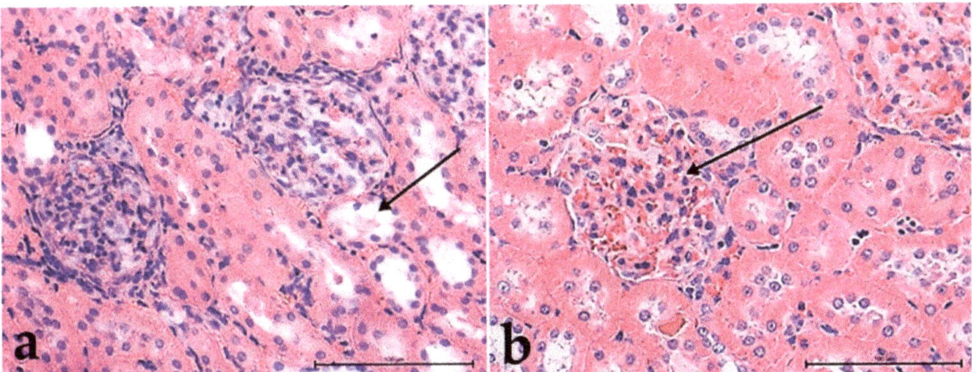

Figure 1. The structure of the kidneys in the control groups: in normal, **I** (**a**) and high-cholesterol diet, **II** (**b**); (**a**) note the lack of the urine in proximal tubules and slightly enlarged distal tubules (black arrow); (**b**) glomerulus filled with blood (black arrow) hematoxylin and eosin (H&E) staining. Mag. 400×. Bar scale 100 µm.

The results of morphological studies indicated in groups **IIa** (Figure 2b), **III** (Figure 2c), **IIIa** (Figure 2d) moderately increased urogenesis (visible intensive glomerular filtration), and slightly greater congestion were present in groups: **II** (Figure 2a), **III** (Figure 2c) and **IIIa** (Figure 2d). In rats fed with a high-cholesterol diet and watered the extract of *Ilex paraguariensis* (group **IIa**), there was some toxic effect in the kidney cortex, visible as the presence of proliferating fibroblasts and lymphocytes near large vessels (Figure 2b).

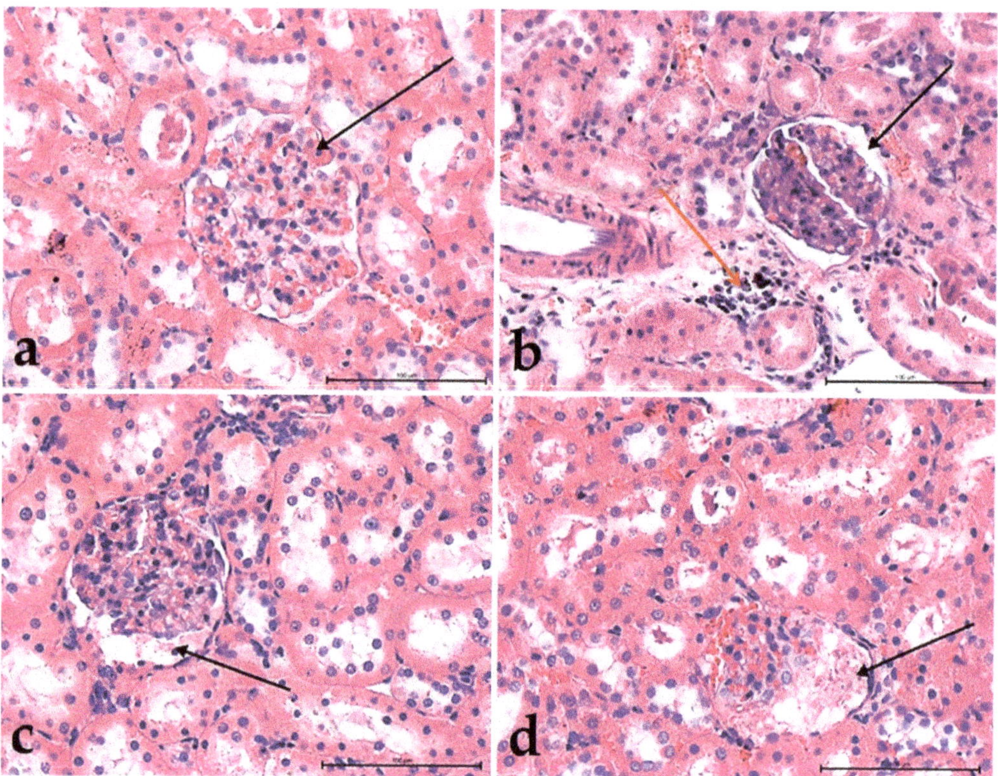

Figure 2. The changes in the structure of the kidneys in groups: **II** (**a**), **IIa** (**b**), **III** (**c**), **IIIa** (**d**). Glomeruli with different urine content (black arrow). Note the presence of leukocytes in the kidney parenchyma in **IIa** group (**b**) (red arrow). H&E staining. Mag. 400×. Bar scale 100 μm.

Among the three fractions tested, the most potent activity enhancing urogenesis with numerous sites showing the arrested vascular flow and increased activity of connective tissue cells were observed in groups: **IV** (Figure 3a), **IVa** (Figure 3b), **Va** (Figure 3d), **VI** (Figure 3e) were the weakest in group **V** (Figure 3c). In group **VIa**, the outlook was similar, but it seems that the proximal and distal tubules were less filled with urine (Figure 3f). The toxic effect was observed as venous stasis with the presence of proliferating fibroblasts. Moreover, extensive lymphocyte infiltrates in large vessels' vicinity were particularly noticeable in group **IVa** (Figure 3b).

3.3. Alcian Blue Staining

Morphological analysis of Alcian blue staining did not show significant differences in polysaccharides' content within the glomeruli in control groups. In both groups, glomeruli were filled with numerous erythrocytes (Figure 4a,b).

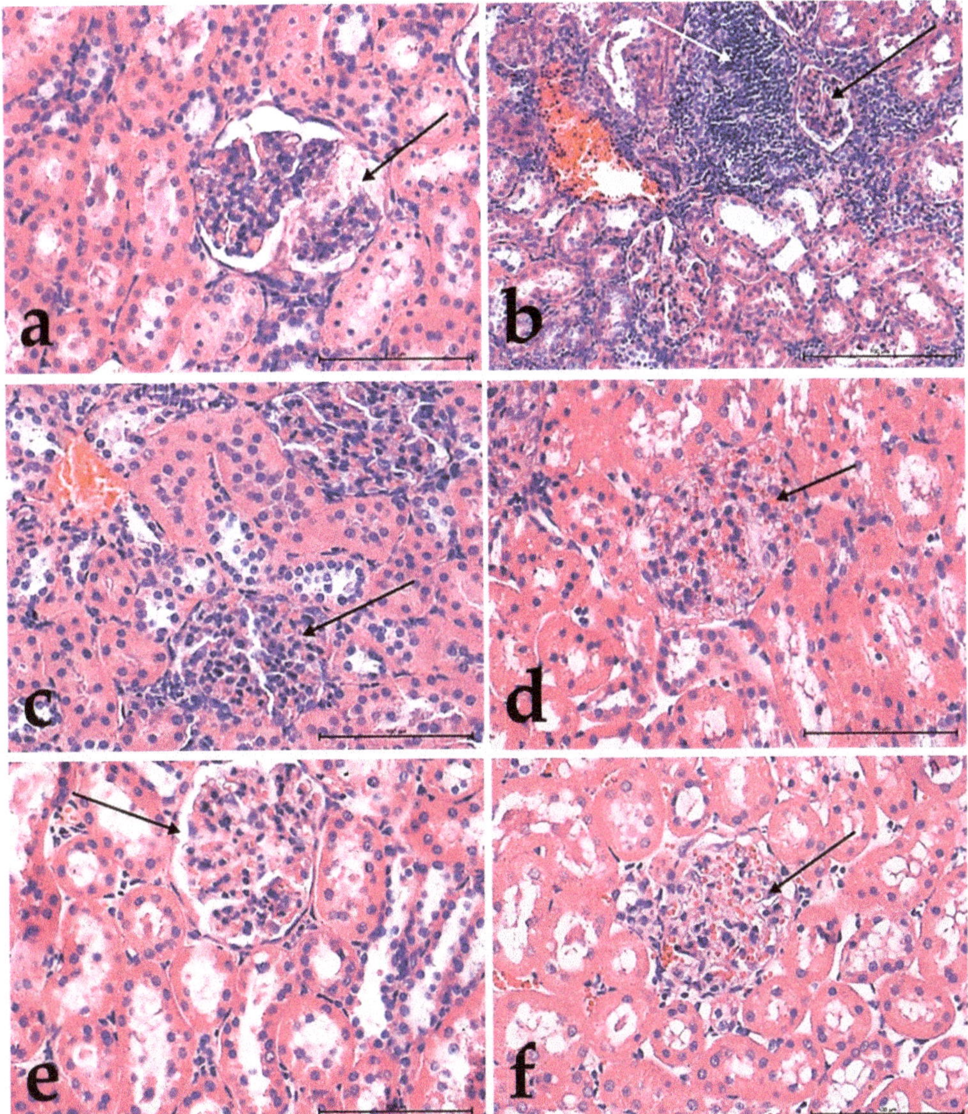

Figure 3. The changes in the structure of the kidneys in groups: **IV (a)**, **IVa (b)**, **V(c)**, **Va (d)**, **VI (e)**, and **VIa (f)**. A glomerulus with different urine and blood content (black arrow). The massive lymphocyte infiltration around the glomerulus in group **IVa (b)** (white arrow) H&E staining. Mag. 400×. Bar scale 100 μm.

The tissue staining with Alcian blue showed that in groups **II** (Figure 5a), **III** (Figure 5c), **IV** (Figure 6a), and **V** (Figure 6c), the active compounds had an influence on the blood-urine barrier, leading to increased urogenesis. The study revealed that cholesterol significantly reduces this effect because the polysaccharide content decrease was observed only for group **VIa** (Figure 6f). In the other groups, no such differences were found (Figures 5b,d and 6b).

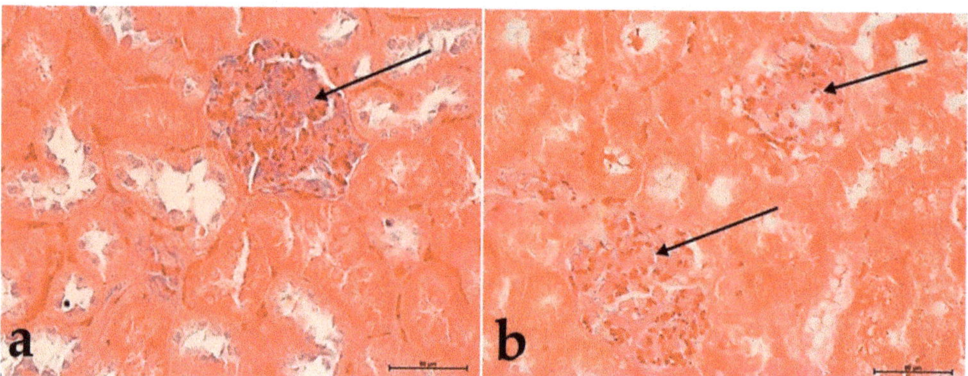

Figure 4. The changes in the kidneys' structure in the controls group: in **I**, animals fed with normal (**a**) and **II**, high-cholesterol diet (**b**). Note the high content of proteoglycans (blue color) in the capillary tuft of the glomerulus (black arrow). Alcian blue staining. Mag. 200×. Bar scale 50 μm.

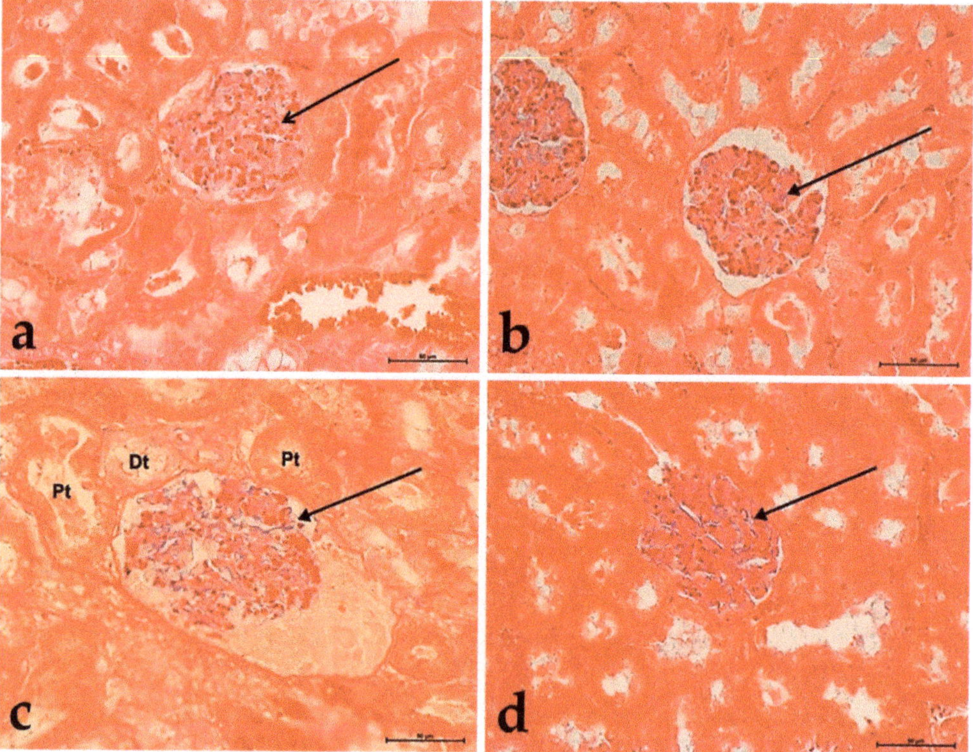

Figure 5. The changes in the structure of the kidneys in animals fed with a normal diet—groups **II** (**a**), **III** (**c**), and in animals fed with high-cholesterol diet—groups **IIa** (**b**), **IIIa** (**d**). A glomerulus with different proteoglycan content (arrow). Note the presence of urine in the glomerulus in group **III** (**c**). Alcian blue staining. Mag. 200×. Bar scale 50 μm.

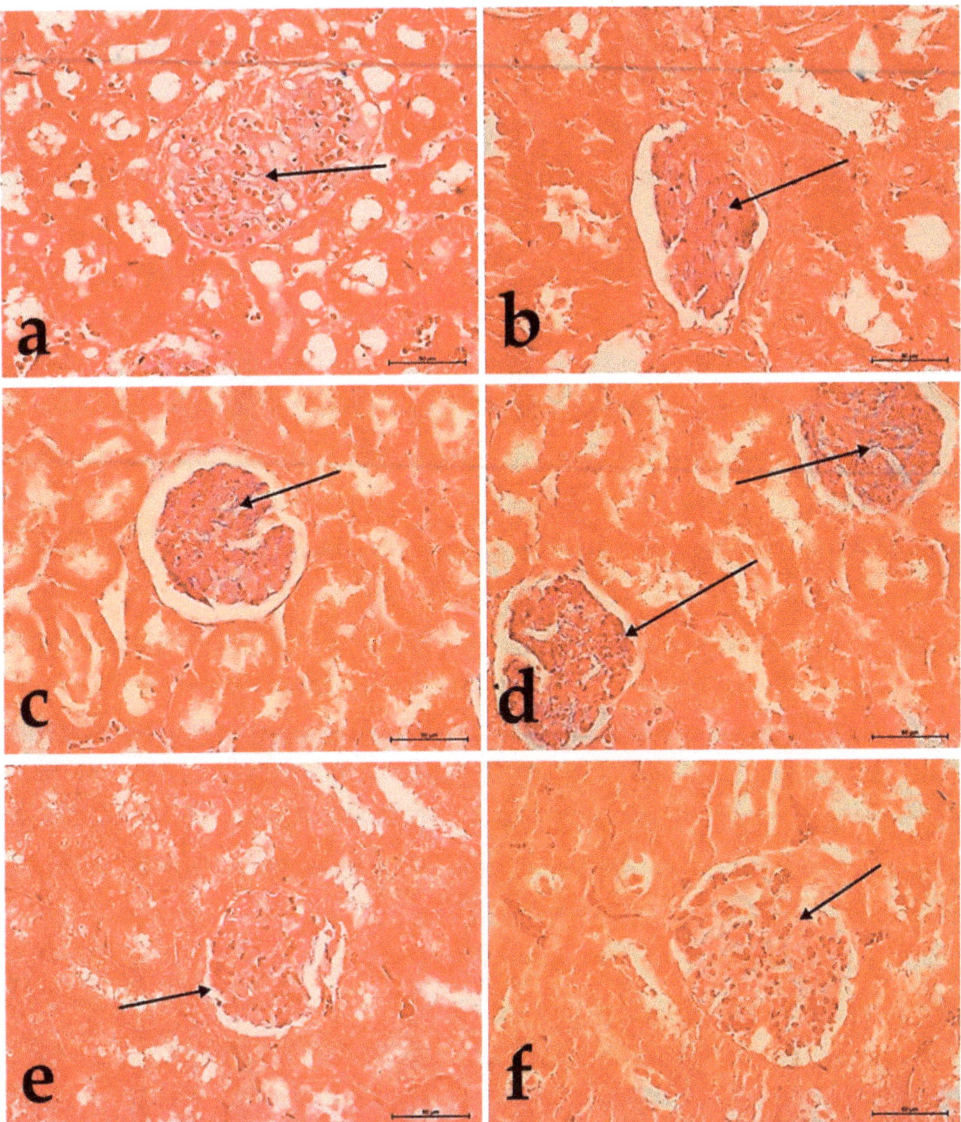

Figure 6. The changes in the proteoglycans content in the kidneys from groups **IV** (**a**), **V** (**c**), and **VI** (**e**) fed with normal diet and from groups **IVa** (**b**), **Va** (**d**), **VIa** (**f**) fed with a high-cholesterol diet. A glomerulus with different urine and blood content (black arrow). Noticeable differences in the lumen of proximal and distal tubules result from different levels of diuresis. Alcian blue staining. Mag. 400×, Bar scale 50 μm.

3.4. Morphometric Analysis

The morphometric analysis of the first nutrient group showed a significant reduction in the content of basal membrane polysaccharides within the glomeruli in groups **II**, **III**, **IV**, and **V** fed with a normal diet. These observations were confirmed for the second group—fed with a high-cholesterol diet in the morphometric study. A slight decrease in the polysaccharide content was observed in group **VI**.

3.4.1. Thickness of the Basement Membrane

The results of the morphometric analysis of the basement membrane's thicknesses in the glomerulus showed that glycoprotein content in the glomerulus is different in subsequent groups (Figure 7). The differences between group **I** and groups **II**, **III**, **IV**, **V** were statistically significant (Figure 7a). The **I**st group in relation to **VI**th is negligible (Figure 7a). In the high cholesterol groups, the content of polysaccharides in the glomerulus was equal in most groups except the **IVa** group. The differences between the **Ia** and **IVa** groups were statistically significant (Figure 7b).

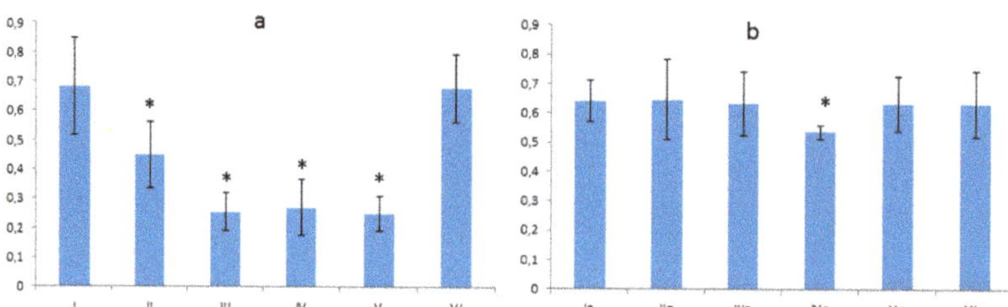

Figure 7. The membrane thickness of the capillary tuft. (**a**) Differences between the control group—I and groups **II**, **III**, **IV**, **V** were statistically significant at $p = 0.05$ (and signed by *); (**b**) Difference between the control group—**Ia** and **IVa** group was statistically significant at $p = 0.05$ (and signed by *). Vertical scale units expressed in μm.

3.4.2. Comparison of the Surface of the Glomerular Capsule to the Capillary Tuft

In animals fed with a regular diet, there was an increase in glomerular capsule size in groups **II** and **III**. Minimal statistically significant growth was observed in group **V** and a decrease in group **IV**. There were no changes in group **VI** (Figure 8a). In the animals fed with a high-cholesterol diet, there was an apparent decrease of the glomerular surface in **IIa**, **IIIa**, **IVa**, and **Va** groups and a slight increase in group **VIa** (Figure 8b). A statistically significant differences compared to the control were reported in groups: **II**, **III**, **IV**, **V**, **IIa**, **IIIa**, **IVa**, **Va**, **VIa**.

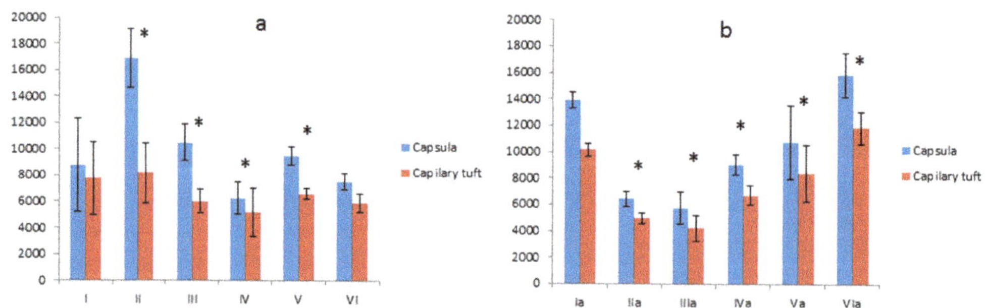

Figure 8. The capsule's surface area and capillary tuft in kidneys of animals fed with regular diet (**a**) and fed with a high-cholesterol diet (**b**). Differences were statistically significant in respect to the control group at $p = 0.05$ (and signed by *). Vertical scale units expressed in μm.

3.4.3. The Ratio between the Size of the Glomerular Capsule and the Capillary Tuft

In this analysis, there is an increase of the glomerular surface capsule in groups: **II**, **III**, **V** in relation to group **I**. In groups **IV** and **VI**, the increases were not statistically significant (Figure 9a). In the second nutritional group, a decrease in this value is generally observed, which in the situation of the slightly increased surface area of group **Ia** indicates a somewhat normalizing effect in group **IIa** and **Va** (more potent) and contained in the components in groups **IIIa**, **IVa**, and **VIa** on vascular tuft (Figure 9b).

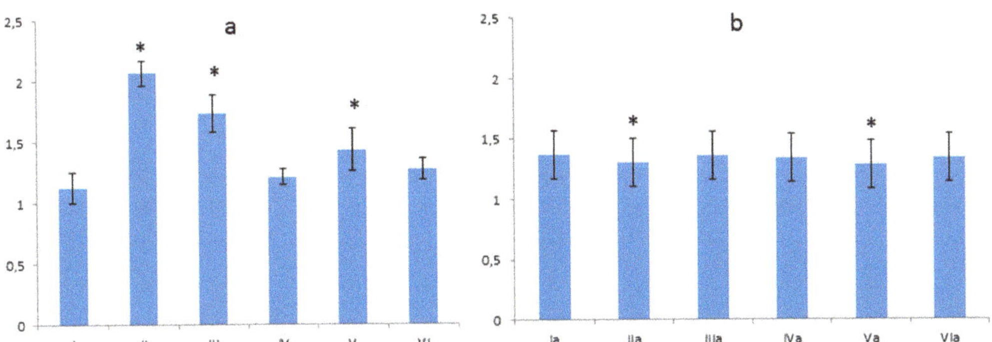

Figure 9. The ratio of the capsule surface to the capillary tuft surface in rats fed with regular diet (**a**). The ratio of the capsule surface to the capillary tuft surface in rats fed with a high-cholesterol diet (**b**). Differences were statistically significant in respect to the control group at $p = 0.05$ (and signed by *). Vertical scale units expressed in μm.

The detailed statistical differences concerning the performed measurements between all groups are presented in Appendix A (Tables A4–A9).

4. Discussion

The water extracts of *I. meservae* leaf contain: water-soluble microelements, carbohydrates and nitrogen-containing metabolites, polar phenolics (caffeoylquinic acids and glycosides of flavonoids), and saponins (glycosides of triterpenoids). Starting the work on testing the safety of prolonged administration of *I. meservae* leaf infusion in animals (rodents), we assumed to observe significant differences among the kidney-targeted effects of administration of standard infusion and the main, easy-separable groups of constituents of *Ilex* leaf (phenolics, saponins, and non-polar terpenoids). From the three assessed fractions, the positive role of polyphenols is widely described, whereas data about saponins as well as terpenoids are limited. However, together they seem to have numerous positive activities. In vivo studies conducted by de Oliveira et al. [20] indicate that *Ilex paraguariensis* plays an important, protective role in obesity and liver function. Peroral administration of water, ethanol, and refined *n*-butanol extracts from its leaves at levels 200, 400, and 800 mg/kg BW/day of (calc. on dry extract) during 30 days resulted in a decrease in serum triglycerides and cholesterol levels and a decrease in the atherosclerotic index in animal models [20]. The results also indicate the potential positive effect of this extract on the cardiovascular system [21]. Some of the pharmacological effects of *I. paraguariensis* are associated with a high content of caffeic acid (and its derivatives), flavonoids, and hydroxylated derivatives of cinnamic acid, which have antioxidant and anti-inflammatory activities [22]. Among the biological activities of Yerba Mate, it is essential to highlight its inhibitory effect on the enzymes involved in the initiation and maintenance of the inflammatory response [23]. Most of the reports concentrate on the positive effects of *I. paraguariensis*, but there are also reports on toxic properties of *Ilex* spp. on human and animal bodies. Therefore, according to de Andrade et al. [24], administration of acute and subchronic toxicity of Yerba Mate dose (2 g/kg body weight of Wistars rats) did not change the macroscopic and histopathological assessment of organs, including liver and kidneys. However, Kataoka et al. [25] found changes in kidneys after green tea administration in rats different from our research. They found that polyphenol intake during lactation by offsprings may play a protective role against a high-fat diet. They also found that polyphenols' high content in a low-fat diet is causing inflammation and fibrosis in kidneys. According to our study, in group feed with polyphenols, increased inflammation in the kidney was associated with increased urination. One possible reason for the toxic effect of polyphenols is the high amounts of chlorogenic acid (CGA) among the polyphenols groups. Ae-Sim Cho et al. found that chlorogenic acid exhibits anti-obesity property and improves lipid metabolism in high-fat diet-induced obese mice [26]. Similar conclusions can be based on Metwally results [27]. This phenomenon's possible mechanism is that CGA promotes fatty acid catabolism by regulating the AMPK pathway [28]. This positive feature of polyphenols in organisms is associated with reducing fat in blood plasma but can be a reason for kidney damage in a low-fat diet, especially at high doses; such correlation was noted by Murakami et al. [29]. Furthermore, the concentration of chlorogenic acid in tea is significantly lower than in Yerba Mate. The following phenolics were detected and counted previously in several European Ilex species and cultivars [12]: a pattern of mono- and dicaffeoylquinic acids accompanied by some flavonoids like quercetin 3-O-rutinoside, kaempferol 3-O-rutinoside, and unspecified quercetin 3-O-hexoside. According to Zwyrzykowska et al. [12], a sum of chlorogenic acids is comparable between *I. paraguariensis* and *I. meserveae* and reaches about 50% of total polyphenols. The differences are visible in dicaffeoyloquinic acid (40 vs. 11 rel%) and flavonoids (3 vs. 30 rel%). The results implicate that extracts should be used in pathological conditions but not in healthy animals. Interestingly, we found that polyphenols have light toxic action on kidneys and induce inflammation process; however, in both extracts (*I. paraguariensis* and *I. meserveae*). only in *I. paraguariensis* did we observe lymphocyte infiltration in the kidney stroma.

Plants of the genus *Ilex* are a rich source of saponins that may have antiproliferative effects and also exhibit anti-hypercholesterolemic, anti-parasitic, and anti-inflammatory effects [30,31]. It is possible that saponins, known to be able to induce cholestasis by liver destruction and act as irritants to the GI (gastrointestinal) tract, may also act synergistically or facilitate the intestinal absorption of co-administered substances and therefore induce additional negative effects on the organism. Another study also reports that *I. paraguariensis* is characterized by water-soluble polysaccharides, mainly arabinogalactans, which have a strong protective effect against gastritis [32]. Moreover, Dartora et al. showed in an animal study that oral administration of rhamnogalacturonan from the leaves of *I. paraguariensis* might be a promising adjuvant for sepsis treatment [33].

There is no comprehensive literature about the tested substances' influence on the filtration barrier status. Initial histological analysis showed that depending on the diet used, the extracts from *I. paraguariensis* and *I. meserveae* are diuretic, and this is due to the different mechanisms of action of the individual components of the extracts. The intercellular content has a significant influence on selective permeability in renal filtration [34]. The glomerular basement membrane comprises proteoglycans—mainly heparan sulfate—and proteins—type IV collagen and laminins [5]. Heparan sulfate proteoglycans in the glomerular basal membrane (HSPGs) are a class of biomolecules with structural and regulatory functions. They are involved in biological processes such as glomerular filtration, cell adhesion, migration, proliferation, and differentiation. In response to numerous cytokines, proteoglycans are degraded by enzymes released by neighboring cells. Therefore, the content decreases during any inflammation process [35]. In our study, increased infiltration by leukocytes was noted in groups **II**, **III**, and **IVa**, and sporadically, single cells were found in other groups. To determine the changes occurring in the glomerulus, morphometric analysis of the size of the glomerular capsule and the capillary tuft was used, as well as the content of polysaccharides within the glomerulus. Our results suggest that increased urogenesis related to the use of *Ilex* extracts and fractions is associated with a decrease in proteoglycans content in the capillary tuft.

5. Conclusions

A standard animal model was used in this study, as well as the cholesterol dose that is usually used in this type of study. It was shown that the synergistic effect of polyphenols and terpenoids in a high-cholesterol diet can be used equally to protect the kidney. The effect of the biologically active compounds used can be determined in the future using a metabolic model, or an animal model of multifunctional aging. In the animal models proposed for the future, the compounds used could affect the already occurring changes in urogenesis, indicating a detrimental or protective effect. The applied dose of active compounds and friction could also be a limiting factor. However, the dose of extracts was applied according to the available literature [19].

Saponin fraction from *Ilex* × *meserveae* seems to have no influence on kidney status at the administered concentrations. However, polyphenols and terpenoids present in dry extracts and the fresh infusions from *Ilex* × *meserveae* and *Ilex paraguariensis* together with co-extracted substances in a normal diet cause a nephrotoxic effect which is decreased by a high-cholesterol diet. This synergistic effect can be used equally to protect the kidney against polyphenols and terpenoids.

Author Contributions: P.K. and R.M.N.: conceptualization, writing—review and editing, investigation, methodology, data curation; A.Z.-W.: writing, investigation, R.K.: methodology, investigation, editing; M.W.: writing—review and editing, data curation, investigation; A.S.: investigation, supervision, project administration. All authors have read and agreed to the published version of the manuscript.

Funding: This research was funded National Science Centre (Poland), grant NCN 2015/19/B/NZ9/02971.

Data Availability Statement: Not applicable.

Acknowledgments: Authors are grateful to Przemysław Bąbelewski (Department of Horticulture, Wrocław University of Environmental and Life Sciences) for assistance in species assessment. The technical assistance of Hanna Czapor-Irzabek (UHPLC-MS; PAEBS, Faculty of Pharmacy, Wroclaw Medical University) is greatly appreciated.

Conflicts of Interest: The authors declare no conflict of interest. The funders had no role in the design of the study; in the collection, analyses, or interpretation of data; in the writing of the manuscript, or in the decision to publish the results.

Appendix A

Appendix A.1. Terpenoid Fraction

Appendix A.1.1. Preparation

To 100 g of dry, ground (<0.05 mm) leaves of *I. meserveae*, 350 mL of chloroform was added, cold-macerated for 24 h, and evaporated on a rotary evaporator. The residue was dissolved in 50 mL of hexane and washed three times with 50 mL of a methanol: water mixture (80:20 with 20 µL 2M hydrochloric acid 37%). After centrifugation and evaporation of hexane extract, the oily green residue was obtained. It was separated on a chromatography column (Kieselgel 60, 230–400 mesh, Merck) into the waxy- and triacylglycerol fractions (eluted with hexane:diethyl ether = 80:1) and the expected terpenoids (eluted with the same solvents in ratio 1:1). After preliminary identification (TLC and GC-MS), it was concentrated in vacuo to give 0.8 g of a white amorphous substance.

Appendix A.1.2. Analysis

The triterpenoid profile was evaluated using a N,O-*bis*-(trimethylsilyl)trifluoroacetamide (BSTFA) derivatization method. GC-MS Shimadzu QP 2020 (Shimadzu, Kyoto, Japan) was used for identification of constituents. 500 µL of pyridine and 50 µL of BSTFA were added to the dry sample (approximately 20 mg). The mixture was transferred to a vial and heated for 25 min at 70 °C. The separation was achieved using a Zebron ZB-5 capillary column (30 m, 0.25 mm, 0.25 µm; Phenomenex, Torrance, CA, USA). GC-MS analysis was performed according to the following parameters: scans were performed from 40 to 1050 *m/z* in electron impact ionization (EI) at 70 eV with a mode of 10 scans per second. Analyses were performed using helium as a carrier gas at a flow rate of 1.0 mL/min at split 1:20, and the following program: 100 °C for 1 min, 2.0 °C/min temperature rise from 100 to 190 °C; 5 °C/min temperature rise from 190 to 300 °C. The injector was maintained at 280 °C, respectively. Compounds were identified using two different analytical methods for comparison; retention times with authentic chemical compounds (Supelco C7-C40 Saturated Alkanes Standard) and mass spectra were obtained, with an available library (Willey NIST 17, fit index above 90%).

All triterpenoids were identified as corresponding trimethylsilylderivatives (TMS).

Table A1. GC-MS profile of triterpenoid fraction of *Ilex meserveae*.

KI exp.	KI lit.	RT [min]	Compound	*Ilex × meserveae* RA [%]
2835	2832	20.65	*all-trans*-Squalene	0.69
3162	3141	24.93	*a*-Tocopherol, TMS derivative	2.21
3361	3370	27.78	(3β)-Olean-18-en-3-ol (TMS)	tr
3370	3344	27.93	β-Sitosterol (TMS)	7.67
3384	3353	28.16	β-Amyrin (TMS)	9.05
3397	3385	28.37	Germanicol (TMS)	2.69
3420	3406	28.79	α-Amyrin (TMS)	28.97
3427	3435	28.93	Lupeol (TMS)	17.34
3508	3523	30.50	Lupeoyl acetate	0.75
3530	3540	30.99	Uvaol, 2-O-TMS	2.35
3563	3560	31.73	(3α)-Lup-20(29)-ene-3,28-diol (O,O-*bis*-TMS)	12.13
3580	3588	32.11	Betulinic acid (O,O-*bis*-TMS)	1.90
3596	3591	32.46	Oleanolic acid (TMS)	1.00
3643	3657	33.60	Ursolic acid (TMS)	5.22
3650	n.a.	33.75	Maslinic acid * (TMS)	6.54

KI exp.—retention index calculated according to linear *n*-alkanes; KI lit.—values of retention indices presented in NIST17 (national institute of standards and technology; Gaithersburg, MD, USA) database; RT—retention time; *—tentatively identified, only based on the mass spectrum, retention index is missed in the database; RA—relative area.

Appendix A.2. Saponin Fraction

Appendix A.2.1. Preparation

60 g of lyophilized *Ilex × meserveae* "Blue Angel" leaves powder was 1-day macerated at room temperature, two times, each with 600 mL 70% methanol. After that, the extracts were combined, filtered through a filter paper, and the solution of 30 g lead (II) acetate trihydrate in 70% methanol (in a total amount of 100 mL) was added to precipitate ballast substances (e.g., phenolics). The extract was centrifuged at 3000 rpm for 15 min (MPW). The residue containing complexes of ballast substances was subjected to an experimental and safe process of its utilization by Paweł Lochyński., Wrocław University of Science and Technology. The possible residues of lead ions in the saponin-rich fraction were precipitated with the addition of diluted sulphuric acid, centrifugation, and further neutralization of the supernatant.

The supernatant was diluted with distilled water to obtain 40% methanol concentration and centrifuged again. Afterward, the second supernatant was applied to a 20 g octadecyl SPE column (J. T. Baker) in two runs. The saponin enriched fraction was recovered from the SPE bed by a minimal volume of pure methanol. Combined saponin fractions from SPE were concentrated with a vacuum evaporator (Büchi) at 40 °C to dryness, and the glassy residue was lyophilized.

One reference Holly (*Ilex aquifolium*) and two reference Mate (*Ilex paraguariensis*) were prepared similarly to *I. meserveae* saponin fraction (Il.6) on an analytical scale. A measure of 100 mg of each plant substance was macerated in 2 mL of 70% methanol for 15 min. in an ultrasonic bath (Bandelin). After that, the solution of 0.05 g lead (II) acetate trihydrate in 70% methanol (in a total amount of 0.5 mL) was added to precipitate ballast substances. The extracts were further treated as described above.

Appendix A.2.2. UHPLC-ESI-MS Conditions of Analysis of Saponins

The UHPLC Ultimate 3000 apparatus (Thermo Fisher Sci.) consisted of a quaternary pump with a vacuum degasser, autosampler, and a thermostated column chamber combined with an ESI-qTOF Compact detector (Bruker Daltonics).

Separations were achieved on Kinetex C-18 (150 × 2.1 mm, 2.6 μm; Phenomenex) with a flow rate of 0.3 mL/min and column temperature of 30 °C. The elution system consisted of the following gradient steps of water (A) and acetonitrile (B) with 0.1% addition of formic acid: 0→1 min. (2→30% B in A), 1→31 min. (30→60% B in A), 31→31.5 min. (60→100% B in A), 31.5→35.5 min. (100% B in A). The extracts were diluted with acetonitrile/water (1:1, v/v) to obtain the final concentrations of 0.02 mg per mL; the injection volumes were 5 μL.

The detector worked in negative mode and was calibrated with the Tunemix^{TM} (Bruker Daltonics) with m/z SD below 1 ppm. A calibration segment was at the beginning of every single run. The mass accuracy of main saponins found in reference Mate extract and verified with literature was within 3 ppm. The main instrumental parameters were as follows: scan range 50–2200 m/z, nebulizer pressure 1.5 bar, dry gas—nitrogen—7.0 L/min, temperature 200 °C, capillary voltage 2.2 kV, ion energy 5 eV, collision energy 10 eV vs. 30 eV, low mass set at 200 m/z. The analysis of the obtained mass spectra was carried out using a Data Analysis (Bruker Daltonics).

Appendix A.2.3. Results of UHPLC-ESI-MS Saponin Profiling

Because no information on saponin profile in *I. meserveae* was found in the literature, its saponin profile was compared with commercially available Mate (*I. paraguariensis*) and holly (*I. aquifolium*). With the information that *I. meserveae* is suspected to be a hybrid of *I. rugosa* and *I. aquifolium* or *I. rugosa* and *I. cornuta*, it was expected that similar compounds as in the mother plant could be found in Il.6.

Comparing saponin profiles of *I. meserveae* and *I. paraguariensis* led to the observation that their pattern in *I. meserveae* is much less complicated than in commercial Mate, as shown in Table A2. In this case, compounds **1–4** were of similar relative intensities. Five of them (**3–5, 9, 13**) were also found in Mate with different concentrations. The most intensive peaks belonged to substance 3 (of molecular weight (MW) 1074 Da) and substance 5 (MW 928 Da), which was found in both Yerba Mate samples on the level of about 10–70% and 9–19% of the relative area (RA). Substances 1 and 2 were absent in the analyzed Yerba Mate.

As we do not possess reference saponins from Mate, we can speculate (by relative retention and accurate molecular mass measurements) that peak 18, the most intensive one in Yerba Mate, could be matesaponin 1 (MW 912 Da), peak 16 or 17 could be matesaponin 2 (MW 1058 Da), peak 3 or 4 could be matesaponin 3 (MW 1074 Da), peak 9 could be matesaponin 4 (MW 1220 Da), and peak 6 could be matesaponin 5 (MW 1382 Da). Chemical Abstracts search for substance 1 revealed five isomeric saponins (MW 1236 Da) isolated from genus *Ilex*, while for substance 2, six isomeric saponins (MW 1074 Da) isolated from the genus *Ilex*. That initial finding showed that isolation and NMR verification of each structure would be necessary to define the substances precisely if needed. However, some additional assumptions were made using interpretations included in a paper of Negrin and co-workers [36].

Table A2. UHPLC-ESI-MS comparison of main saponin profiles of *I. meserveae* saponin-rich fraction (Il.6) and reference commercial Mate samples (Il.1a and Il.1b).

No.	Compound Number Reported in [36]	[M–H]⁻ Measured [m/z]	Proposed Formula	Il.1a		Il.1b		Il.6	
				RT [min]	RA [%]	RT [min]	RA [%]	RT [min]	RA [%]
1.	—	1235.61	$C_{59}H_{96}O_{27}$	—	—	—	—	6.78	73.40
2.	39 [a]	1073.56	$C_{53}H_{86}O_{22}$	—	—	—	—	7.40	70.53
3.	39 [a]	1073.56	$C_{53}H_{86}O_{22}$	8.07	60.97	8.03	10.70	8.06	100.00
4.	39 [a]	1073.56	$C_{53}H_{86}O_{22}$	8.10	20.01	8.07	5.07	8.10	11.53
5.	40	927.50	$C_{47}H_{76}O_{18}$	8.24	18.63	8.22	8.24	8.26	99.22
6.	41	1381.68	$C_{65}H_{106}O_{31}$	8.48	11.11	8.47	2.78	—	—
7.	43	1131.56	$C_{55}H_{88}O_{24}$	8.70	10.70	—	—	—	—
8.	45	1101.55	$C_{54}H_{86}O_{23}$	8.85	14.20	—	—	—	—
9.	47	1219.61	$C_{59}H_{96}O_{26}$	9.10	59.09	9.11		9.11	13.99
9a.	48	911.50	$C_{47}H_{76}O_{17}$	9.17	5.10	—	—	9.21	31.68
10.	49	927.50	$C_{47}H_{76}O_{18}$	9.45	18.28	9.42	8.86	9.45	10.62
11.	56 [b]	1235.58	$C_{62}H_{92}O_{25}$	10.13	4.05	—	—	—	—
12.	56 [b]	1235.58	$C_{62}H_{92}O_{25}$	10.40	9.40	—	—	—	—
13.	55	1057.56	$C_{53}H_{86}O_{21}$	10.78	27.96	10.72	11.11	10.82	48.36
14.	56 [b]	1235.58	$C_{62}H_{92}O_{25}$	11.08	4.35	—	—	—	—
15.	57	1115.57	$C_{55}H_{88}O_{23}$	11.23	20.73	11.17	7.61	—	—
16.	62	1057.56	$C_{53}H_{86}O_{21}$	11.85	100.00	11.78	94.64	—	—
17.	64	1057.56	$C_{53}H_{86}O_{21}$	12.09	68.77	12.04	54.22	—	—
18.	66 [c]	911.50	$C_{47}H_{76}O_{17}$	12.49	93.17	12.44	100.00	—	—
19.	66 [c]	911.50	$C_{47}H_{76}O_{17}$	12.68	11.98	12.63	7.92	—	—
20.	66 [c]	911.50	$C_{47}H_{76}O_{17}$	12.91	20.45	12.85	13.00	—	—
21.	72	1085.56	$C_{54}H_{86}O_{22}$	13.63	8.95	13.56	2.80	—	—
22.	77	895.51	$C_{47}H_{76}O_{16}$	14.34	20.19	14.28	18.10	—	—
23.	80	895.51	$C_{47}H_{76}O_{16}$	14.63	6.43	14.55	5.83	—	—
24.	83	953.52	$C_{49}H_{78}O_{18}$	14.86	30.53	14.77	25.58	—	—
25.	82 [d]	1219.59	$C_{59}H_{96}O_{26}$	16.24	3.25	16.21	3.55	—	—
26.	82 [d]	1219.59	$C_{59}H_{96}O_{26}$	16.54	2.84	16.53	2.26	—	—
27.	108 [e]	895.51	$C_{47}H_{76}O_{16}$	20.36	3.68	20.36	4.03	—	—
28.	108 [e]	895.51	$C_{47}H_{76}O_{16}$	20.50	2.34	20.50	1.83	—	—
29.	110	749.45	$C_{41}H_{66}O_{12}$	22.18	3.33	22.15	3.42	—	—

RT—retention time, RA—the relative area of peak, when the area of the largest one is calculated as 100%. The same letters in the column referring to [36] means the same possible assignments of detected compounds.

Table A3. Database- and MS/MS-based identification of main saponins present in *Ilex paraguariensis* (Il.1), *I. aquifolium* (Il.5), and *I. meserveae* (Il.6).

Compound	Source	MS/MS Interpretation	Probable Identification
1 Rt = 6.8 min; calc. [M–H]⁻ = 1235.6061, err. 0.5 ppm; neutral formula: $C_{59}H_{96}O_{27}$	Il.5 Il.6	911 [M–(Hex+Hex)–H]⁻ 765 [M–(Hex+Hex)–dxHex–H]⁻ 749 [M–(Hex+Hex)–Hex–H]⁻ 731 [M–(Hex+Hex)–Hex–18–H]⁻ 603 [M–(Hex+Hex)–Hex–dxHex–H]⁻ 471 [M–(Hex+Hex)–Hex–(dxHex+Pen)–H]⁻	kudinoside N (SA)
2 Rt = 7.4 min; calc. [M–H]⁻ = 1073.5538, err. −0.1 ppm; neutral formula: $C_{53}H_{86}O_{22}$	Il.6	749 [M–(Hex+Hex)–H]⁻ 731 [M–(Hex+Hex)–18–H]⁻ 453 [M–(Hex+Hex+18)–(dxHex+Pen)–H]⁻	latifoloside L (PA) matesaponin 3 (UA)

Table A3. *Cont.*

Compound	Source	MS/MS Interpretation	Probable Identification
3/4 Rt = 8.0 min/8.1 min calc. [M–H]⁻ = 1073.5538, err. −0.3 ppm; neutral formula: $C_{53}H_{86}O_{22}$	Il.1 Il.5 Il.6	911 [M–Hex–H]⁻ 765 [M–Hex–dxHex–H]⁻ 749 [M–Hex–Hex–H]⁻ 731 [M–Hex–Hex–18–H]⁻ 603 [M–Hex–(Hex+dxHex)–H]⁻ 471 [M–Hex–(Hex+dxHex)–Pen–H]⁻	latifoloside L (PA) matesaponin 3 (UA)
5 Rt = 8.3 min; calc. [M–H]⁻ = 927.4959, err. 0.2 ppm; neutral formula: $C_{47}H_{76}O_{18}$	Il.1 Il.5 Il.6	765 [M–Hex–H]⁻ 603 [M–Hex–Hex–H]⁻ 471 [M–Hex–Hex–Pen–H]⁻	ilexoside XV (SA) ilexoside II (PA) ilekudinoside E (PA) ilexsaponin B3 (IG-B)
9 Rt = 9.1 min; calc. [M–H]⁻ = 1219.6117, err. −0.2 ppm; neutral formula: $C_{59}H_{96}O_{26}$	Il.1 Il.5 Il.6	895 [M–(Hex+Hex)–H]⁻ 749 [M–(Hex+Hex)–dxHex–H]⁻ 733 [M–(Hex+Hex)–Hex–H]⁻ 715 [M–(Hex+Hex)–Hex–18–H]⁻ 587 [M–(Hex+Hex)–(Hex+dxHex)–H]⁻ 569 [M–(Hex+Hex)–(Hex+dxHex)–18–H]⁻ 455 [M–(Hex+Hex)–(Hex+dxHex)–Pen–H]⁻	matesaponin 4 (UA)
13 RT = 10.8 min; calc. [M–H]⁻ = 1057.5589, err. 0.1 ppm; neutral formula: $C_{53}H_{86}O_{21}$	Il.1 Il.5 Il.6	733 [M–(Hex+Hex)–H]⁻ 587 [M–(Hex+Hex)–dxHex–H]⁻ 455 [M–(Hex+Hex)–dxHex–Pen–H]⁻	matesaponin 2 (UA) or isomer
16 RT = 11.85 min; calc. [M–H]⁻ = 1057.5589, err. 0.5 ppm; neutral formula: $C_{53}H_{86}O_{21}$	Il.1	895 [M–Hex–H]⁻ 749 [M–Hex–dxHex–H]⁻ 733 [M–Hex–Hex–H]⁻ 715 [M–Hex–Hex–18–H]⁻ 587 [M–Hex–(Hex+dxHex)–H]⁻ 569 [M–Hex–(Hex+dxHex)–18–H]⁻ 455 [M–Hex–(Hex+dxHex)–Pen–H]⁻	matesaponin 2 (UA) ilekudinoside A (OA)
17 RT = 12.09 min; calc. [M–H]⁻ = 1057.5589, err. −0.5 ppm; neutral formula: $C_{53}H_{86}O_{21}$	Il.1	895 [M–Hex–H]⁻ 749 [M–Hex–dxHex–H]⁻ 733 [M–Hex–Hex–H]⁻ 715 [M–Hex–Hex–18–H]⁻ 587 [M–Hex–(Hex+dxHex)–H]⁻ 569 [M–Hex–(Hex+dxHex)–18–H]⁻ 455 [M–Hex–Hex–dxHex–Pen–H]⁻	ilekudinoside A (OA)
18 RT = 12.49 min; calc. [M–H]⁻ = 911.5010, err. −1.6 ppm; neutral formula: $C_{47}H_{76}O_{17}$	Il.1	749 [M–Hex–H]⁻ 587 [M–Hex–Hex–H]⁻ 569 [M–Hex–Hex–18–H]⁻ 455 [M–Hex–Hex–Pen–H]⁻	matesaponin 1 (UA) guaiacin B (OA)
24 RT = 14.86min; calc. [M–H]⁻ = 953.5115, err. −1.1 ppm; neutral formula: $C_{49}H_{78}O_{18}$	Il.1	749 [M–Hex(Ac)–H]⁻ 731 [M–Hex(Ac)–18–H]⁻ 629 [M–Hex–Hex–H]⁻ 587 [M–Hex(Ac)–Hex–H]⁻ 569 [M–Hex(Ac)–Hex–18–H]⁻ 455 [M–Hex–Hex–Pen–H]⁻	*I. amara* saponin (UA), 77-52-1 *I. amara* saponin (OA), 508-02-1

Probable identification based on (1) saponin database of *Aquifoliaceae*, (2) MS/MS fragmentation pathway. Aglycones: IG-B—ilexgenin B, OA—oleanolic acid, PA—pomolic acid, SA—siaresinolic acid, UA—ursolic acid. MS/MS loss of sugar unit: dxHex—deoxyhexose, Hex—hexose, Hex(Ac)—acetylhexose, Pen—pentose.

Table A4. The membrane thickness of capillary tuft in rats fed with normal diet. Statistical differences between all experimental groups; * significant, 0 insignificant.

Group No	I	II	III	IV	V	VI
I	-	*	*	*	*	0
II	*	-	*	*	*	*
III	*	*	-	0	0	*
IV	*	*	0	-	0	*
V	*	*	0	0	-	*
VI	0	*	*	*	*	-

Table A5. The membrane thickness of capillary tuft in rats fed with high cholesterol diet. Statistical differences between all experimental groups; * significant, 0 insignificant.

Group No	Ia	IIa	IIIa	IV	Va	VIa
Ia	-	0	0	*	0	0
IIa	0	-	0	*	0	0
IIIa	0	0	-	*	0	0
IVa	*	*	*	-	*	*
V	0	0	0	*	-	0
VIa	0	0	0	*	0	-

Table A6. The capsule's surface area and capillary tuft in kidneys of animals fed with regular diet. Statistical differences between all experimental groups; * significant, 0 insignificant.

Group No	I	II	III	IV	V	VI
I	-	*	*	*	*	0
II	*	-	*	*	*	*
III	*	*	-	*	0	*
IV	*	*	*	-	*	*
V	*	*	0	*	-	*
VI	0	*	*	*	*	-

Table A7. The capsule's surface area and capillary tuft in kidneys of animals in high cholesterol diet. Statistical differences between all experimental groups; * significant, 0 insignificant.

Group No	Ia	IIa	IIIa	IV	Va	VIa
Ia	-	*	*	*	*	*
IIa	*	-	0	*	*	*
IIIa	*	0	-	*	*	*
IVa	*	*	*	-	0	*
V	*	*	*	0	-	*
VIa	*	*	*	*	*	-

Table A8. The ratio of the capsule surface to the capillary tuft surface in rats fed with regular diet. Statistical differences between all experimental groups; * significant, 0 insignificant.

Group No	I	II	III	IV	V	VI
I	-	*	*	0	*	0
II	*	-	*	*	*	*
III	*	*	-	*	*	*
IV	0	*	*	-	*	0
V	*	*	*	*	-	0
VI	0	*	*	0	0	-

Table A9. The ratio of the capsule surface to the capillary tuft surface in rats fed with a high-cholesterol diet. Statistical differences between all experimental groups; * significant, 0 insignificant.

Group No	Ia	IIa	IIIa	IV	Va	VIa
Ia	-	*	0	0	*	0
IIa	*	-	*	*	0	*
IIIa	0	*	-	0	*	0
IVa	0	*	0	-	*	0
V	*	0	*	*	-	*
VIa	0	*	0	0	*	-

References

1. Keservani, R.K.; Kesharwani, R.K.; Vyas, N.; Jain, S.; Raghuvanshi, R.; Sharma, A.K. Nutraceutical and functional food as future food: A review. *Der Pharm. Lett.* **2010**, *2*, 106–116.
2. Cogoi, L.; Giacomino, M.S.; Pellegrino, N.; Anesini, C.; Filip, R. Nutritional and phytochemical study of *Ilex paraguariensis* fruits. *J. Chem.* **2013**, *2013*, 750623. [CrossRef]
3. Bracesco, N.; Sanchez, A.; Contreras, V.; Menini, T.; Gugliucci, A. Recent advances on *Ilex paraguariensis* research: Minireview. *J. Ethnopharmacol.* **2011**, *136*, 378–384. [CrossRef]
4. Siri-Tarino, P.W. Effects of diet on high-density lipoprotein cholesterol. *Curr. Atheroscler. Rep.* **2011**, *13*, 453–460. [CrossRef] [PubMed]
5. Gao, H.; Long, Y.; Jiang, X.; Liu, Z.; Wang, D.; Zhao, Y.; Li, D.; Sun, B.-l. Beneficial effects of yerba mate tea (*Ilex paraguariensis*) on hyperlipidemia in high-fat-fed hamsters. *Exp. Gerontol.* **2013**, *48*, 572–578. [CrossRef] [PubMed]
6. Arçari, D.P.; Bartchewsky, W.; dos Santos, T.W.; Oliveira, K.A.; Funck, A.; Pedrazzoli, J.; de Souza, M.F.; Saad, M.J.; Bastos, D.H.; Gambero, A. Antiobesity effects of yerba mate extract (*Ilex paraguariensis*) in high-fat diet-induced obese mice. *Obesity* **2009**, *17*, 2127–2133. [CrossRef]
7. He, L.; Hao, L.; Fu, X.; Huang, M.; Li, R. Severe hypertriglyceridemia and hypercholesterolemia accelerating renal injury: A novel model of type 1 diabetic hamsters induced by short-term high-fat/high-cholesterol diet and low-dose streptozotocin. *BMC Nephrol.* **2015**, *16*, 51. [CrossRef]
8. Gluba-Brzozka, A.; Franczyk, B.; Rysz, J. Cholesterol Disturbances and the Role of Proper Nutrition in CKD Patients. *Nutrients* **2019**, *18*, 2820. [CrossRef]
9. Bravo, L.; Goya, L.; Lecumberri, E. LC/MS characterization of phenolic constituents of mate (*Ilex paraguariensis* St. Hil.) and its antioxidant activity compared to commonly consumed beverages. *Food Res. Int.* **2007**, *40*, 393–405. [CrossRef]
10. Balzan, S.; Hernandes, A.; Reichert, C.L.; Donaduzzi, C.; Pires, V.A.; Junior, A.G.; Junior, E.L.C. Lipid-lowering effects of standardized extracts of *Ilex paraguariensis* in high-fat-diet rats. *Fitoterapia* **2013**, *86*, 115–122. [CrossRef]
11. Anesini, C.; Turner, S.; Cogoi, L.; Filip, R. Study of the participation of caffeine and polyphenols on the overall antioxidant activity of mate (*Ilex paraguariensis*). *LWT-Food Sci. Technol.* **2012**, *45*, 299–304. [CrossRef]
12. Zwyrzykowska, A.; Kupczyński, R.; Jarosz, B.; Szumny, A.; Kucharska, A.Z. Qualitative and quantitative analysis of polyphenolic compounds in *Ilex* sp. *Open Chem.* **2015**, *13*, 1303–1312. [CrossRef]
13. Salmon, A.H.; Neal, C.R.; Harper, S.J. New aspects of glomerular filtration barrier structure and function: Five layers (at least) not three. *Curr. Opin. Nephrol. Hypertens.* **2009**, *18*, 197–205.
14. Jarad, G.; Miner, J.H. Update on the glomerular filtration barrier. *Curr. Opin. Nephrol. Hypertens.* **2009**, *18*, 226–232. [CrossRef]
15. Miner, J.H. The glomerular basement membrane. *Exp. Cell Res.* **2012**, *318*, 973–978. [CrossRef]
16. Jarad, G.; Cunningham, J.; Shaw, A.S.; Miner, J.H. Proteinuria precedes podocyte abnormalities inLamb2−/−mice, implicating the glomerular basement membrane as an albumin barrier. *J. Clin. Investig.* **2006**, *116*, 2272–2279. [CrossRef]
17. de Morais, E.C.; Stefanuto, A.; Klein, G.A.; Boaventura, B.C.; de Andrade, F.; Wazlawik, E.; di Pietro, P.F.; Maraschin, M.; da Silva, E.L. Consumption of yerba mate (*Ilex paraguariensis*) improves serum lipid parameters in healthy dyslipidemic subjects and provides an additional LDL-cholesterol reduction in individuals on statin therapy. *J. Agric. Food Chem.* **2009**, *57*, 8316–8324. [CrossRef]
18. Włodarczyk, M.; Szumny, A.; Gleńsk, M. Lanostane-type saponins from *Vitaliana primuliflora*. *Molecules* **2019**, *24*, 1606. [CrossRef]
19. de Resende, P.E.; Verza, S.G.; Kaiser, S.; Gomes, L.F.; Kucharski, L.C.; Ortega, G.G. The activity of mate saponins (*Ilex paraguariensis*) in intra-abdominal and epididymal fat, and glucose oxidation in male Wistar rats. *J. Ethnopharmacol.* **2012**, *144*, 735–740. [CrossRef] [PubMed]
20. de Oliveira, E.; Lima, N.; Conceição, E.; Peixoto-Silva, N.; Moura, E.; Lisboa, P. Treatment with *Ilex paraguariensis* (yerba mate) aqueous solution prevents hepatic redox imbalance, elevated triglycerides, and microsteatosis in overweight adult rats that were precociously weaned. *Braz. J. Med. Biol. Res.* **2018**, *51*, e7342. [CrossRef] [PubMed]
21. Gebara, K.S.; Gasparotto Junior, A.; Palozi, R.A.; Morand, C.; Bonetti, C.I.; Gozzi, P.T.; de Mello, M.R.; Costa, T.A.; Cardozo Junior, E.L. A randomized crossover intervention study on the effect a standardized mate extract (*Ilex paraguariensis* A. St.-Hil.) in men predisposed to cardiovascular risk. *Nutrients* **2021**, *13*, 14. [CrossRef] [PubMed]

22. Pasquali, T.R.; Roman, S.S.; Dal Pra, V.; Cansian, R.L.; Mossi, A.J.; Oliveira, V.J.; Mazutti, M.A. Acute toxicity and anti-inflammatory effects of supercritical extracts of *Ilex paraguariensis*. *Afr. J. Pharm. Pharmacol.* **2011**, *5*, 1162–1169.

23. Görgen, M.; Turatti, K.; Medeiros, A.R.; Buffon, A.; Bonan, C.D.; Sarkis, J.J.; Pereira, G.S. Aqueous extract of *Ilex paraguariensis* decreases nucleotide hydrolysis in rat blood serum. *J. Ethnopharmacol.* **2005**, *97*, 73–77. [CrossRef]

24. De Andrade, F.; de Albuquerque, C.A.C.; Maraschin, M.; da Silva, E.L. Safety assessment of yerba mate (*Ilex paraguariensis*) dried extract: Results of acute and 90 days subchronic toxicity studies in rats and rabbits. *Food Chem. Toxicol.* **2012**, *50*, 328–334. [CrossRef]

25. Kataoka, S.; Norikura, T.; Sato, S. Maternal green tea polyphenol intake during lactation attenuates kidney injury in high-fat-diet-fed male offspring programmed by maternal protein restriction in rats. *J. Nutr. Biochem.* **2018**, *56*, 99–108. [CrossRef]

26. Cho, A.-S.; Jeon, S.-M.; Kim, M.-J.; Yeo, J.; Seo, K.-I.; Choi, M.-S.; Lee, M.-K. Chlorogenic acid exhibits anti-obesity property and improves lipid metabolism in high-fat diet-induced-obese mice. *Food Chem. Toxicol.* **2010**, *48*, 937–943. [CrossRef]

27. Metwally, D.M.; Alajmi, R.A.; El-Khadragy, M.F.; Yehia, H.M.; AL-Megrin, W.A.; Akabawy, A.M.; Amin, H.K.; Moneim, A.E.A. Chlorogenic acid confers robust neuroprotection against arsenite toxicity in mice by reversing oxidative stress, inflammation, and apoptosis. *J. Funct. Foods* **2020**, *75*, 104202. [CrossRef]

28. Sudhakar, M.; Sasikumar, S.J.; Silambanan, S.; Natarajan, D.; Ramakrishnan, R.; Nair, A.J.; Kiran, M.S. Chlorogenic acid promotes development of brown adipocyte-like phenotype in 3T3-l1 adipocytes. *J. Funct. Foods* **2020**, *74*, 104161. [CrossRef]

29. Murakami, A. Dose-dependent functionality and toxicity of green tea polyphenols in experimental rodents. *Arch. Biochem. Biophys.* **2014**, *557*, 3–10. [CrossRef]

30. Hu, T.; He, X.-W.; Jiang, J.-G.; Xu, X.-L. Efficacy evaluation of a Chinese bitter tea (*Ilex latifolia* Thunb.) via analyses of its main components. *Food Funct.* **2014**, *5*, 876–881. [CrossRef] [PubMed]

31. Arçari, D.P.; Bartchewsky Jr, W.; dos Santos, T.W.; Oliveira, K.A.; de Oliveira, C.C.; Gotardo, É.M.; Pedrazzoli Jr, J.; Gambero, A.; Ferraz, L.F.; Carvalho, P.d.O. Anti-inflammatory effects of yerba mate extract (*Ilex paraguariensis*) ameliorate insulin resistance in mice with high fat diet-induced obesity. *Mol. Cell. Endocrinol.* **2011**, *335*, 110–115. [CrossRef] [PubMed]

32. Maria-Ferreira, D.; Dartora, N.; da Silva, L.M.; Pereira, I.T.; de Souza, L.M.; Ritter, D.S.; Iacomini, M.; Werner, M.F.D.; Sassaki, G.L.; Baggio, C.H. Chemical and biological characterization of polysaccharides isolated from *Ilex paraguariensis* A. St.-Hil. *Int. J. Biol. Macromol.* **2013**, *59*, 125–133. [CrossRef] [PubMed]

33. Dartora, N.; de Souza, L.M.; Paiva, S.M.M.; Scoparo, C.T.; Iacomini, M.; Gorin, P.A.J.; Rattmann, Y.D.; Sassaki, G.L. Rhamnogalacturonan from *Ilex paraguariensis*: A potential adjuvant in sepsis treatment. *Carbohydr. Polym.* **2013**, *92*, 1776–1782. [CrossRef]

34. Scott, R.P.; Quaggin, S.E. The cell biology of renal filtration. *J. Cell Biol.* **2015**, *209*, 199–210. [CrossRef] [PubMed]

35. Rops, A.L.W.M.M.; van der Vlag, J.; Lensen, J.F.M.; Wijnhoven, T.J.M.; van den Heuvel, L.P.W.J.; van Kuppevelt, T.H.; Berden, J.H.M. Heparan sulfate proteoglycans in glomerular inflammation. *Kidney Int.* **2004**, *65*, 768–785. [CrossRef] [PubMed]

36. Negrin, A.; Long, C.; Motley, T.J.; Kennelly, E.J. LC-MS metabolomics and chemotaxonomy of caffeine-containing Holly (*Ilex*) species and related taxa in the *Aquifoliaceae*. *J. Agric. Food Chem.* **2019**, *67*, 5687–5699. [CrossRef]

MDPI

St. Alban-Anlage 66

4052 Basel

Switzerland

Tel. +41 61 683 77 34

Fax +41 61 302 89 18

www.mdpi.com

Foods Editorial Office

E-mail: foods@mdpi.com

www.mdpi.com/journal/foods

9 783036 530000